Building Vibration and Soil Dynamics

Building Vibration and Soil Dynamics

Guest Editors
Chao He
Wenbo Tu
Chao Zou
Yunlong Guo

Basel • Beijing • Wuhan • Barcelona • Belgrade • Novi Sad • Cluj • Manchester

Guest Editors

Chao He
College of Transportation
Engineering
Tongji University
Shanghai
China

Wenbo Tu
School of Transportation
Engineering
East China Jiaotong
University
Nanchang
China

Chao Zou
School of Civil and
Transportation Engineering
Guangdong University
of Technology
Guangzhou
China

Yunlong Guo
School of Engineering
University of Birmingham
Birmingham
United Kingdom

Editorial Office
MDPI AG
Grosspeteranlage 5
4052 Basel, Switzerland

This is a reprint of the Special Issue, published open access by the journal *Buildings* (ISSN 2075-5309), freely accessible at: www.mdpi.com/journal/buildings/special_issues/98M5MMIHH1.

For citation purposes, cite each article independently as indicated on the article page online and using the guide below:

Lastname, A.A.; Lastname, B.B. Article Title. *Journal Name* **Year**, *Volume Number*, Page Range.

ISBN 978-3-7258-2812-8 (Hbk)
ISBN 978-3-7258-2811-1 (PDF)
https://doi.org/10.3390/books978-3-7258-2811-1

© 2024 by the authors. Articles in this book are Open Access and distributed under the Creative Commons Attribution (CC BY) license. The book as a whole is distributed by MDPI under the terms and conditions of the Creative Commons Attribution-NonCommercial-NoDerivs (CC BY-NC-ND) license (https://creativecommons.org/licenses/by-nc-nd/4.0/).

Contents

About the Editors vii

Preface ix

Lingshan He and Ziyu Tao
Building Vibration Measurement and Prediction during Train Operations
Reprinted from: *Buildings* 2024, *14*, 142, https://doi.org/10.3390/buildings14010142 1

Qiangqiang Shua, Kexing Liu, Jingkai Li and Wenbo Tu
Time Domain Nonlinear Dynamic Analysis of Vertically Loaded Tapered Pile in Layered Soils
Reprinted from: *Buildings* 2024, *14*, 445, https://doi.org/10.3390/buildings14020445 15

Jinbao Yao, Zhaozhi Wu, Xiaofeng Cao, Nianping Wu and Nan Zhang
Derivation and Application of Analytical Coupling Loss Coefficient by Transfer Function in Soil–Building Vibration
Reprinted from: *Buildings* 2024, *14*, 1933, https://doi.org/10.3390/buildings14071933 31

Xinwei Luo, Xuan Jiang, Qingsong Feng, Wenlin Hu, Qinming Tu and Yanming Chen
Analysis of Train-Induced Vibration Transmission and Distribution Characteristics in Double-Layer Metro Depot
Reprinted from: *Buildings* 2024, *14*, 1702, https://doi.org/10.3390/buildings14061702 54

Puyu Zhan, Suduo Xue, Xiongyan Li, Guojun Sun and Ruisheng Ma
Seismic Assessment of Large-Span Spatial Structures Considering Soil–Structure Interaction (SSI): A State-of-the-Art Review
Reprinted from: *Buildings* 2024, *14*, 1174, https://doi.org/10.3390/buildings14041174 70

Zhijun Zhang, Xiaozhen Li, Xun Zhang, Guihong Xu and Anjie Wu
Measurements and Evaluation of Road Traffic-Induced Micro-Vibration in a Workshop Equipped with Precision Instruments
Reprinted from: *Buildings* 2024, *14*, 1142, https://doi.org/10.3390/buildings14041142 103

Xiangrong Guo and Shipeng Wang
Research on the Dynamic Response of the Multi-Line Elevated Station with "Integral Station-Bridge System"
Reprinted from: *Buildings* 2024, *14*, 758, https://doi.org/10.3390/buildings14030758 122

Yuhong Xie, Zhou Cao and Jian Yu
Effect of Soil Anisotropy on Ground Motion Characteristics
Reprinted from: *Buildings* 2023, *13*, 3017, https://doi.org/10.3390/buildings13123017 139

Guifeng Zhao, Meng Wang, Ying Liu and Meng Zhang
Dynamic Response of Transmission Tower-Line Systems Due to Ground Vibration Caused by High-Speed Trains
Reprinted from: *Buildings* 2023, *13*, 2884, https://doi.org/10.3390/buildings13112884 156

Banglu Xi, Guangzi Li and Xiaochuan Chen
Experimental Study on Bearing Behavior and Soil Squeezing of Jacked Pile in Stiff Clay
Reprinted from: *Buildings* 2023, *13*, 2609, https://doi.org/10.3390/buildings13102609 186

About the Editors

Chao He

Dr. Chao He undertook research at McMaster University during the period from 2017 to 2020 and was enlisted in Tongji University's "Young Talent Program" in 2020. Currently, he serves as a Distinguished Researcher at the School of Transportation Engineering, Tongji University. His research primarily concentrates on the dynamics of railway transportation systems and computational methods for modeling the interaction between soil and structural components.

Wenbo Tu

Dr. Wenbo Tu is a Professor, affiliated with East China Jiaotong University. He was enlisted in the 7th China Association for Science and Technology Young Talent Support Program. He holds the positions of Leading Academic and Technical Lead in Jiangxi Province and is recognized as a high-level talent in Jiangxi Province.

Chao Zou

Dr. Chao Zou, an Associate Professor at Guangdong University of Technology, focuses on the dynamics of vibration and noise in railway transportation, road traffic, and Transit-Oriented Development (TOD). He has contributed to numerous studies assessing the environmental impacts of vibration and noise from subway vehicle depots in cities such as Guangzhou, Shenzhen, Wuhan, Foshan, Hangzhou, Nanjing, and Changsha.

Yunlong Guo

Dr. Yunlong Guo is working on AI and digital twin-based automated technology to achieve railway circularity. His research focuses on railway geotechnics and railway structure monitoring with a focus on cutting-edge methods (DIC, PIV, GPR, image analysis) and maglev-derived system development. He also works on applying new materials and designs to railway tracks, e.g., recycled rubber chips, recycled ballast, and sleeper design, and railway inspections using ground-penetrating radar and track-inspecting trains.

Preface

The growing demand for sustainable transportation has led to a significant interest in developing rail transit networks for both intra-and intercity travel. This shift towards more environmentally friendly modes of transport is driven by the need to reduce carbon emissions, alleviate traffic congestion, and promote efficient urban mobility. Rail systems are often viewed as a viable solution due to their ability to move large numbers of passengers with lower energy consumption compared with road vehicles.

However, alongside these advancements in rail infrastructure, train-induced vibrations that are transmitted through soils to nearby buildings have become widely recognized as environmental concerns. These vibrations can stem from various sources, such as the movement of trains on tracks, wheel–rail interactions, and even operational factors such as acceleration and braking. The transmission of these vibrations through different soil types can lead to significant negative influences on nearby structures, sensitive equipment housed within those structures, and the comfort levels of residents living in proximity to railway lines.

In addition to train-induced vibrations, natural disasters such as earthquakes pose serious threats that can have substantial physical impacts on buildings. Earthquakes generate ground motions that may compromise structural integrity, leading not only to economic losses but also to potential human casualties and injuries. The unpredictability associated with seismic events necessitates robust engineering solutions that are aimed at enhancing buildings' resilience against such forces.

Consequently, this reprint will focus on several critical aspects related to these phenomena: it will explore advanced modeling methods that accurately simulate the propagation characteristics of both train-induced vibrations and seismic waves as they interact with various soil types and building foundations. Understanding how these vibrational forces propagate is essential for designing effective mitigation strategies that minimize their impact.

Furthermore, this reprint will delve into innovative techniques for vibration mitigation, which may include passive damping systems or active control measures that are designed specifically for rail transit environments. It will also consider best practices in construction design aiming at improving structural resilience against the dynamic loads caused by both trains and seismic activity.

By addressing these topics comprehensively—ranging from theoretical models through practical applications—this reprint aims not only to contribute valuable insights into current research but also provide guidance for engineers and policymakers who are involved in urban planning and infrastructure development related to sustainable transportation systems.

Chao He, Wenbo Tu, Chao Zou, and Yunlong Guo
Guest Editors

Article

Building Vibration Measurement and Prediction during Train Operations

Lingshan He [1] and Ziyu Tao [2,*]

[1] Guangzhou Urban Planning & Design Survey Research Institute Co., Ltd., Guangzhou 510000, China; zacharyhe@foxmail.com
[2] Department of Civil and Environmental Engineering, The Hong Kong Polytechnic University, Hong Kong 999077, China
* Correspondence: zi-yu.tao@polyu.edu.hk

Abstract: Urban societies face the challenge of working and living in environments filled with vibration caused by transportation systems. This paper conducted field measurements to obtain the characteristics of vibration transmission from soil to building foundations and within building floors. Subsequently, a prediction method was developed to anticipate building vibrations by considering the soil and structure interaction. The rigid foundation model was simplified into a foundation–soil system connected via spring damping, and the building model is based on axial wave transmission within the columns and attached floors. Building vibrations were in response to measured input vibration levels at the ground and were validated through field measurements. The influence of different building heights on soil and structure vibration propagation was studied. The results showed that the predicted vibrations match well with the measured vibrations. The proposed prediction model can reasonably predict the building vibration caused by train operations. The closed-form method is an efficient tool for predicting floor vibrations prior to construction.

Keywords: train-induced vibration; soil–structure interaction; prediction model; vibration measurement

Citation: He, L.; Tao, Z. Building Vibration Measurement and Prediction during Train Operations. *Buildings* **2024**, *14*, 142. https://doi.org/10.3390/buildings14010142

Academic Editor: Shaohong Cheng

Received: 21 December 2023
Revised: 29 December 2023
Accepted: 5 January 2024
Published: 6 January 2024

Copyright: © 2024 by the authors. Licensee MDPI, Basel, Switzerland. This article is an open access article distributed under the terms and conditions of the Creative Commons Attribution (CC BY) license (https://creativecommons.org/licenses/by/4.0/).

1. Introduction

Transport-oriented development (TOD) has been widely promoted and applied in urban rail transit infrastructure, including lines, stations, and metro depots, improving the development layout of the subway, community, and industry [1,2]. This procedure allows the surrounding properties to fully enjoy the dividends of rail transit and further shortens the distance between rail transit and buildings. However, as the distance between the vibration source and the building is reduced, the impact of the train operation on the building's vibration is further intensified, which has become one of the core problems to be controlled in the development of rail transit.

The rapid and accurate prediction of long-term environmental vibration caused by train operation and effective guidance for comprehensive vibration reduction and isolation design are the basis for promoting the rapid and benign development of rail transit properties [3,4]. To accurately calculate the building vibration effects, it is necessary to find out the propagation mechanism of vibration in the vibration source, soil, and building. The dynamic interaction between soil and structure is the key factor in determining vibration propagation.

Most existing models usually ignore the coupling between the vibration source and the building [5–7]. In structural design, the boundary condition is usually a fixed base, but in practice, the foundation soil is not completely rigid. Considering the subsoil, foundation structure, and superstructure as a unified dynamic system, the actual stiffness of the subsoil is introduced by considering the deformation ability, and the natural frequency of the system will be reduced [8]. Therefore, considering the dynamic coupling between foundation soil and structure has practical significance in the vibration response of building structures [9–11]. Kuo et al. [12] characterized the dynamic coupling between soil and structure through coupling loss, and defined it as the difference in vibration levels inside and outside the building. Francois et al. [13] studied the influence of dynamic coupling between soil and structure on adjacent buildings and showed that assuming that the building base and the incident wave field movement were consistent could distort calculated vibrations. Coulier et al. [14] showed that when the distance between the vibration source and the building is smaller than the wavelength in the soil, the coupling effect between the soil and the shallow foundation of the building is very significant, and the interaction with the pile foundation is expected to be greater.

The dynamic coupling between soil and structure is investigated through the global analysis method or substructure method [15–17]. The global analysis method takes the structure and foundation soil as a complete dynamic system. However, due to the complexity of the problem, this method can only be realized via numerical simulation. Its disadvantage is that the modeling workload and calculation costs are large, making it challenging for application in large-scale engineering practice. The substructure method models the structure and the foundation independently and calculates the substructure model separately to solve the dynamic response of the superstructure by introducing the boundary conditions of the contact between the soil and the foundation, which offers the advantages of reduced computing memory requirements and higher efficiency [18]. Based on the substructure method, Fiala et al. [19] studied the vibration and radiated noise response of adjacent frame structures caused by train operations. The dynamic coupling between soil and structure was considered by coupling the impedance function of soil in the building base as the boundary condition. Lopes et al. [20] studied the response of soil stiffness to ground frame structure vibrations caused by tunnel train operations under the consideration of the dynamic coupling of soil and structure through numerical simulation. The dynamic coupling of soil and structure was simulated via the boundary element method and the simplified lumped parameter method, respectively. Considering the dynamic interaction between soil and structure is a crucial factor in accurately predicting structural vibration response. Ignoring this effect may result in overly conservative prediction outcomes.

Furthermore, the key to the method of calculating the environmental vibration of rail transit is to improve calculation accuracy, which requires a balance between time and accuracy [21–24]. The use of physical transfer functions to evaluate vibration effects is reliable and effective, and Connolly et al. [25] synthesized thousands of vibration data records from seven European countries to confirm this view. The physical properties of soil mass are the most important factors affecting vibration calculation. However, it is difficult to obtain a large number of the field test data needed to establish physical transfer functions in engineering practice, so the transfer function method has not been widely used in environmental vibration prediction. In the past decade, numerical modeling has been widely used in the field of environmental vibration [26]. Traditional numerical models need to integrate three components: vibration source, vibration transmission path, and building structure [27,28], which can usually be divided into two parts: the vehicle–track–soil coupling model and soil–structure coupling model. To find a balance between prediction accuracy and prediction time cost, scholars from various countries have conducted a lot of model simplification and algorithm optimization work, such as the finite-boundary element model [29], 2.5D finite element model [30,31], and finite difference model [32]. Due to the uncertainty of many parameters in numerical models, it is necessary to conduct repeated checks of the model through field test data [33] to calibrate the model parameters.

To solve this problem, a hybrid model vibration prediction method has gradually been developed that combines field testing and vibration theory techniques to avoid errors caused by model simplification. This approach also mitigates the issue of numerical models requiring an excessive number of calculation parameters, and providing a new idea for the environmental vibration prediction of rail transit [34]. Guo et al. [35] established a theoretical model of the building structure for vibration prediction based on the dynamic stiffness matrix method, allowing the calculation of the frequency domain response. The validity of the model was verified by using the measured data of the parking lot of Nanjing University Town. Sanayei et al. [36,37] described vertical vibration wave conduction by assembling single, independent structural column components and attached floor components, and established a one-dimensional impedance theoretical prediction model with measured building foundation vibration as input. The impedance model was verified via a four-story building scale model composed of aluminum columns and MDF in laboratory experiments and field tests at the Boston Hynes Convention Center. Zou et al. [38] extended the one-dimensional impedance model to a two-dimensional impedance model, taking into account the propagation of axial waves and bending waves in frame columns and shear walls. Later, to consider the influence of the transfer structure on the vibration propagation rule, Zou et al. [39] extended the two-dimensional impedance model to a three-dimensional impedance model under the condition of multiple-load input [40]. The model takes into account the different forms of vibration wave source in the transfer structure more reasonably, which is especially suitable for large high-rise buildings.

The current practice shows that the vibration response of buildings is affected by the type of building foundation and soil layer parameters, and the coupling effect of soil and structure affects the accuracy of the vibration prediction method. In this paper, the characteristics of vibration propagation loss between soil and structure are illustrated through field tests, the influence of train operation on building vibration is predicted by establishing a prediction model, and the influence of different building heights on the vibration propagation of soil and structure is studied. The objective is to develop a methodology involving ground vibration measurements for obtaining the ground vibration response, thereby enabling an accurate prediction of building vibration response during mitigation measure design prior to construction. This approach utilizes ground vibration as a model input, mitigating uncertainties associated with vibration propagation in tunnel and soil layers.

2. Field Measurement

2.1. Measurement Setup

The building that is close to the tunnel of Nanjing Metro Line 3 was selected to carry out the measurements. The horizontal distance between the building and the tunnel is about 40 m, and the buried depth is 17 m. The spatial relationship between the building and the tunnel is shown in Figure 1, and the photo of the building is shown in Figure 2.

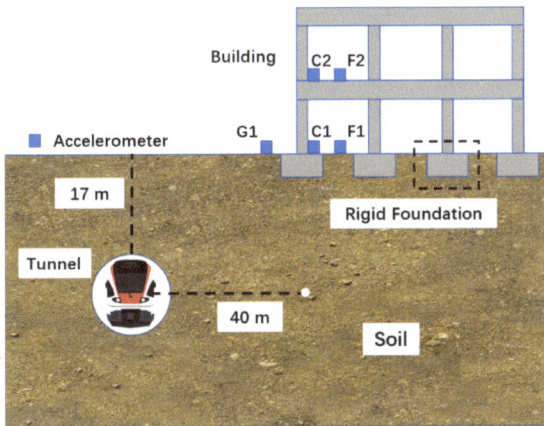

Figure 1. Spatial relationship between the building and the tunnel.

Figure 2. Measured 2-story building.

According to the geological drilling survey data, the soil is divided into four layers: plain fill, clay, muck, and silty clay. Table 1 lists the physical parameters of the soil layer.

Table 1. Physical parameters of the soil layer.

Name	Thickness (m)	Density (kg/m³)	Poisson's Ratio	Modulus of Elasticity (MPa)
Plain fill	2.1	16	0.45	120
Clay	1.3	18.4	0.46	130
Muck	8.4	17.5	0.48	300
Silty clay	10.6	19.3	0.45	380

The building is a reinforced concrete-frame structure with 2 floors and a height of 8 m, which relies on structural columns for load-bearing with a size of 0.4 × 0.4 m². The building foundation is a rigid foundation with a base size of 1.2 × 1.2 m². Table 2 lists the dimensions and dynamics parameters of the building structure, where h, t, w, E, and ρ are the height, thickness, width, elastic modulus, density of the structural components, respectively. The damping ratio and Poisson's ratio of the reinforced concrete structural components are 0.02 and 0.2, respectively.

Table 2. Physical parameters of the building structure.

Structural Component	h (m)	t (m)	w (m)	E (Gpa)	ρ (kg/m³)
Column	4	0.4	0.4	32.5	2500
Floor slab	-	0.1	-	30	2500
Rigid foundation	-	1.2	1.2	32.5	2500

A total of 4 measuring points were arranged inside the building and measuring point was arranged outside the building to collect the vibration acceleration caused by train operations, as shown in Figure 1. In total, 20 train pass-by events were recorded, with the trains traveling at approximately 50 km/h. The sampling frequency was set to 512 Hz. The JM3873 wireless acquisition system was selected as the accelerometer which is made by the Jing Ming Technology Yangzhou China was a working day with a relatively low pedestrian flow. During the measurements, the weather conditions were characterized by clear skies and low wind speeds.

Figure 3. Accelerometer.

2.2. Measurement Results

The time history of acceleration of each measuring point caused by train operations is separated from the records of continuous synchronous testing, and the typically measured time history of a train pass-by event is shown in Figure 4.

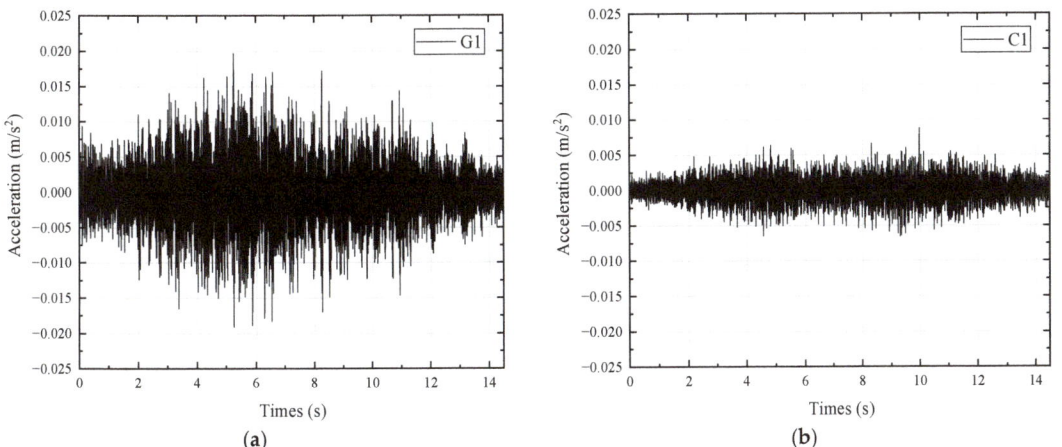

Figure 4. Measured typical time history of acceleration. (**a**) Ground G1; (**b**) 1st-floor C1.

The duration of ground and building vibrations caused by train operation is about 14.5 s. The ground vibration with a peak acceleration of about 0.02 m/s^2 is greater than the column vibration with a peak acceleration of about 0.008 m/s^2.

The measured acceleration is calculated and converted into a 1/3 octave band to obtain the vertical vibration acceleration level. The acceleration levels are presented in the form of an envelope diagram to show the influence range caused by different train pass-by events, as shown in Figure 5.

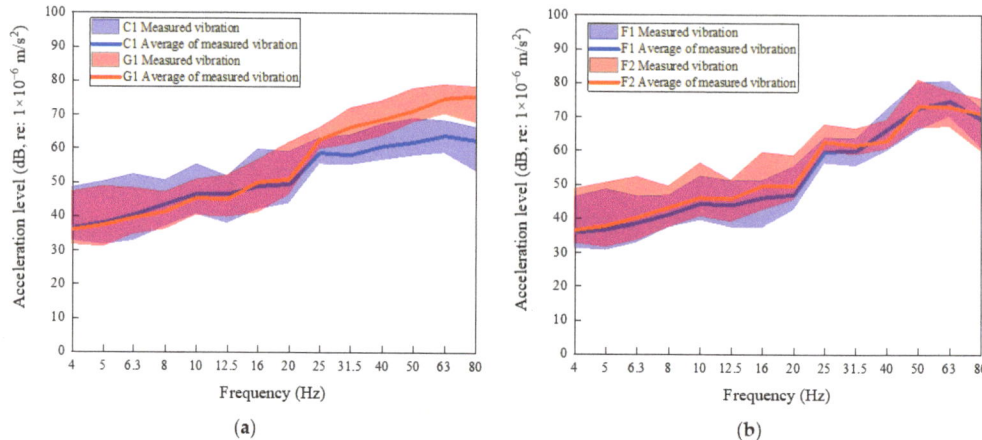

Figure 5. Measured acceleration levels of a 1/3 octave band frequency. (**a**) Comparison of ground vibration and column vibration on 1st floor; (**b**) comparison of floor vibration on 1st floor and 2nd floor.

Comparing the location of G1 and C1 in Figure 5a, the vibration loss of vibration propagating from the soil to the structure is more obvious, and it tends to decrease significantly at a high frequency, while the attenuation of middle and low frequencies is not obvious. The dividing line is at 25 Hz, and the vibration acceleration level above 25 Hz decreases significantly.

Comparing the location of F1 and F2, with the increase in building height, the difference in vibration acceleration in the frequency domain is not obvious. In the propagation process, the characteristic frequencies are consistent, which is because the floor height is small and the building system tends to be excited as a whole, and the difference in vibration acceleration between the first and second floors is small. In addition, due to the small mass and volume of the building foundation compared with those of the soil, it can be regarded as the interaction between the building structure as a whole and the soil.

3. Vibration Prediction and Validation

To predict the building vibrations, this paper presents a hybrid prediction method based on the measured ground vibration and the prediction model of a building structure that considers the soil–structure interaction. The purpose is that the vibration response of the new building can be calculated via the prediction model of a building structure after the measured vibrations of ground soil surface are obtained. The advantage of this method is that tunnels and different soil layers do not need to be considered in the model, because the foundation of the building is buried in the topsoil, and the vibration of the topsoil surface is obtained via measurement. Hence, a prerequisite of this method is to obtain the measured data of ground vibration.

This model can greatly improve the efficiency of solving the coupling loss between the rigid foundation, soil, and structure. Firstly, the coupling model at the interface between a rigid foundation and soil is simplified as a spring-damping coupling component. The soil layer, where the rigid foundation is located, is considered homogeneous soil, and its parameters aligned with those of the surface soil layer. Subsequently, by adding the building structure to the top of the foundation and considering the influence of the building structure on the coupling loss, the coupling loss of the soil–foundation system under the influence of the building structure is obtained. Finally, the building foundation vibration is solved according to the in situ measured ground vibration and soil–foundation coupling loss, and the building response is solved based on the building prediction model. The model diagram is shown in Figure 6.

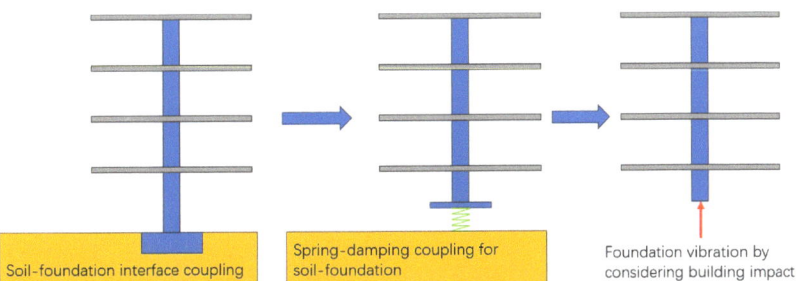

Figure 6. Schematic diagram of vibration prediction model for rigid foundation and building.

3.1. Prediction Model

3.1.1. Soil–Structure Interaction Model

For a rigid foundation, where the contact area between the foundation and soil is small, the dynamic interaction between them is weaker than the inertial interaction of the superstructure, and the influence of soil layer interface on the soil–structure interaction is lower. Therefore, the rigid foundation model can be simplified into a foundation–soil system connected via spring damping. The empirical formula based on boundary element research by Auersch [41] can describe the dynamic behavior of the shallow foundation and soil, and its expression is as follows:

$$k \approx 3.4 G_s \sqrt{A_s} \tag{1}$$

$$c \approx 1.6 \rho_s v_s A_s \tag{2}$$

where A_s is the area of the foundation; G_s, ρ_s, and v_s are the shear modulus, density, and shear wave velocity of soil, respectively. k and c are the stiffness and damping of the foundation, respectively.

Due to the large proportion of building quality being in the building–rigid foundation system, foundation modeling cannot be conducted at the base–structure node without ignoring the influence of the building. For the load-bearing column, the dynamical behavior can be described by the following governing equation:

$$-E_c A_c u''(x_s) = k(u - u_0) + c(\dot{u} - \dot{u}_0) \tag{3}$$

where primes and dots denote spatial and time derivatives. u, E_c, and A_c are the vertical displacement, modulus of elasticity, and area of the column, respectively. u_0 is a free-field displacement.

Its solution can be expressed by the following equation:

$$\frac{u_c}{u_0} = \frac{\cos a\zeta}{\cos a} \frac{1}{1 + iq \tan a} \tag{4}$$

where u_c is column displacement. i is equal to $\sqrt{-1}$. $\zeta = x/H$ is the position relative to building height, H. x is the distance from the roof. $a = 2\pi f H / v_L$ is the normalized frequency parameter. f is the frequency. $v_L = E_c / \rho_c$ is the wave speed of the column. ρ_c is the density of the column. The expression for parameter q is defined as follows:

$$q = \frac{\sqrt{E_c \rho_c} A_c}{1.6 \sqrt{G_s \rho_s} A_s} \tag{5}$$

3.1.2. Building Model

The building model mainly considers the propagation of axial vibration waves in the building, specifically vertical vibrations. Usually, the vibration caused by train operations propagates upward through vertical load-bearing structures, such as structural columns. Since medium- and high-frequency vibration waves can effectively attenuate in the floor, their influence is limited to a short distance from the junction of the vertical load-bearing structure and the floor. Consequently, vibrations have a minimal impact on the mutual propagation of each vertical load-bearing structure. Thus, the propagation of vibration in a building can be considered upward propagation along a single vertical bearing structure with the attached floor, and the vibration of the floor is the sum of the vibration energy carried by each vertical bearing structure. The vibration propagation model can only consider a single vertical bearing structure and each floor. For each component of the building structure, the rod and the plate element can be used to simulate the structural column and the floor slab.

The load-bearing column is simulated as a finite-length rod element with two nodes at a specific frequency, and both ends are connected to adjacent floors. Considering the axial wave propagation in the load-bearing column, the impedance matrix representing the relationship between force and velocity is shown as follows:

$$z_c = \frac{E_c(1+i\eta_c)A_c\beta_c}{iw\sin(\beta_c l_c)} \begin{bmatrix} \cos(\beta_c l_c) & -1 \\ -1 & \cos(\beta_c l_c) \end{bmatrix} \quad (6)$$

where $\beta_c = \omega\sqrt{\frac{\rho_c}{E_c(1+i\eta_c)}}$ is the axial wavenumber, ω is the circular frequency, and l_c is the length of the column. E_c and η_c are the elastic modulus and loss factor of the column.

The floor slab and the vertical load-bearing structure are assumed to be rigidly connected and are simulated as infinite plate elements with the same terms under a specific frequency of point load. The impedance expression of the floor is as follows:

$$z_f = 8\sqrt{\frac{\rho_f E_f(1+j\eta_f)h_f^4}{12(1-v_f^2)}} \quad (7)$$

where ρ_f, h_f, v_f, E_f, and η_f are the material density, thickness, Poisson's ratio, elastic modulus, and loss factor of the floor, respectively.

3.1.3. System Assembly

The system model is composed by summing the forces and reactions of each component on the node in the vertical direction. Conceptually, this is achieved by considering the force acting on a rigid massless block with an unknown vertical vibration at each node, where the force of each component is achieved by acting on a different surface of the mass block. If there is a vertical external force, it can be applied at each node of the floor slab. The train excitation input is characterized by the force at the bottom column of building on the ground. Hence, to obtain this exciting force, it is necessary to obtain the ground vibration measured on site, and use Equation (4) to obtain the building column vibration at the bottom of the building. Using this building column vibration as the input of the building model, the vibration of each floor of the building can be obtained by solving the following formula:

$$F = ZV \quad (8)$$

where F and V are the external forces and velocity vectors of each floor, respectively; Z is

3.2. Model Validation

3.2.1. Transmission Ratio from Soil to Structure

To better compare the effects of soil–structure coupling loss, all time domain signals from ground measuring point G1 and indoor measuring point C4 are analyzed via fast Fourier transform, and the transmission ratio of vibration from soil to building is calculated. On this basis, the proposed model is used to calculate the transmission ratio of vibration from soil to building, and the comparison between the calculated coupling loss and the measured coupling loss is shown in Figure 7a. The coupling loss curve exhibits a peak value around 10–20 Hz, likely attributed to the formant arising from the overall vibration of the building structure and soil. The coupling loss curve moves from a low frequency to a high frequency, and vibration acceleration attenuation moves from a small to large trend. The predicted coupling loss curves closely align with the measured results.

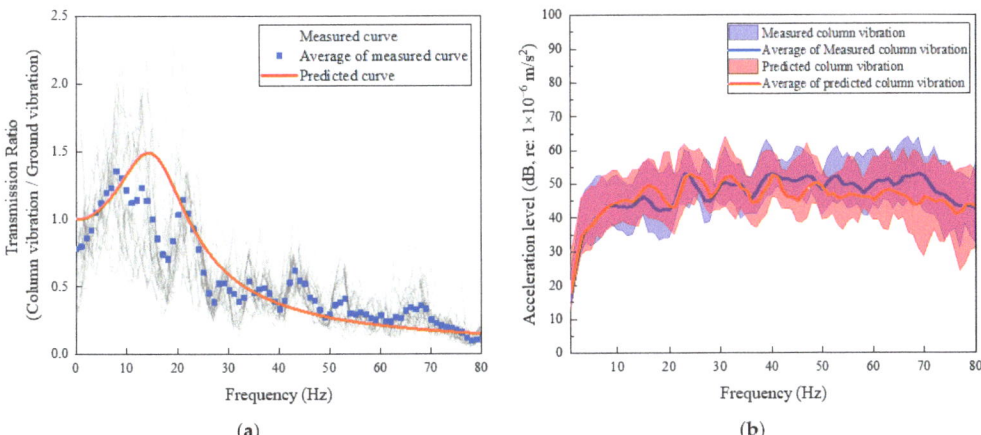

Figure 7. Comparison of measured vibration and predicted vibration from soil to column. (**a**) Transmission ratio; (**b**) column vibration on 1st floor.

By combining the predicted coupling loss curve with the measured ground vibration, the predicted foundation vibration can be obtained, as shown in Figure 7b. The predicted column vibration on the 1st floor is close to the measured vibration. It can be considered that the soil and structure coupling loss can be effectively predicted based on the measured ground vibration and the proposed prediction model.

3.2.2. Building Vibration

For the indoor vibration prediction of the building, measuring points C2 and F2 are selected to represent the column vibration and center of floor vibration, respectively. The envelope diagram and average value of measured and predicted vibrations are compared, as shown in Figure 8.

The predicted vibrations match well with the measured vibrations, and the envelope graph remains consistent across all train pass-by events. The largest discrepancy between the measured average vibration and the predicted average vibration for the column is observed at 80 Hz, exhibiting a difference of 9.2%. Similarly, in terms of floor vibration, the most significant deviation between the measured and predicted average vibrations occurs at 25 Hz, with a disparity of 4.8%. It can be considered that the proposed prediction model can reasonably predict the building vibration caused by train operations.

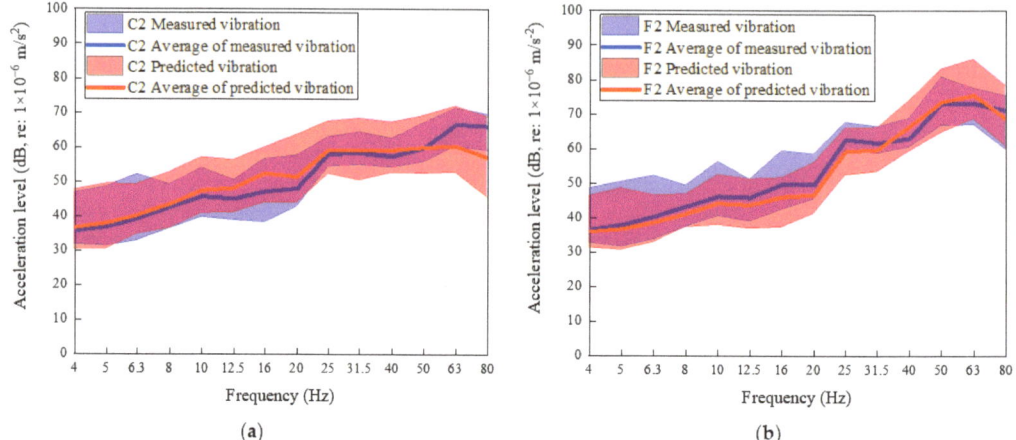

Figure 8. Comparison of measured vibration and predicted vibration within the building (**a**) at the column; (**b**) at the floor center.

4. Discussion

Based on the proposed model, the influence of different building heights on soil–structure interaction is studied by changing the number of building floors. When considering different numbers of floors, since rigid foundations are usually used for shorter buildings, the effects of building height on soil–structure coupling loss are calculated by selecting floors 2, 5, and 8, as shown in Figure 9.

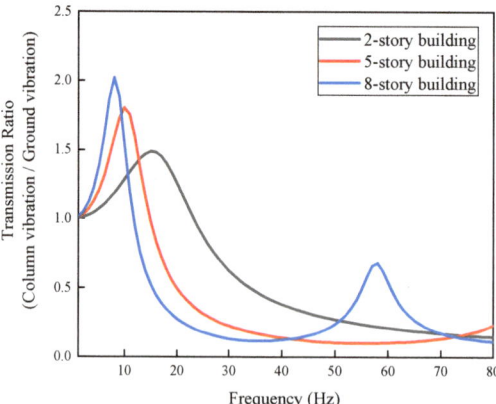

Figure 9. Transmission ratio of soil and structure.

With the increase in the number of floors, the interaction between soil and structure shows an obvious inertia effect. The peak value of the coupling loss curve appears from 20 Hz for a two-story building to 10 Hz for an eight-story building due to the building mass increases. This resonance is attributed to the eigenfrequency of the rigidly founded building, this being approximately 30 Hz, resulting in a 60 Hz elastic building resonance where the foundation and roof vibrate and amplify in an anti-phase [41]. This frequency is

Figure 10 compares the difference between the vibration of each floor inside the building and the ground vibration. The vibration of the two-story building shows a gradually increasing trend from the ground floor to the upper floor, but the increase is very small. In the range of 4–20 Hz, the indoor vibrations are greater than the ground vibration, and the vibration amplification peak is at 16 Hz with a value of about 3 dB. In the range greater than 20 Hz, the ground vibration is greater than indoor vibrations, and the attenuation peak is reached at 80 Hz with about 17 dB.

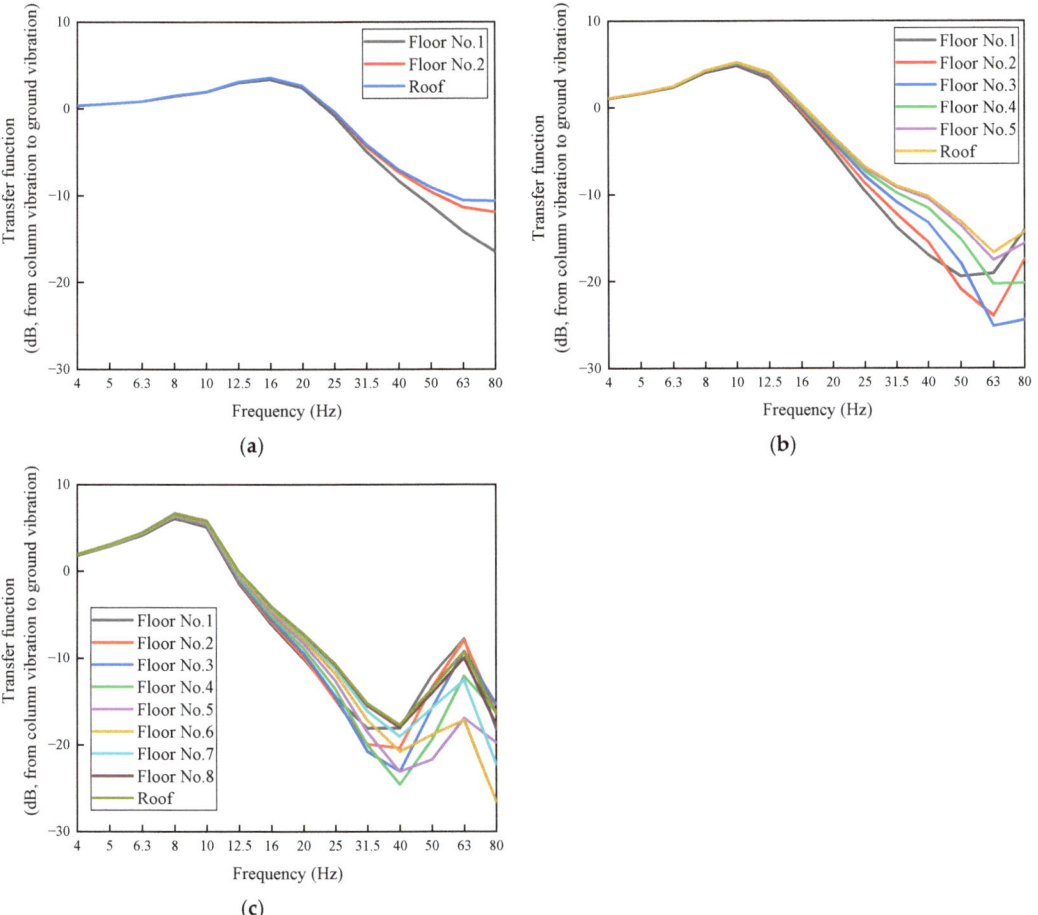

Figure 10. Difference between each floor vibration inside the building and ground vibration. (**a**) Two-story building; (**b**) five-story building; (**c**) eight-story building.

For the five-story building, the propagation law in the building changes, and the overall performance is that the vibration acceleration level of the first floor to the fifth floor decreases first and then increases in the frequency band of 50–80 Hz, and in the range of 4–50 Hz, the acceleration level of the top floor is greater than that of the first floor. In the range of 4–16 Hz, the indoor vibration is greater than the ground vibration with a peak value of 4 dB at 10 Hz. In a range greater than 16 Hz, the ground vibration is greater than the indoor vibration.

For the eight-story building, the low-frequency peak is further increased, and the middle- and high-frequency attenuation capacity is further enhanced. For the propagation in the building, with the increase in the number of floors, the propagation law shows that the vibration acceleration level decreases first and then increases from the 1st floor to the 8th floor. In the range of 4–12.5 Hz, the indoor vibrations are greater than the ground vibration. In range greater than 12.5 Hz, the ground vibration is greater than the indoor vibrations.

In frame structure buildings, the axial impedance of the load-bearing column far exceeds the bending wave impedance of the floor. Impedance differences appear at the junction of columns and floors on each floor of the building, resulting in a discontinuity of axial impedance. Due to this impedance discontinuity between the high-impedance column and the low-impedance floor, only a small amount of axial upward transmission energy is transferred to the floor. The axial movement of the column in the vertical direction generates floor-bending waves, which propagate outward from the junction of the column and the floor to the outer diameter and are finally dissipated by floor damping. Therefore, in the above analysis of three different building heights, when the number of floors is small, the axial impedance of the column is much larger than that of the floor in the building system. In the propagation process, the vibration propagates along the axis, and the damping of the column itself is not enough to attenuate the vibration, resulting in the vibration energy propagating to the top layer and then being superimposed with the incident wave, resulting in the amplification of the vibration along each layer. As the number of floors increases to a certain height, the damping of the entire building system and the dissipation of each floor will partially attenuate the vibration propagating in the middle floor, so that the vibration decreases first and then increases with building height.

The proposed method also has a limitation; namely, its applicability is restricted to the vibration prediction in newly constructed buildings along operational subway lines, as it necessitates obtaining ground vibrations through field measurements. Furthermore, this method is exclusively suitable for buildings with rigid or shallow foundations due to the previous research on deep foundations revealing distinct soil–structural coupling losses in deep foundation systems, such as pile foundations [42,43].

5. Conclusions

Aiming at the problem of building vibration caused by train operations, a building vibration prediction method considering the coupling loss of soil and structure is established in this paper. Through the field test of the ground and a two-story building that is close to the vibration source, the vibration propagation law in the ground through the building foundation and within the building is obtained, and the validity of the model is verified via measurement. The propagation rules of vibration in buildings with different floors are studied, and the following conclusions are obtained:

(1) The predicted vibration is in good agreement with the measured vibration, and the established prediction model can reasonably predict the vibration of a building with a rigid foundation caused by train operation once the ground vibration can be obtained via measurement.
(2) The coupling loss is obvious as the vibration propagates from the soil to the rigid foundation, and the vibration tends to decrease significantly with an increase in frequency.
(3) For s rigid building foundation, the coupling effect of soil and structure causes a low-frequency building–soil resonance and a high-frequency amplitude attenuation. This effect becomes more pronounced with higher building heights, leading to a more significant change in high-frequency amplitude.

Author Contributions: Conceptualization, L.H. and Z.T.; methodology, Z.T.; validation, Z.T.; formal analysis, L.H.; investigation, L.H.; resources, L.H.; data curation, L.H.; writing—original draft preparation, L.H.; writing—review and editing, Z.T.; visualization, Z.T.; supervision, Z.T.; project administration, Z.T.; funding acquisition, Z.T. All authors have read and agreed to the published version of the manuscript.

ber 51908139.

Data Availability Statement: The data presented in this study are available on request from the corresponding author.

Acknowledgments: The graduate students of Zixiong Shen, Jiahao Hu, Yitao Qiu, and Xuming Li are appreciated for helping us to complete the field measurements.

Conflicts of Interest: Author Lingshan He was employed by the Guangzhou Urban Planning & Design Survey Research Institute Co., Ltd. The remaining authors declare that the research was conducted in the absence of any commercial or financial relationships that could be construed as a potential conflict of interest.

References

1. Zou, C.; Wang, Y.; Moore, J.A.; Sanayei, M. Train-induced field vibration measurements of ground and over-track buildings. *Sci. Total Environ.* **2017**, *575*, 1339–1351. [CrossRef]
2. Zou, C.; Wang, Y.; Wang, P.; Guo, J. Measurement of ground and nearby building vibration and noise induced by trains in a metro depot. *Sci. Total Environ.* **2015**, *536*, 761–773. [CrossRef]
3. He, C.; Zhou, S.; Guo, P. Mitigation of Railway-Induced Vibrations by Using Periodic Wave Impeding Barriers. *Appl. Math. Model.* **2022**, *105*, 496–513. [CrossRef]
4. Liu, Q.; Li, X.; Zhang, X.; Zhou, Y.; Chen, Y.F. Applying Constrained Layer Damping to Reduce Vibration and Noise from a Steel-Concrete Composite Bridge: An Experimental and Numerical Investigation. *J. Sandw. Struct. Mater.* **2020**, *22*, 1743–1769. [CrossRef]
5. Li, X.; Chen, Y.; Zou, C.; Chen, Y. Train-Induced Vibration Mitigation Based on Foundation Improvement. *J. Build. Eng.* **2023**, *76*, 107106. [CrossRef]
6. Liu, Q.; Thompson, D.J.; Xu, P.; Feng, Q.; Li, X. Investigation of Train-Induced Vibration and Noise from a Steel-Concrete Composite Railway Bridge Using a Hybrid Finite Element-Statistical Energy Analysis Method. *J. Sound Vib.* **2020**, *471*, 115197. [CrossRef]
7. Liu, W.; Liang, R.; Zhang, H.; Wu, Z.; Jiang, B. Deep Learning Based Identification and Uncertainty Analysis of Metro Train Induced Ground-Borne Vibration. *Mech. Syst. Signal Process.* **2023**, *189*, 110062. [CrossRef]
8. He, C.; Zhou, S.; Di, H.; Guo, P.; Xiao, J. Analytical Method for Calculation of Ground Vibration from a Tunnel Embedded in a Multi-Layered Half-Space. *Comput. Geotech.* **2018**, *99*, 149–164. [CrossRef]
9. Colaço, A.; Barbosa, D.; Alves Costa, P. Hybrid Soil-Structure Interaction Approach for the Assessment of Vibrations in Buildings Due to Railway Traffic. *Transp. Geotech.* **2022**, *32*, 100691. [CrossRef]
10. Zhou, S.; He, C.; Guo, P.; Yu, F. Dynamic Response of a Segmented Tunnel in Saturated Soil Using a 2.5-D FE-BE Methodology. *Soil Dyn. Earthq. Eng.* **2019**, *120*, 386–397. [CrossRef]
11. Colaço, A.; Alves Costa, P.; Amado-Mendes, P.; Calçada, R. Vibrations Induced by Railway Traffic in Buildings: Experimental Validation of a Sub-Structuring Methodology Based on 2.5D FEM-MFS and 3D FEM. *Eng. Struct.* **2021**, *240*, 112381. [CrossRef]
12. Kuo, K.A.; Papadopoulos, M.; Lombaert, G.; Degrande, G. The Coupling Loss of a Building Subject to Railway Induced Vibrations: Numerical Modelling and Experimental Measurements. *J. Sound Vib.* **2019**, *442*, 459–481. [CrossRef]
13. François, S.; Pyl, L.; Masoumi, H.R.; Degrande, G. The Influence of Dynamic Soil–Structure Interaction on Traffic Induced Vibrations in Buildings. *Soil Dyn. Earthq. Eng.* **2007**, *27*, 655–674. [CrossRef]
14. Coulier, P.; Lombaert, G.; Degrande, G. The Influence of Source–Receiver Interaction on the Numerical Prediction of Railway Induced Vibrations. *J. Sound Vib.* **2014**, *333*, 2520–2538. [CrossRef]
15. Auersch, L. Response to Harmonic Wave Excitation of Finite or Infinite Elastic Plates on a Homogeneous or Layered Half-Space. *Comput. Geotech.* **2013**, *51*, 50–59. [CrossRef]
16. Auersch, L. Dynamic Behavior of Slab Tracks on Homogeneous and Layered Soils and the Reduction of Ground Vibration by Floating Slab Tracks. *J. Eng. Mech.* **2012**, *138*, 923–933. [CrossRef]
17. Auersch, L. Wave Propagation in the Elastic Half-Space Due to an Interior Load and Its Application to Ground Vibration Problems and Buildings on Pile Foundations. *Soil Dyn. Earthq. Eng.* **2010**, *30*, 925–936. [CrossRef]
18. Clot, A.; Arcos, R.; Romeu, J. Efficient three-dimensional building-soil model for the prediction of ground-borne vibrations in buildings. *J. Struct. Eng.* **2017**, *143*, 04017098. [CrossRef]
19. Fiala, P.; Degrande, G.; Augusztinovicz, F. Numerical Modelling of Ground-Borne Noise and Vibration in Buildings Due to Surface Rail Traffic. *J. Sound Vib.* **2007**, *301*, 718–738. [CrossRef]
20. Lopes, P.; Alves Costa, P.; Calçada, R.; Silva Cardoso, A. Influence of Soil Stiffness on Building Vibrations Due to Railway Traffic in Tunnels: Numerical Study. *Comput. Geotech.* **2014**, *61*, 277–291. [CrossRef]
21. Papadopoulos, M.; François, S.; Degrande, G.; Lombaert, G. The Influence of Uncertain Local Subsoil Conditions on the Response of Buildings to Ground Vibration. *J. Sound Vib.* **2018**, *418*, 200–220. [CrossRef]
22. Edirisinghe, T.L.; Talbot, J.P. The Significance of Source-Receiver Interaction in the Response of Piled Foundations to Ground-Borne Vibration from Underground Railways. *J. Sound Vib.* **2021**, *506*, 116178. [CrossRef]
23. Galvín, P.; Mendoza, D.L.; Connolly, D.P.; Degrande, G.; Lombaert, G.; Romero, A. Scoping Assessment of Free-Field Vibrations Due to Railway Traffic. *Soil Dyn. Earthq. Eng.* **2018**, *114*, 598–614. [CrossRef]
24. López-Mendoza, D.; Romero, A.; Connolly, D.P.; Galvín, P. Scoping Assessment of Building Vibration Induced by Railway Traffic. *Soil Dyn. Earthq. Eng.* **2017**, *93*, 147–161. [CrossRef]

25. Connolly, D.P.; Alves Costa, P.; Kouroussis, G.; Galvin, P.; Woodward, P.K.; Laghrouche, O. Large Scale International Testing of Railway Ground Vibrations across Europe. *Soil Dyn. Earthq. Eng.* **2015**, *71*, 1–12. [CrossRef]
26. He, C.; Zhou, S.; Guo, P.; Di, H.; Zhang, X. Modelling of Ground Vibration from Tunnels in a Poroelastic Half-Space Using a 2.5-D FE-BE Formulation. *Tunn. Undergr. Space Technol.* **2018**, *82*, 211–221. [CrossRef]
27. Zhou, S.; He, C.; Di, H.; Guo, P.; Zhang, X. An Efficient Method for Predicting Train-Induced Vibrations from a Tunnel in a Poroelastic Half-Space. *Eng. Anal. Bound. Elem.* **2017**, *85*, 43–56. [CrossRef]
28. Ma, M.; Liu, W.; Qian, C.; Deng, G.; Li, Y. Study of the Train-Induced Vibration Impact on a Historic Bell Tower above Two Spatially Overlapping Metro Lines. *Soil Dyn. Earthq. Eng.* **2016**, *81*, 58–74. [CrossRef]
29. Galvín, P.; Domínguez, J. Experimental and Numerical Analyses of Vibrations Induced by High-Speed Trains on the Córdoba–Málaga Line. *Soil Dyn. Earthq. Eng.* **2009**, *29*, 641–657. [CrossRef]
30. Liang, X.; Yang, Y.B.; Ge, P.; Hung, H.-H.; Wu, Y. On Computation of Soil Vibrations Due to Moving Train Loads by 2.5D Approach. *Soil Dyn. Earthq. Eng.* **2017**, *101*, 204–208. [CrossRef]
31. Cheng, G.; Feng, Q.; Sheng, X.; Lu, P.; Zhang, S. Using the 2.5D FE and Transfer Matrix Methods to Study Ground Vibration Generated by Two Identical Trains Passing Each Other. *Soil Dyn. Earthq. Eng.* **2018**, *114*, 495–504. [CrossRef]
32. Verbraken, H.; Lombaert, G.; Degrande, G. Verification of an Empirical Prediction Method for Railway Induced Vibrations by Means of Numerical Simulations. *J. Sound Vib.* **2011**, *330*, 1692–1703. [CrossRef]
33. Yang, J.; Zhu, S.; Zhai, W.; Kouroussis, G.; Wang, Y.; Wang, K.; Lan, K.; Xu, F. Prediction and Mitigation of Train-Induced Vibrations of Large-Scale Building Constructed on Subway Tunnel. *Sci. Total Environ.* **2019**, *668*, 485–499. [CrossRef]
34. Lopes, P.; Ruiz, J.F.; Alves Costa, P.; Medina Rodríguez, L.; Cardoso, A.S. Vibrations inside Buildings Due to Subway Railway Traffic. Experimental Validation of a Comprehensive Prediction Model. *Sci. Total Environ.* **2016**, *568*, 1333–1343. [CrossRef]
35. Guo, T.; Cao, Z.; Zhang, Z.; Li, A. Numerical Simulation of Floor Vibrations of a Metro Depot under Moving Subway Trains. *J. Vib. Control* **2018**, *24*, 4353–4366. [CrossRef]
36. Sanayei, M.; Zhao, N.; Maurya, P.; Moore, J.A.; Zapfe, J.A.; Hines, E.M. Prediction and Mitigation of Building Floor Vibrations Using a Blocking Floor. *J. Struct. Eng.* **2012**, *138*, 1181–1192. [CrossRef]
37. Sanayei, M.; Kayiparambil, P.A.; Moore, J.A.; Brett, C.R. Measurement and Prediction of Train-Induced Vibrations in a Full-Scale Building. *Eng. Struct.* **2014**, *77*, 119–128. [CrossRef]
38. Zou, C.; Moore, J.A.; Sanayei, M.; Wang, Y. Impedance Model for Estimating Train-Induced Building Vibrations. *Eng. Struct.* **2018**, *172*, 739–750. [CrossRef]
39. Zou, C.; Moore, J.A.; Sanayei, M.; Wang, Y.; Tao, Z. Efficient Impedance Model for the Estimation of Train-Induced Vibrations in over-Track Buildings. *J. Vib. Control* **2021**, *27*, 924–942. [CrossRef]
40. Zou, C.; Moore, J.A.; Sanayei, M.; Tao, Z.; Wang, Y. Impedance Model of Train-Induced Vibration Transmission Across a Transfer Structure into an Overtrack Building in a Metro Depot. *J. Struct. Eng.* **2022**, *148*, 04022187. [CrossRef]
41. Auersch, L. Building Response Due to Ground Vibration - Simple Prediction Model Based on Experience with Detailed Models and Measurements. *Int. J. Acoust. Vib.* **2010**, *15*, 101. [CrossRef]
42. Tao, Z.-Y.; Zou, C.; Yang, G.-R.; Wang, Y.-M. A semi-analytical method for predicting train-induced vibrations considering train-track-soil and soil-pile-building dynamic interactions. *Soil Dyn. Earthq. Eng.* **2023**, *167*, 107822. [CrossRef]
43. Li, X.; Chen, Y.; Zou, C.; Wu, J.; Shen, Z.; Chen, Y. Building coupling loss measurement and prediction due to train-induced vertical vibrations. *Soil Dyn. Earthq. Eng.* **2023**, *167*, 107644. [CrossRef]

Disclaimer/Publisher's Note: The statements, opinions and data contained in all publications are solely those of the individual author(s) and contributor(s) and not of MDPI and/or the editor(s). MDPI and/or the editor(s) disclaim responsibility for any injury to people or property resulting from any ideas, methods, instructions or products referred to in the content.

Article

Time Domain Nonlinear Dynamic Analysis of Vertically Loaded Tapered Pile in Layered Soils

Qiangqiang Shua [1,2], Kexing Liu [3], Jingkai Li [4] and Wenbo Tu [3,*]

[1] School of Architectural Engineering, Sichuan University of Arts and Science, Dazhou 635000, China
[2] Building Environment Engineering Technology Research Center in Dazhou, Sichuan University of Arts and Science, Dazhou 635000, China
[3] School of Transportation Engineering, East China Jiaotong University, Nanchang 330013, China
[4] China Huaxi Engineering Design & Construction Co., Ltd., Chengdu 610000, China
* Correspondence: wenbotu@ecjtu.edu.cn

Citation: Shua, Q.; Liu, K.; Li, J.; Tu, W. Time Domain Nonlinear Dynamic Analysis of Vertically Loaded Tapered Pile in Layered Soils. *Buildings* **2024**, *14*, 445. https://doi.org/10.3390/buildings14020445

Academic Editors: Humberto Varum and Fabrizio Gara

Received: 26 December 2023
Revised: 23 January 2024
Accepted: 30 January 2024
Published: 6 February 2024

Copyright: © 2024 by the authors. Licensee MDPI, Basel, Switzerland. This article is an open access article distributed under the terms and conditions of the Creative Commons Attribution (CC BY) license (https://creativecommons.org/licenses/by/4.0/).

Abstract: A simplified model is proposed for predicting the nonlinear dynamic response of vertically loaded tapered piles in the time domain, in which the tapered pile is divided into several frustum segments and the four-spring is used for the simulation of the soil–pile interaction. The differential equations for the tapered pile are given and solved by the finite difference method. The vertical dynamic response of a typical tapered pile is investigated, and the consistency of the computational results compared with the finite element results convincingly verifies the reliability of the proposed simplified model. Then, recommended segment numbers, considering the computational efficiency and accuracy requirements for the dynamic analysis of tapered piles, are given. And parametric studies are also carried out to investigate the effect of soil and pile parameters on the nonlinear dynamic response of the tapered pile. The results show that soil nonlinearity significantly affects the vertical dynamic characteristics of the tapered pile. And the tapered pile shows better dynamic characteristics than the cylindrical pile with the same volume and pile length. In addition, the properties of the soil along the upper part of the tapered pile have a more considerable effect on the dynamic response of the tapered pile. These results help to further improve the theory of nonlinear dynamic response analysis of tapered piles and promote its widespread application in engineering practice.

Keywords: tapered pile; time domain; vertical load; dynamic response

1. Introduction

Tapered piles are a novel type of pile that originated in the former Soviet Union in the 1970s. Compared to cylindrical piles of the same volume, calculated by the theoretical or numerical method used in engineering practices [1,2], the load-carrying capacity is increased by 0.5~2.5 times [3–6]. Due to its advantages over traditional pile foundations mentioned above, it is also used in structural foundations, slope retaining, and railway or highway weak foundation treatment projects [7–9]. Many researchers have investigated the vertical and lateral bearing capacity of the tapered pile. However, a limited understanding of the tapered pile–soil dynamic interaction mechanisms has affected the widespread use of tapered piles. Sudhendu and Ghosh [10] proposed a simplified analytical method for the dynamic characteristics of the vertically loaded tapered piles and gave the factors affecting the dimensionless stiffness and damping coefficients of the pile. Xie and Vaziri [11] established a mathematical model for the vertical harmonic vibration of tapered piles and verified it by field tests. Wu et al. [12] and Wang et al. [13] divided the tapered pile into several continuous stepped cylindrical segments and then solved the vertical vibration differential equation using the Rayleigh–Love bar theory to obtain the dynamic impedance of the pile. Nevertheless, all the above analyses were based on the assumption that the soil is a uniform elastic medium and that the soil is assumed to be an elastic material.

Actually, soil is distributed in a layered pattern and will show nonlinear characteristics when subjected to a large load. Limited studies have been conducted to calculate the nonlinear dynamic response of tapered piles considering the nonlinear interaction between the soil and the pile, and the effects of the soil stratification on the dynamic characteristics of the vertically loaded tapered pile were also not considered [14–17].

On the other hand, the tapered pile is a typical pile whose cross-section changes linearly with pile length, and the interaction between the soil and tapered pile is improved by a wedge-shaped surface, as shown in Figure 1. However, the current theoretical and analytical models mentioned above almost always treat tapered piles approximately as stepped piles, the inclined surface is treated as a vertical surface, and the effects of the taper angle are not fully considered for a dynamic analysis of tapered piles [18–20]. Hu et al. [21] provide a solution to calculate the dynamic impedance of the vertically harmonic loaded tapered pile, considering the effects of the tapered angle in accordance with the elastic dynamic Winkler theory. However, the soil was considered an elastic body, and this approach cannot be further extended to the nonlinear dynamic response analysis. This means that it is impossible to obtain the time-dependent response of the foundation, and at the same time, it is impossible to obtain the dynamic response of the foundation in the case of complex loads, such as seismic loads, wind loads, etc. Obviously, the current understanding of the bearing characteristics of tapered piles is still insufficient, which will prevent its development and application.

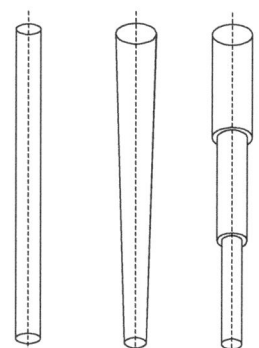

Cylindrical pile   Tapered pile   Stepped pile

Figure 1. Illustrations of a cylindrical pile, tapered pile and stepped pile.

To address the shortcomings of the current research analysis of tapered piles under vertical loads, this paper focuses on investigating the nonlinear dynamic response of tapered piles, considering the taper angle effect. Firstly, a pile–soil interaction model based on the nonlinear Winkler dynamic theory was established, and the nonlinear vertical dynamic characteristics of the tapered pile were obtained by solving the vertical vibration differential equation using the finite difference method and the Newmark-β method. Secondly, the effect of the number of tapered pile segments on the calculation accuracy of the nonlinear dynamic response is discussed. Finally, parametric studies were conducted to explore the effect of the tapered pile and soil parameters on the dynamic characteristics of the vertically loaded tapered pile. These results help to further improve the theory of nonlinear dynamic response analysis of tapered piles and promote its widespread application in engineering practices.

2.1. Vertical Vibration Analysis Model

As shown in Figure 1, the vibration properties of a tapered pile subjected to vertical harmonic load in layered soil are established based on the dynamic Winkler theory [22,23], where the taper angle, pile length and pile top radius of the tapered pile are θ, L and r_0, respectively. And the assumptions below are introduced to facilitate the analysis:

(1) It is assumed that the pile is elastic, the surface of the pile is always boned to the soil, the pile deformation exists only in the y-z plane and the displacement perpendicular to the pile is neglected, as shown in Figure 2.
(2) It is assumed that the soil is distributed in layers, and the height of the pile discretization is the same as that of the soil discretization. The pile–soil interaction was simulated as a continuous series of separate soil springs and dampers distributed around the pile shaft.
(3) It is assumed that the vertical harmonic load acts on the pile top, and the vertical vibration mode of the pile also exhibits a harmonic vibration mode.
(4) It is assumed that each layer of soil is isotopically homogeneous. The soil around the pile is assumed to be a plane strain model with no forces on the surface of the soil.

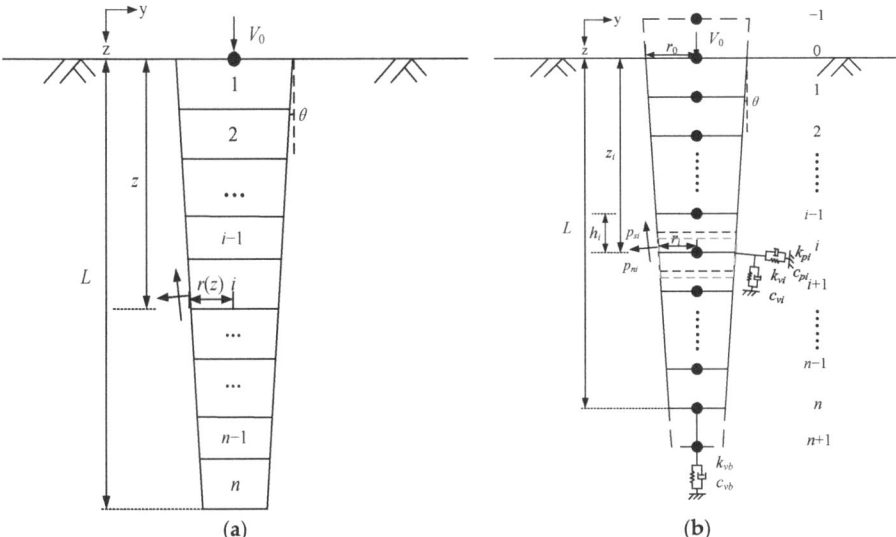

Figure 2. Vertical vibration analysis model of the tapered pile: (**a**) pile differential element; (**b**) discretized by the finite differential method.

As illustrated in Figure 2a, the tapered pile is divided into n frustum segments along the pile shaft According to the dynamic equilibrium condition, the differential equation of vertical vibration of ith segment of the tapered pile, neglecting the internal damping coefficients of the tapered pile, can be expressed as follows:

$$m(z)\frac{\partial^2 w(z,t)}{\partial t^2} + c\frac{\partial w(z,t)}{\partial t} + kw(z,t) = EA(z)\frac{\partial^2 w(z,t)}{\partial z^2} + E\frac{dw(z,t)}{dz}\frac{dA(z)}{dz} = 0 \quad (1)$$

where E is the elastic modulus of pile, k and c are the stiffness and damping coefficient of soil, $m(z)$ and $A(z)$ are pile mass per unit length and cross-sectional area at depth z, $w(z,t)$ is the vertical displacement of pile at depth z and time t.

To solve Equation (1), the equation is discretized by the finite difference method, and the model can be illustrated as in Figure 2b. The height, radius and vertical displacement of the i_{th} segment are h_i, r_i and w_i, respectively. The soil surrounding the tapered pile is modeled as a continuous distribution of springs and dampers to model the soil–pile dynamic interaction, where the springs k_{vi} and dampers c_{vi} are perpendicular to the normal direction of the pile foundation, and the spring k_{pi} and damper c_{pi} are parallel to the normal direction of the pile foundation. The tangential force p_{ti} and normal force p_{ni} along the tapered pile will be generated at the ith segment of the pile when the pile experiences a vertical dynamic load $V = V_0 e^{i\omega t}$ with circular frequency ω. This is different from the cylindrical pile due to the different form of the pile–soil contact surface. The vibration equilibrium equation at each point of a vertically loaded tapered pile can be derived at equidistant points z_i, z_{i-1} and z_{i+1} [24], as follows:

$$\frac{E_{p0}A_{p0}}{h_i^2}[2w_0 - 2w_1] + k_{vp0}w_0 = \frac{E_{p0}A_{p0}}{h_i^2}(1 - \frac{ah_i}{az+b})\frac{2h_i p_0}{E_{p0}A_{p0}}, (i = 0) \tag{2}$$

$$\frac{E_{pi}A_{pi}}{h_i^2}[-(1-(\frac{ah_i}{az+b}))w_{i-1} + 2w_i - (1+\frac{ah_i}{az+b})w_{i+1}] + k_{vpi}w_i = 0, (i = 1 \sim n-1) \tag{3}$$

$$\frac{E_{pn}A_{pn}}{h_i^2}[-2w_{n-1} + (2+(1+\frac{ah_i}{az+b})\frac{2h_i k_{bz}}{E_{pn}A_{pn}})w_n] + k_{vpn}w_n = 0, (i = n) \tag{4}$$

$$k_{vpi} = k_{vi}(\cos\theta)^2 + k_{pi}(\sin\theta)^2, \tag{5}$$

$$c_{vpi} = c_{vi}(\cos\theta)^2 + c_{pi}(\sin\theta)^2, \tag{6}$$

where E_{pi}, m_{pi} and A_{pi} are the elastic modulus, mass and cross-sectional area of the ith frustum pile segment, respectively. The dynamic stiffness and damping coefficients of the springs and dampers along the tapered pile shaft are to be determined as [25–27]

$$k_{vi} \approx 0.6 E_{si}\left(1 + \frac{1}{2}\sqrt{a_{0i}}\right), \tag{7}$$

$$c_{vi} \approx 2\beta_{si}\frac{k_{vi}}{\omega} + \pi \rho_{si} V_{si} d_i a_{0i}^{-\frac{1}{4}}, \tag{8}$$

$$k_{pi} \approx 1.2 E_{si}, \tag{9}$$

$$c_{pi} \approx 2\beta_{si}\frac{k_{pi}}{\omega} + 6\rho_{si} V_{si} d_i a_{0i}^{-\frac{1}{4}}, \tag{10}$$

where E_{si}, V_{si}, ρ_{si} and β_{si} are the soil elastic modulus, shear wave velocity, density and damping ratio of the ith soil segment. D_i is the diameter of the ith pile segment. $A_{0i} = \omega d_i / V_{si}$ is the dimensionless frequency.

The pile tip segment can be approximated as a rigid mass block embedded in an elastic foundation. Excluding the springs along the pile shaft, the dynamic stiffness and damping coefficients of the spring at the pile tip can be obtained from the vertical vibration solution given by Lysmer and Richart [28]:

$$k_{vb} = \frac{4 G_b r_b}{1 - v_b}, \tag{11}$$

$$c_{vb} = \frac{3.4 r_b \sqrt{G_b \rho_b}}{1 - v_b}, \tag{12}$$

where k_{vb} and c_{vb} are the stiffness and damping coefficients of soil at the pile tip, respectively. G_b, v_b and ρ_b are the soil shear modulus, Poisson's ratio and soil density at the pile tip, respectively.

Combining the finite difference method and the Newmark-β method, the dynamic

$$\{\Delta w\} = \left(\frac{E_{pi}A_{pi}}{h_i^2}[I_{pl}] + b_0[M_p] + b_1[C_v] + [K_{vp}]\right)^{-1}\{\Delta \vec{F}_v\} = [\check{K}_v]^{-1}\{\Delta \vec{F}_v\}, \tag{13}$$

$$\{\Delta \tilde{F}_v\} = [\Delta F_{v0} \quad \Delta F_{v1} \quad \cdots \quad \Delta F_{vi} \quad \cdots \quad \Delta F_{v(n-1)} \quad \Delta F_{vn}]^T, \tag{14}$$

$$\Delta F_{vi} = \Delta F_v + m_i[b_2 \dot{w}(t) + b_3 \ddot{w}(t)] + c_{vi}[b_4 \dot{w}(t)], \tag{15}$$

where $[\tilde{K}_v]$ is the vertical equivalent total stiffness matrix, $\{\Delta w\}$ and $\{\Delta \tilde{F}_v\}$ are the increments of the vertical displacement and vertical equivalent load vectors. β_0, b_1, b_2, b_3, b_4 are the Newmark parameters, $[M_p]$, $[K_{vp}]$ and $[C_v]$ are the mass, dynamic stiffness and damping matrices of the tapered pile, respectively. $[I_{pl}]$ is the coefficient matrix generated by the computational process, given as

$$[I_{pl}] = \begin{bmatrix} 2 & -2 & & & & \\ -(1-A_1) & 2 & -(1+A_1) & & & \\ & & \ddots & & & \\ & & -(1-A_i) & 2 & -(1+A_i) & \\ & & & \ddots & & \\ & & & -(1-A_{n-1}) & 2 & -(1+A_{n-1}) \\ & & & & & 2+(1+A_n)\frac{2k_{bz}}{h_i} \end{bmatrix}, \tag{16}$$

where $A_i = \frac{-\tan\theta h_i}{-\tan\theta z_i + r_0}$, z_i is the length of the ith pile node to the ground surface.

2.2. Nonlinear Analysis Model

Actually, the soil surrounding the pile will exhibit typical nonlinear characteristics when the soil is subjected to a vertical dynamic load. A hyperbolic nonlinear model with the Masing criterion was introduced to simulate the nonlinear characteristics of soil, as shown in Figure 3 [29–31].

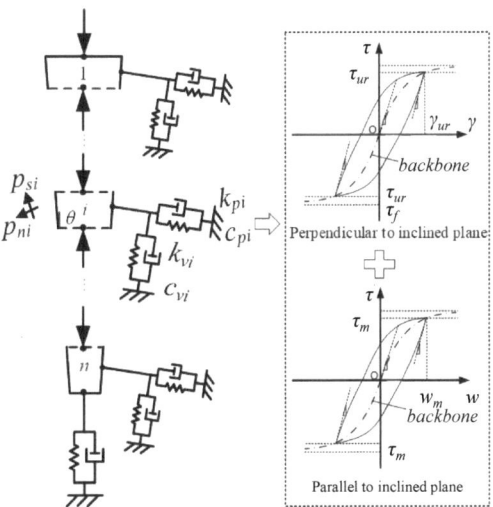

Figure 3. Nonlinear analysis model of the tapered pile.

For the soil parallel to the inclined plane of the tapered pile, the stress–displacement hyperbolic nonlinear model is given as [29,30]

$$\tau \pm \tau_m = \frac{w \pm w_m}{a + \frac{|w-w_m|}{2b}}, \tag{17}$$

where τ_m is the yield shear stress that corresponds to the displacement w_m. a is the initial flexibility factor at the soil–pile interface and b is the reciprocal of the asymptotic coefficient of the shear stress–displacement curve.

$$a = \frac{r_0 \ln|r_m/r_0|}{G_s}, \tag{18}$$

$$b = \frac{0.8}{\sigma'_n \tan \delta}, \tag{19}$$

where r_m is the radius of influence of soil displacement near the pile, δ is the effective friction angle at the soil–pile interface and σ'_n is the effective stress acting on the pile.

For the soil perpendicular to the inclined plane of the tapered pile, the stress–strain hyperbolic nonlinear model was used to model the soil nonlinearity, given as [31,32]

$$\tau_d \pm \tau_{ur} = \frac{\gamma_d \pm \gamma_{ur}}{\frac{1}{G_0} + \frac{|\gamma_d - \gamma_{ur}|}{2\tau_f}}, \tag{20}$$

$$\gamma_d = \frac{(1+v)y}{5r_0}, \tag{21}$$

where τ_d and γ_d are the current shear stress and shear strain of the soil. T_f is the shear stress at failure, and τ_{ur} is the yield shear stress corresponding to the shear strain γ_{ur}. G_0, y and r_0 are the initial small strain shear modulus of soil, the lateral displacement and the radius of the pile, respectively.

For the soil at the pile tip, the stress–strain hyperbolic nonlinear model with the Masing criterion was introduced [33,34]

$$\sigma_d \pm \sigma_m = \frac{\varepsilon_d \pm \varepsilon_m}{\frac{1}{E_0} + \frac{|\varepsilon_d - \varepsilon_m|}{2\sigma_f}}, \tag{22}$$

where σ_d and ε_d are the normal stress–strain of soil. Σ_m is the yield normal stress corresponding to the vertical strain ε_m. σ_f is the ultimate normal stress. E_0 is the initial small strain elastic modulus of soil.

Finally, in combination with the nonlinear model of the soil mentioned above, the nonlinear dynamic response of the vertically loaded tapered pile can be obtained by Equation (13) according to the initial displacement using the iterative method [35].

3. Validation and Sensitivity Analysis

3.1. Validation

The dynamic responses of vertically loaded tapered piles in homogeneous elastic soil are given in Bryden et al. [36]. The density, shear modulus and Poisson's ratio of soil are 1800 kg/m³, 12.5 MPa and 0.25, respectively. A mass block of 5000 kg is placed on top of the pile. The elastic modulus and density of the pile are assumed to be 20 GPa and 2400 kg/m³. The equivalent radius and pile length of the tapered pile are 0.1 m and 5.0 m. The equivalent radius is defined as

$$r_{eq}^2 = \frac{1}{3}(r_0^2 + r_0 r_b + r_b^2), \tag{23}$$

where r_b is the pile radius at the pile tip.

Figure 4 shows the stiffness and damping coefficients of the tapered pile with a taper angle $\theta = 1.5°$ calculated using the simplified proposed analysis method. The results are almost identical to those calculated by Bryden et al. [36], who give the variation curves of under different taper angles. On the other hand, to compare the dynamic characteristics of the floating pile and end-bearing pile, the ratio of V_b/V_s is set as = 10,000 and 1 for end-bearing and floating pile, respectively [37]. On the other hand, Ghazavi [19] obtained the normalized amplitude of the tapered pile supporting a rigid foot under vertical harmonic loading and investigated the effect of different taper angles on the vertical dynamic response of the tapered pile when the equivalent radius is the same. Here, the normalized amplitude is given as

$$A_w = \frac{\omega^2}{\sqrt{\left(\frac{k_{vp}}{M_t} - \omega^2\right)^2 + \left(\frac{\omega c_{vp}}{M_t}\right)^2}}, \quad (24)$$

where M_t and ω are the footing mass and the circular frequency of the excitation load. K_{vp} and c_{vp} are the dynamic stiffness and damping of the tapered pile.

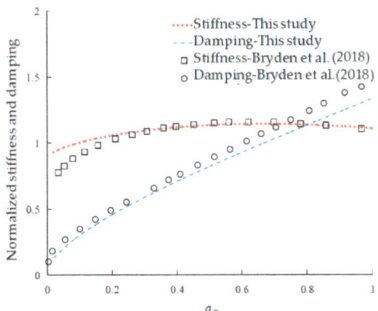

Figure 4. Comparison with the vertical dynamic stiffness and damping coefficients of the tapered pile when the tapered angle θ = 1.5° in Ref. [36].

Then, the normalized dynamic vertical responses of the tapered pile with various taper angles are compared in Figure 5. The results show that the theoretical results of this paper agree well with Bryden et al. [36], which verifies the rightness of the simplified model. It can be seen that the normalized amplitude decreases with the increase in the taper angle for the floating pile and end-bearing pile, and the resonant frequency is nearly constant for the floating pile while the resonance frequency of end-bearing piles decreases significantly. This is due to the difference in soil properties between the pile tip and the pile shaft. Compared to the dynamic response of a vertically loaded tapered pile with θ = 0°, the dynamic amplitude of the tapered pile was approximately reduced by 20% for a tapered pile with θ = 1.5°, which also shows the ability of the taper angle θ to improve the dynamic characteristics of the pile.

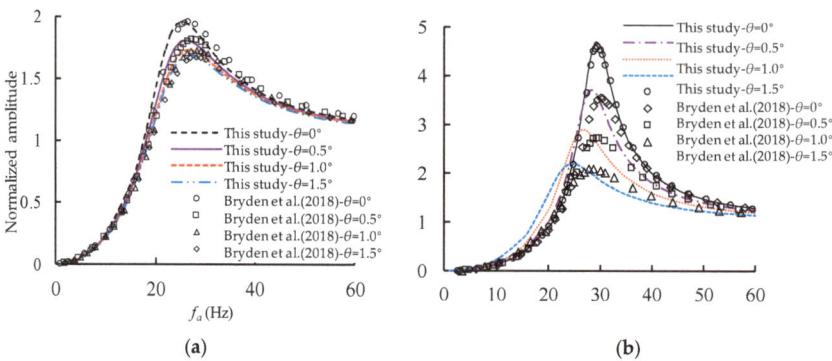

Figure 5. Comparison of the vertical dynamic responses of tapered pile with Ref. [36]: (**a**) floating pile; (**b**) end-bearing pile.

3.2. Sensitivity Analysis

The calculation accuracy of the finite difference method is closely related to the number of pile frustum segments, namely the length of the calculation element [38,39]. Sensitivity analysis was performed for typically tapered piles with length-to-radius ratios ranging from $L/r_b = 30$ to 70 and 110. The dynamic harmonic load V_0 is set at 100 kN. The parameters of the tapered pile and soil are listed in Table 1 [40,41].

Table 1. Properties of the tapered pile for different slenderness ratios.

	Parameter	Value
Tapered pile	Pile length L Equivalent radius r_{eq} Elastic modulus E_p Density ρ_p	2.5 0.1 m 20 GPa 2400 kg/m^3
Soil	Shear wave velocity V_s Shear modulus G_s Density ρ_s Poisson's ratio v	82.5 m/s 12.5 MPa 1800 kg/m^3 0.25

Figure 6 shows the time history curves of vertical displacements of tapered piles at different length-to-diameter ratios. It can be found that the dynamic response of the tapered pile will tend to a stable value as the number of pile segments n increases. Therefore, for the tapered pile with L/r_b = 30–110, the accuracy requirements and computational efficiency can be met when n is taken as 200 [42,43]. Comparing the vertical displacements under different L/r_b ratios, it can be seen that when the radius of the pile tip is the same, the longer the pile length is, the smaller the vertical dynamic response of the pile is, and the resonance frequency increases with the increase in the L/r_b ratio. This is due to the vertical stiffness of the pile increasing with the increase in pile length, which means that the ability to resist vertical deformation is better. In the meantime, there is a typical shift in the displacement at the first cycle, which is also consistent with the results of El Naggar and Bentley [44]. The reason may be attributed to the increased nonlinearity of the soil and the large iterative time step in the dynamic analysis process.

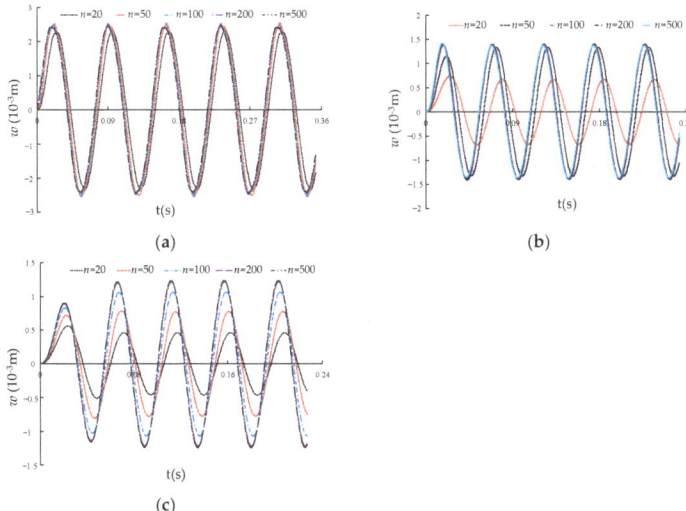

Figure 6. The time history curves of vertical displacements of tapered piles for different L/r_b ratios at pile top: (**a**) $L/r_b = 30$ (resonant frequency $f_a = 14$ Hz); (**b**) $L/r_b = 70$ (resonant frequency $f_a = 19$ Hz); (**c**) $L/r_b = 110$ (resonant frequency $f_a = 22$ Hz).

4. Parameter Discussion

4.1. Effect of Taper Angle on the Nonlinear Dynamic Response of the Tapered Pile

The taper angle is the main factor affecting the dynamic response of the tapered pile [36,45]. In this section, the dynamic responses of tapered piles with four different taper angles, 0°, 1°, 2° and 3°, are calculated, and the other parameters are the same as in Table 1.

Figure 7 shows the nonlinear dynamic response of a vertically loaded tapered pile under different taper angles. The results show that the resonance frequency of the tapered pile increases as the taper angle increases, and the nonlinear dynamic responses of the tapered pile in the vertical direction increase with a decreasing taper angle, which is consistent with the static bearing characteristics. The same pattern can be seen in the hysteresis curves of loading–displacement. However, since the reasonable selection of the pile parameters of tapered piles is beneficial to the reduction in construction costs, the dynamic response characteristics of vertically loaded tapered piles are of more concern when the pile volume is the same.

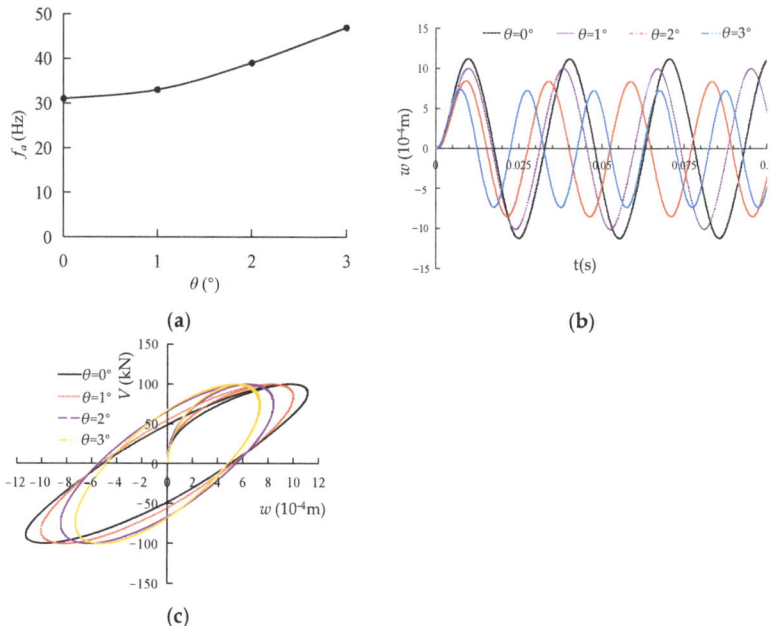

Figure 7. Nonlinear dynamic characteristics of the vertically loaded tapered pile for different taper angles with constant pile tip radius and pile length: (**a**) resonant frequency; (**b**) time history curves of vertical displacement at pile top; (**c**) hysteresis curves of loading and vertical displacement.

In order to keep the pile volume constant, two different cases of varying the diameter of the tapered pile and varying the length of the tapered pile are considered. Therefore, the resonant frequency, time history curves and hysteresis curves of the constant volume taper piles under various taper angles are shown in Figures 8 and 9. The results show that the nonlinear dynamic response of the vertically loaded tapered pile increases with a decreasing taper angle. And it can also be found that the resonance frequency of the tapered pile increases and the dynamic response of the constant volume tapered pile decreases more significantly when keeping the pile tip radius constant as compared to keeping the pile length constant. It is shown that better vertical dynamic performance can be obtained when the tapered pile keeps the radius of the pile tip constant while varying the length of the pile. This phenomenon was also found in the hysteresis curves of loading–displacement. With the increase in the taper angle, the slope of the hysteresis loop gradually increases and the amplitude of the vertical dynamic response gradually decreases. The area and shape of the hysteresis loop also tend to be more and more elliptical, which indicates that the nonlinear influence between the pile and soil is gradually reduced. This can provide engineering guidance for the design of tapered piles. It is worth noting that, although the larger the taper angle, the smaller the dynamic response of the vertically loaded tapered pile, the constructability of the pile should also be considered in engineering practice.

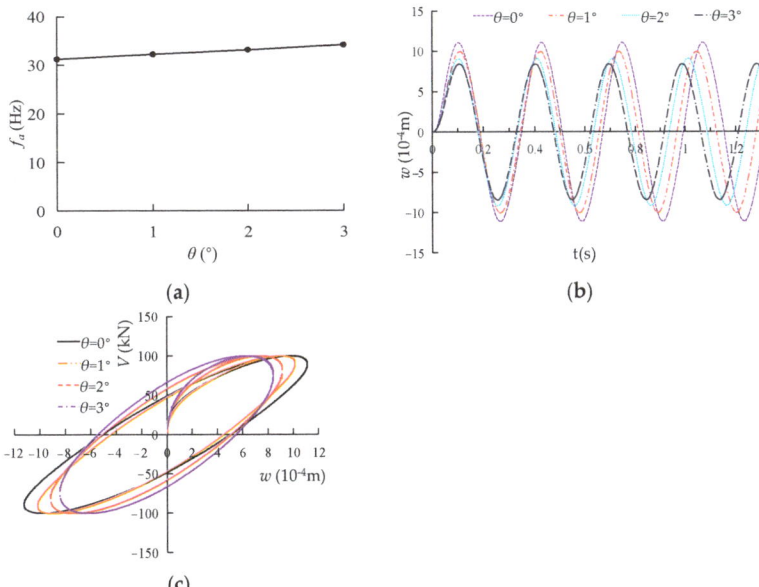

Figure 8. Nonlinear dynamic characteristics of the vertically loaded tapered pile for different taper angles with constant volume and pile tip radius: (**a**) resonant frequency; (**b**) time history curves of vertical displacement at pile top; (**c**) hysteresis curves of loading and vertical displacement.

4.2. Effect of Soil Elastic Modulus on the Nonlinear Dynamic Response of the Tapered Pile

The soil properties are also important factors influencing the dynamic behavior of tapered piles [46,47]. Three different pile–soil elastic modulus ratios, E_p/E_s = 500, 1000 and 2000, are considered for further analysis of the effect of soil parameters on the nonlinear dynamic response of the vertically loaded tapered pile. The taper angle of the pile θ is set at 1°, and the other parameters can be found in Table 1.

The resonant frequency, time history curves and hysteresis curves of loading and vertical displacement of the tapered piles with different elastic modulus ratios E_p/E_s are depicted in Figure 10. The results show that the nonlinear dynamic response of the vertically loaded tapered pile increases as the elastic modulus ratio E_p/E_s increases. The amplitude of the dynamic response increases by 267% when the elastic modulus ratios E_p/E_s are increased from 500 to 2000. This is mainly attributed to the fact that the decrease in soil modulus surrounding the pile leads to a decrease in the frictional resistance of the tapered pile. The hysteresis curve shows a counterclockwise inclination with an increase in soil modulus. This is because the increase in the soil modulus leads to an increase in the soil constraint on the pile, which also leads to a reduction in the degree of nonlinearity between the tapered pile and the soil.

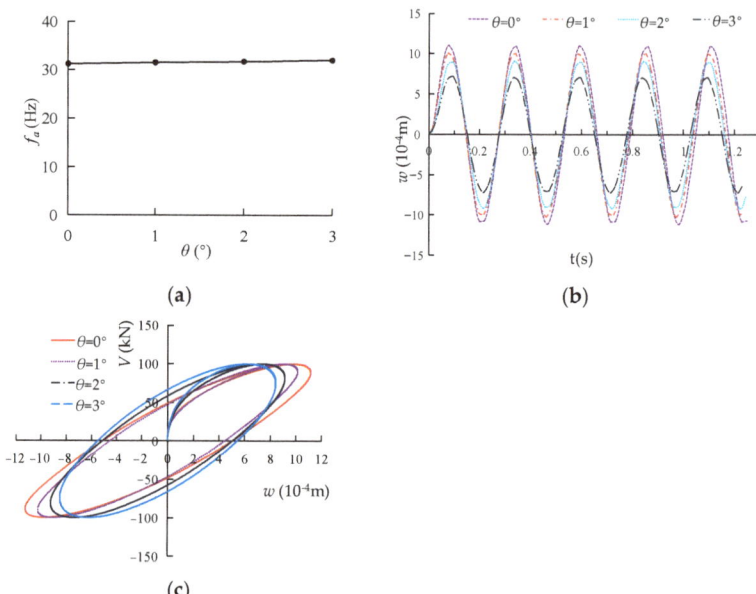

Figure 9. Nonlinear dynamic characteristics of the vertically loaded tapered pile for different taper angles with constant volume and pile length: (**a**) resonant frequency; (**b**) time history curves of vertical displacement at pile top; (**c**) hysteresis curves of loading and vertical displacement.

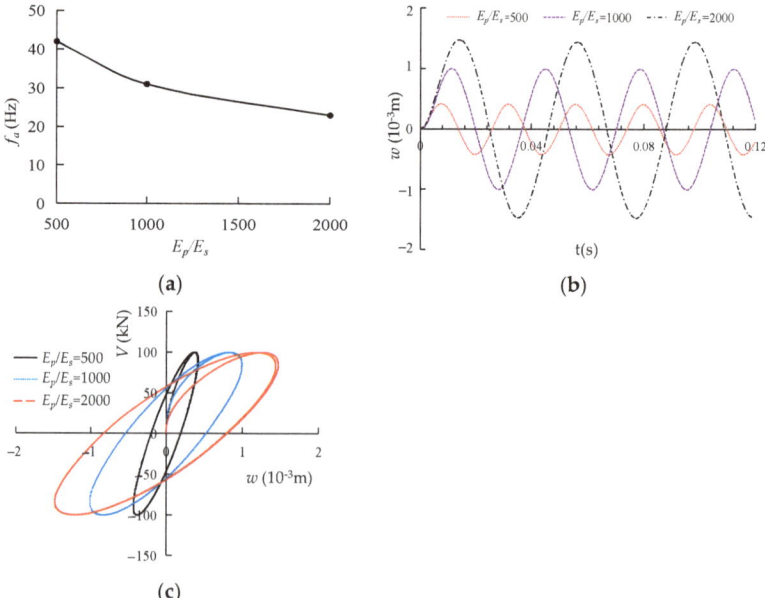

Figure 10. Nonlinear dynamic characteristics of the vertically loaded tapered pile at different pile–soil modulus ratios: (**a**) resonant frequency; (**b**) time history curves of vertical displacement at pile top; (**c**) hysteresis curves of loading and vertical displacement.

4.3. Effect of Soil Stratification on the Nonlinear Dynamic Response of the Tapered Pile

It should be noted that the former analysis of the pile is considered to be a foundation embedded in a uniform elastic half-space, while the soil in engineering practice shows layer distribution [48,49]. To clarify the effect of soil stratification on the dynamic response of the tapered pile, both two and three soil layers are discussed in this section.

Considering the soil is divided into two layers: the shear velocity of the upper soil layer V_{su} = 150 m/s remains unchanged, and the shear velocity of the lower soil layer V_{sd} gradually increases from 100 m/s to 200 m/s; the shear velocity of the lower soil layer V_{sd} = 150 m/s remains unchanged, and the shear velocity of the upper soil layer V_{su} gradually increases from 100 m/s to 200 m/s. The taper angle of the pile θ is set at 1°, and the other parameters are the same as in Table 1. The resonant frequency and time history curves of the constant volume tapered piles when considering the soil stratification are shown in Figures 11 and 12. The results show that the resonant frequency of the tapered pile increases as the shear velocity of the upper and lower soil layers increases, while the vertical dynamic response of the tapered pile decreases. Obviously, the increase in the shear wave velocity of the upper and lower soil layers will lead to an increasing ability of the tapered pile to resist vertical deformation. However, the changes in dynamic properties caused by changing the properties of the upper soil layer are more obvious, which indicates that the upper soil properties play an important role in the nonlinear dynamic response of the vertically loaded tapered pile.

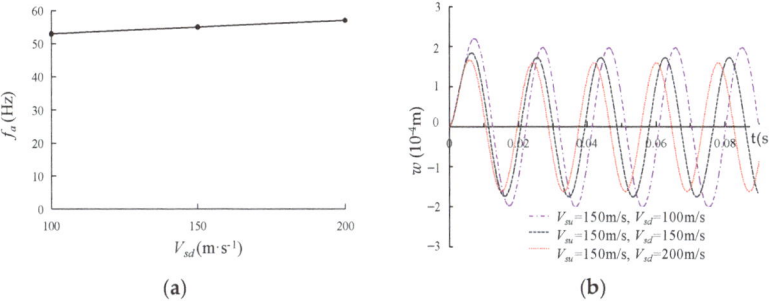

Figure 11. Nonlinear dynamic characteristics of the vertically loaded tapered pile at different shear wave velocities of lower soil: (a) resonant frequency; (b) time history curves of vertical displacement at pile top.

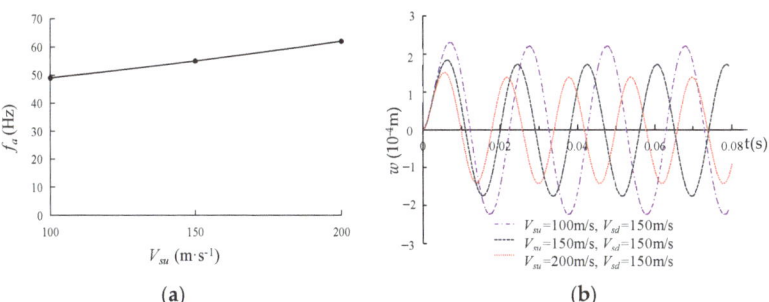

Figure 12. Nonlinear dynamic characteristics of the vertically loaded tapered pile at different shear wave velocities of upper soil: (a) resonant frequency; (b) time history curves of vertical displacement at pile top.

To further analyze the influence of soil stratification, soil divided into three layers is considered: the shear velocity of the upper and lower soil layers $V_{su} = V_{sd}$ = 150 m/s remains unchanged, and the shear velocity of the middle soil layer V_{sm} gradually increases from 100 m/s to 200 m/s. The nonlinear dynamic response of the vertically loaded tapered pile is shown in Figure 13. It is found that the vertical displacement at the top of the tapered pile decreases gradually as the shear modulus of the middle soil layer increases, which also implies that the overall dynamic stiffness of the vertically loaded tapered pile is increasing. However, the change in resonant frequency and vertical displacement of the pile top is not significant, which is similar to changing the properties of the lower soil.

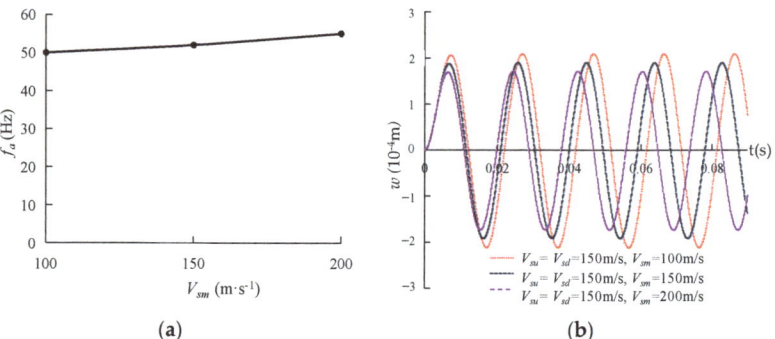

Figure 13. Nonlinear dynamic characteristics of the vertically loaded tapered pile at different shear wave velocities of middle soil: (**a**) resonant frequency; (**b**) time history curves of vertical displacement at pile top.

5. Conclusions

In this paper, the dynamic responses of the vertically loaded tapered piles are investigated based on the nonlinear dynamic Winkler model with six springs, and the influences of pile and soil parameters on the dynamic responses of vertically loaded tapered piles are discussed. The main conclusions are drawn as follows:

(1) The simplified analytical model proposed in this study can effectively obtain the nonlinear dynamic properties of the vertically loaded tapered piles, and it can ensure high computational accuracy when the pile–soil segment n is equal to 200, which also greatly reduces the computational cost.

(2) The taper angle and soil elastic modulus are the principal factors affecting the nonlinear dynamic response of vertically loaded tapered piles. The dynamic response of the tapered pile decreases with an increase in the taper angle or a decrease in the pile–soil elastic modulus ratio.

(3) For the constant volume tapered pile, the vertical dynamic response decreases more significantly when keeping the pile end radius constant as compared to when keeping the pile length constant.

(4) The soil stratification has a great influence on the nonlinear dynamic characteristics of vertically loaded tapered piles, especially the properties of the upper soil layer along the pile shaft.

Author Contributions: Conceptualization, W.T.; Methodology, W.T.; Software, Q.S. and K.L.; Validation, K.L., J.L. and W.T.; Formal analysis, K.L.; Data curation, W.T.; Writing—original draft preparation, Q.S., K.L. and J.L.; Writing—review and editing, W.T.; Visualization, J.L. and W.T.; Supervision, W.T.; Funding acquisition, Q.S. and W.T. All authors have read and agreed to the published version of the manuscript.

Funding: This work was financially supported by the National Natural Science Foundation of China (52268069), the Open Project of the Building Environment Engineering Technology Research Center in Dazhou (SDJ2022ZC-02) and Jiangxi Provincial Cultivation Program for Academic and Technical

Data Availability Statement: Data are available upon reasonable request.

Conflicts of Interest: Author Jingkai Li was employed by the company China Huaxi Engineering Design & Construction Co., Ltd. The remaining authors declare that the research was conducted in the absence of any commercial or financial relationships that could be construed as a potential conflict of interest.

References

1. Conte, E.; Pugliese, L.; Troncone, A.; Vena, M. A simple approach for evaluating the bearing capacity of piles subjected to inclined loads. *Int. J. Geomech.* **2021**, *21*, 04021224. [CrossRef]
2. Tamboura, H.H.; Isobe, K.; Ohtsuka, S. End bearing capacity of a single incompletely end-supported pile based on the rigid plastic finite element method with non-linear strength property against confining stress. *Soils Found.* **2022**, *62*, 101182. [CrossRef]
3. Wei, J.; El Naggar, M.H. Experimental study of axial behaviour of tapered piles. *Can. Geotech. J.* **1998**, *35*, 641–654. [CrossRef]
4. Liu, J.; Wang, Z.H. Experimental study on the bearing capacity of wedge pile. *J. Tianjin Univ. (Sci. Technol.)* **2002**, *35*, 257–260. (In Chinese)
5. Jiang, J.P.; Gao, G.Y.; Gu, B.H. Comparison of belled pile, tapered pile and equal diameter pile. *Chin. J. Geotech. Eng.* **2003**, *25*, 764–766. (In Chinese)
6. El Naggar, M.H.; Sakr, M. Cyclic response of axially loaded tapered piles. *Int. J. Phys. Model. Geotech.* **2002**, *2*, 1–12.
7. Hataf, N.; Shafaghat, A. Optimizing the bearing capacity of tapered piles in realistic scale using 3D finite element method. *Geotech. Geol. Eng.* **2005**, *33*, 1465–1473. [CrossRef]
8. Vali, R.; Mehrinejad Khotbehsara, E.; Saberian, M.; Li, J.; Mehrinejad, M.; Jahandariet, S. A three-dimensional numerical comparison of bearing capacity and settlement of tapered and under-reamed piles. *Int. J. Geotech. Eng.* **2019**, *13*, 236–248. [CrossRef]
9. Shafaghat, A.; Khabbaz, H. Recent advances and past discoveries on tapered pile foundations: A review. *Geomech. Geoengin.* **2022**, *17*, 455–484. [CrossRef]
10. Sudhendu, S.; Ghosh, D.P. Vertical vibration of tapered piles. *J. Geotech. Eng. ASCE* **1986**, *112*, 290–302.
11. Xie, J.; Vaziri, H.H. Vertical vibration of nonuniform piles. *J. Eng. Mech. ASCE* **1991**, *117*, 1105–1118. [CrossRef]
12. Wu, W.B.; Wang, K.H.; Dou, B. Vertical dynamic response of a viscoelastic tapered pile embedded in layered foundation. *J. Vib. Shock* **2013**, *32*, 120–127. (In Chinese)
13. Wang, K.H.; Tong, W.F.; Xiao, C.; Wu, B.J. Study on dynamic response of tapered pile and model test. *J. Hunan Univ. (Nat. Sci.)* **2019**, *46*, 94–102. (In Chinese)
14. Michaelides, O.; Gazetas, G.; Bouckovalas, G.; Chrysikou, E. Approximate non-linear dynamic axial response of piles. *Géotechnique* **1998**, *48*, 33–53. [CrossRef]
15. Singh, S.; Patra, N.R. Free and forced vibration analyses of tapered piles under axial harmonic Loads. *Int. J. Geomech.* **2021**, *21*, 04021023. [CrossRef]
16. Padrón, L.A.; Aznárez, J.J.; Maeso, O. Dynamic analysis of piled foundations in stratified soils by a BEM–FEM model. *Soil. Dyn. Earthq. Eng.* **2008**, *28*, 333–346. [CrossRef]
17. He, C.; Zhou, S.H.; Guo, P.J.; Di, H.G.; Zhang, X.H. Analytical model for vibration prediction of two parallel tunnels in a full-space. *J. Sound Vib.* **2018**, *423*, 306–321. [CrossRef]
18. Liu, J.; He, J.; Wu, Y.P.; Yang, Q.G. Load transfer behaviour of a tapered rigid pile. *Géotechnique* **2012**, *62*, 649–652. [CrossRef]
19. Ghazavi, M. Analysis of kinematic seismic response of tapered piles. *Geotech. Geol. Eng.* **2007**, *25*, 37–44. [CrossRef]
20. Wang, J.; Zhou, D.; Zhang, Y.G.; Cai, W. Vertical impedance of a tapered pile in inhomogeneous saturated soil described by fractional viscoelastic model. *Appl. Math. Model.* **2019**, *75*, 88–100. [CrossRef]
21. Hu, Y.; Tu, W.B.; Gu, X.Q. A simple approach for the dynamic analysis of a circular tapered pile under axial harmonic vibration. *Buildings* **2023**, *13*, 999. [CrossRef]
22. Huang, M.S.; Tu, W.B.; Gu, X.Q. Time domain nonlinear lateral response of dynamically loaded composite caisson-piles foundations in layered cohesive soils. *Soil. Dyn. Earthq. Eng.* **2018**, *106*, 113–130. [CrossRef]
23. El Naggar, M.H.; Novak, M. Non-linear model for dynamic axial pile response. *J. Geotech. Eng.* **1994**, *120*, 308–329. [CrossRef]
24. Zhang, Y. A finite difference method for fractional partial differential equation. *Appl. Math. Comput.* **2009**, *215*, 524–529. [CrossRef]
25. Gazetas, G.; Makris, N. Dynamic pile-soil-pile interaction. Part I: Analysis of axial vibration. *Earthq. Eng. Struct. Dyn.* **1991**, *20*, 115–132. [CrossRef]
26. Makris, N.; Gazetas, G. Dynamic pile-soil-pile interaction. Part II: Lateral and seismic response. *Earthq. Eng. Struct. Dyn.* **1992**, *21*, 145–162. [CrossRef]
27. Gazetas, G.; Fan, K.; Kaynia, A. Dynamic response of pile groups with different configurations. *Soil Dyn. Earthq. Eng.* **1993**, *12*, 239–257. [CrossRef]
28. Lysmer, J.; Richart Jr, F.E. Dynamic response of footings to vertical loading. *J. Soil Mech. Found. Div.* **1966**, *92*, 65–91. [CrossRef]
29. Hardin, B.O.; Drnevich, V.P. Shear modulus and damping in soils: Design equations and curves. *J. Soil Mech. Found. Div.* **1972**, *98*, 667–692. [CrossRef]

30. Zhang, Q.Q.; Zhang, Z.M.; He, J.Y. A simplified approach for settlement analysis of single pile and pile groups considering interaction between identical piles in multilayered soils. *Comput. Geotech.* **2010**, *37*, 969–976. [CrossRef]
31. Tu, W.B.; Huang, M.S.; Gu, X.Q.; Chen, H.P.; Liu, Z.H. Experimental and analytical investigations on nonlinear dynamic response of caisson-pile foundations under horizontal excitation. *Ocean Eng.* **2020**, *208*, 107431. [CrossRef]
32. Kagawa, T.; Kraft, L.M. Lateral load-deflection relationships of piles subjected to dynamic loadings. *Soils Found.* **1980**, *20*, 19–36. [CrossRef] [PubMed]
33. Tu, W.B.; Gu, X.Q.; Chen, H.P.; Fang, T.; Geng, D.X. Time domain nonlinear kinematic seismic response of composite caisson-piles foundation for bridge in deep water. *Ocean Eng.* **2021**, *235*, 109398. [CrossRef]
34. Yu, J.; Huang, M.S.; Li, S.; Leung, C.F. Load-displacement and upper-bound solutions of a loaded laterally pile in clay based on a total-displacement-loading EMSD method. *Comput. Geotech.* **2017**, *83*, 64–76. [CrossRef]
35. Daftardar-Gejji, V.; Jafari, H. An iterative method for solving nonlinear functional equations. *J. Math. Anal. Appl.* **2006**, *316*, 753–763. [CrossRef]
36. Bryden, C.; Arjomandi, K.; Valsangkar, A. Dynamic Axial Stiffness and Damping Parameters of Tapered Piles. *Int. J. Geomech.* **2018**, *18*, 06018014. [CrossRef]
37. Ghazavi, M. Response of tapered piles to axial harmonic loading. *Can. Geotech. J.* **2008**, *45*, 1622–1628. [CrossRef]
38. Troncone, A.; Pugliese, L.; Parise, A.; Conte, E. A simple method to reduce mesh dependency in modelling landslides involving brittle soils. *Géotech. Lett.* **2022**, *12*, 167–173. [CrossRef]
39. Baek, J.; Schlinkman, R.T.; Beckwith, F.N.; Chen, J.S. A deformation-dependent coupled Lagrangian/semi-Lagrangian meshfree hydromechanical formulation for landslide modeling. *Adv. Model. Simul. Eng. Sci.* **2022**, *9*, 1–35. [CrossRef]
40. Dehghanpoor, A.; Ghazavi, M. Response of tapered piles under lateral harmonic vibrations. *Geomate J.* **2012**, *2*, 261–265. [CrossRef]
41. Singh, S.; Patra, N.R. Lateral dynamic response of tapered pile embedded in a cross-anisotropic medium. *J. Earthq. Eng.* **2022**, *26*, 5826–5847. [CrossRef]
42. Cai, Y.Y.; Yu, J.; Zhen, C.T.; Qi, Z.B.; Song, B.X. Analytical solution for longitudinal dynamic complex impedance of tapered pile. *Chin. J. Geotech. Eng.* **2011**, *33*, 392–398. (In Chinese)
43. Wu, W.B.; Wang, K.H.; Wu, D.H.; Ma, B.N. Study of dynamic longitudinal impedance of tapered pile considering lateral inertial effect. *Chin. J. Rock Mech. Eng.* **2011**, *30*, 3618–3625. (In Chinese)
44. El Naggar, M.H.; Bentley, K.J. Dynamic analysis for laterally loaded piles and dynamic p-y curves. *Can. Geotech. J.* **2000**, *37*, 1166–1183. [CrossRef]
45. Gao, L.; Wang, K.H.; Xiao, S.; Wu, J.T.; Wang, N. Vertical impedance of tapered piles considering the vertical reaction of surrounding soil and construction disturbance. *Mar. Georesour. Geotechnol.* **2017**, *35*, 1068–1076. [CrossRef]
46. Kodikara, J.K.; Moore, I.D. Axial response of tapered piles in cohesive frictional ground. *J. Geotech. Eng.* **1993**, *119*, 675–693. [CrossRef]
47. Khan, M.K.; El Naggar, M.H.; Elkasabgy, M. Compression testing and analysis of drilled concrete tapered piles in cohesive-frictional soil. *Can. Geotech. J.* **2008**, *45*, 377–392. [CrossRef]
48. Li, H.; He, C.; Gong, Q.M.; Zhou, S.H.; Li, X.X.; Zou, C. TLM-CFSPML for 3D dynamic responses of a layered transversely isotropic half-space. *Comput. Geotech.* **2024**, *168*, 106131. [CrossRef]
49. He, C.; Zhou, S.H.; Di, H.G.; Guo, P.J.; Xiao, J.H. Analytical method for calculation of ground vibration from a tunnel embedded in a multi-layered half-space. *Comput. Geotech.* **2018**, *99*, 149–164. [CrossRef]

Disclaimer/Publisher's Note: The statements, opinions and data contained in all publications are solely those of the individual author(s) and contributor(s) and not of MDPI and/or the editor(s). MDPI and/or the editor(s) disclaim responsibility for any injury to people or property resulting from any ideas, methods, instructions or products referred to in the content.

Article

Derivation and Application of Analytical Coupling Loss Coefficient by Transfer Function in Soil–Building Vibration

Jinbao Yao, Zhaozhi Wu *, Xiaofeng Cao, Nianping Wu and Nan Zhang

School of Civil Engineering, Beijing Jiaotong University, Beijing 100044, China; jbyao@bjtu.edu.cn (J.Y.); 20121020@bjtu.edu.cn (X.C.); 21121122@bjtu.edu.cn (N.W.); nzhang@bjtu.edu.cn (N.Z.)
* Correspondence: 19125893@bjtu.edu.cn

Abstract: Vibrations generated by railways may undergo amplification or reduction while traversing the foundations, floors, and spans of adjacent structures. This fluctuation in the vibration intensity, identified as a building's coupling loss, is commonly considered in vibration forecasts through the utilization of universal frequency-independent adjustment parameters. This article employs a theoretical analytical approach to investigate the propagation characteristics of Rayleigh waves in elastic foundation soil, as well as the variations at the contact surface of buildings' foundations. Analytical expressions for the coupling loss coefficient are derived to explore the displacement transfer relationship in the soil–structure interaction. To accurately and efficiently analyze the proposed buildings and site, the entire vibration propagation system is decoupled into substructure systems for independent analytical calculations. Theoretical analytical methods are utilized to obtain the displacement transfer functions between the soil and the structures through the refraction and transmission of waves. From a theoretical perspective, a thorough understanding of the interaction between soil and buildings is achieved. The influence of various variables related to railways and foundations on the building responses is analyzed. By comparing with measured data, the correctness of the analytical form of the coupling loss coefficient is validated, filling a gap in the literature due to the lack of analytical research on displacement transfer losses in soil–structure interactions.

Keywords: coupling loss; Rayleigh wave; soil vibration; vibration isolation; theoretical study

Citation: Yao, J.; Wu, Z.; Cao, X.; Wu, N.; Zhang, N. Derivation and Application of Analytical Coupling Loss Coefficient by Transfer Function in Soil–Building Vibration. *Buildings* **2024**, *14*, 1933. https://doi.org/10.3390/buildings14071933

Academic Editor: Fabrizio Gara

Received: 19 May 2024
Revised: 15 June 2024
Accepted: 21 June 2024
Published: 25 June 2024

Copyright: © 2024 by the authors. Licensee MDPI, Basel, Switzerland. This article is an open access article distributed under the terms and conditions of the Creative Commons Attribution (CC BY) license (https://creativecommons.org/licenses/by/4.0/).

1. Introduction

The propagation of vibrations generated by railway systems can significantly impact the structural integrity and comfort levels of adjacent buildings. These vibrations, which may amplify or attenuate as they travel through soil and building foundations, are quantified by the coupling loss coefficient, a critical factor when predicting and managing building responses to environmental vibrations.

In the 1960s and 1970s, the gradual development of high-speed railways emerged as a convenient and efficient transportation infrastructure in developed countries. A significant focus of scholarly inquiry during this period involved the examination of the environmental vibration responses induced by rail transit. In 1850, Professor Krylov et al. [1–3] in the United Kingdom pioneered the establishment of a ground vibration prediction model, basing it on the dynamic loads derived from vehicle loads. Utilizing the classical Green function equation, their work delved into the study of ground vibration responses resulting from rail transit. In 1970, Lang and Kurzweil [4] contributed a predictive Equation for train-induced vibrations through extensive research, correlating the vibration levels and distances with low-frequency vibrations. In 1988, Takemiya Hirokazu [5] employed a quasi-static method to analyze the vibration response of the Shinkansen track system caused by high-speed trains. This comprehensive analysis included the examination of viaduct and pile foundation vibrations. Takemiya studied the resistance surrounding the vibration source and the influence of layered soil on elastic wave propagation in the foundation soil, and then proposed a damping scheme. By comparing three different arrangements encompassing diverse materials, damping effects, and wave damping zones, the optimal damping effect was determined.

In 2020, Yue Jianyong et al. [6] employed a test method to measure the tunnel and surrounding building accelerations before and after subway vibration reduction. This facilitated the analysis and evaluation of subway vibration reduction effects. Building on this, a finite element and infinite element coupling numerical simulation method was employed to simulate the vibration response of surrounding buildings during subway operations. The resulting vibration characteristics were then compared with measured data. Fang Lei's work in 2020 [7] involved simulating soil–structure interaction through the derivation of the stiffness coefficients between soil and building. Additionally, an enhancement to the prediction model's accuracy was achieved through the application of the random forest algorithm. In 2021, Colaco et al. [8–10] predicted the vibration generation and propagation in the track–ground–building system using the 2.5DFEM-MFS method for tracks and the 3DFEM method for buildings. In the same year, Sadeghi et al. [11–13] established and verified a vibration acoustic model for buildings based on field test results. The building parameters were categorized based on the structural and non-structural characteristics, and a model parameter analysis was conducted to study the influence of the building structural and acoustic parameters on structural noise. The derived prediction model facilitated establishing the functional relationship between train-induced vibration acceleration and buildings. In 2021, Yasser E. Ibrahim et al. [14] conducted a detailed three-dimensional finite element analysis of a 10-story reinforced concrete frame structure based on a raft, utilizing ABAQUS 2020. The study, employing a moving point load to simplify the train load, investigated the effects of the train speed and the distance between the train and the building on the vibration response of the foundation structure. The research concluded that the use of open ditches and filled foam ditches mitigated the vibration response induced by trains.

Presently, the numerical simulation method necessitates simplifications and assumptions within the overall finite element model, compromising the accuracy of the calculation results. Although the test methods yield accurate and reliable results, they are time-consuming and expensive. The test method is also susceptible to external factors, such as the environmental conditions, sensor instability, and background noise, introducing interference to the test data and thereby imposing limitations on the derived prediction equations. In light of these considerations, this study derives the coupling loss coefficient of an elastic soil foundation through theoretical analysis.

2. Theoretical Study of the Transfer Function Method

During the operation of a high-speed train, the vibrations generated by the interaction between the train and the track propagate through the roadbed to the surrounding soil and buildings, giving rise to a vibration transmission system involving the train, subgrade, foundation soil, building foundation, and the building itself, as depicted in Figure 1. The ISO14837-1 standard [15] defines the frequency domain function expression $A(f)$ for building vibration, consisting of the source term $S(f)$, the propagation term $P(f)$, and the receiving term $R(f)$. The magnitude of the building vibration $A(f)$ is mathematically represented as $A(f) = S(f)P(f)R(f)$. Consequently, employing the transfer function method, the entire vibration propagation system is decomposed into sub-structural systems for independent analysis and calculation (Equation (1)). These sub-structural systems include the train–track–site soil system (refer to Figure 2), soil–structure dynamic interaction system (refer to Figure 3), and building structural system (refer to Figure 4).

$$U_b(f) = U_0^{soil}(f) \cdot C_l(f) \cdot F_a(f) \tag{1}$$

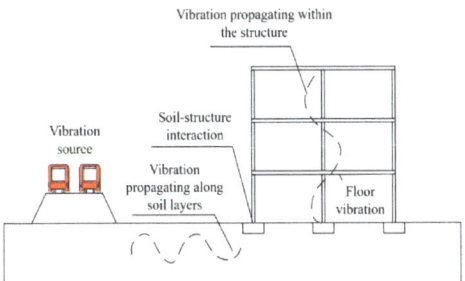

Figure 1. Vibration propagation diagram induced by a high-speed train.

Figure 2. Train–track–site soil system.

Figure 3. Soil–structure dynamic interaction system.

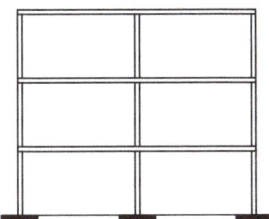

Figure 4. Building structure system.

Equation (1) represents the calculation equation for the transfer function method. Examining the equation, it becomes evident that the vibration response, U_0^{soil}, of the free field comprises the vibration source term, $S(f)$, and the propagation term, $P(f)$. Simultaneously, the receiving term, $R(f)$, comprises the coupling loss coefficient, C_l, and the floor amplification coefficient, F_a. Notably, the floor amplification factor is solely determined by the structural form and material characteristics of the building. To elaborate, once the inherent properties, such as the structural form of the building, are established, the floor amplification factor is determined accordingly. Consequently, the acceptance term, $R(f)$, is solely dependent on the coupling loss coefficient, C_l. This paper focuses on the investigation of the reflection and transmission law of Rayleigh waves propagating through the building foundation. The relationship between incident waves, reflected waves, and transmitted waves is thoroughly analyzed. Subsequently, the coupling loss coefficient of the soil–structure transfer function is derived at the contact surface between the soil and the structure. This approach eliminates the need for a complex modeling process and measurement methods, ensuring both efficiency and accuracy in the calculation process.

2.1. Propagation Law of Rayleigh Waves

The propagation of waves through elastic materials is subject to geometric damping, where, following the principle of the conservation of energy, the amplitude of the wave gradually decreases with increasing wave intensity and diffusion. Miller [16] was the first to characterize the wave field under a concentrated load. Importantly, the acceptance term, $R(f)$, is exclusively determined by the coupling loss coefficient, C_l [17].

Consequently, in incorporating the soil–structure interaction into the process of resolving the vibration response of a building, the floor amplification factor is determined subsequent to establishing the building's form. The determination of the floor amplification factor is integral to comprehensively understanding and applying the soil–structure interaction in the context of solving building vibration responses. On the free interface, when the incidence angle of the SV wave is substantial, a Rayleigh surface waveform manifests on the free interface, propagating along it with its amplitude diminishing as the distance from the interface increases. While there are alternative methods for generating Rayleigh surface waves, in a half space containing layered media, considering the free surface boundary conditions, interlayer continuity, and amplitude limits of an infinite domain, two distinct types of special solutions to waves emerge. One is confined solely within the half space containing layered media, while the other extends beyond the plane. Both categories of special solutions propagate along the free surface, with their amplitudes gradually decreasing to zero at greater depths. These special solutions are identified as Rayleigh surface waves and Love surface waves.

The decay of the vibration is influenced by the properties of the medium and increases with the vertical distance. When the vertical distance is infinite, the displacement tends toward zero, indicating that Rayleigh waves exclusively propagate on the soil surface [18]. While surface waves propagate along the free surface, the energy of body waves not only travels along the free surface but also diffuses deeply within the medium. Consequently, the amplitude of surface waves decays at a much slower rate than that of body waves. Given that Rayleigh waves contribute to over two-thirds of environmental vibration during the propagation process, an examination of the attenuation characteristics of Rayleigh waves aids in understanding the laws governing vibration propagation. Consequently, this paper exclusively focuses on studying the propagation and attenuation of Rayleigh waves.

2.2. Equation Derivation of Coupling Loss Coefficients

Currently, there is a deficiency in the theoretical research on the propagation of Rayleigh waves in discontinuous media, both domestically and internationally. Previous efforts have predominantly relied on numerical methods, yielding specific outcomes. The determination of the amplitude and angle of reflected and refracted waves generated by P-waves and S-waves at the discontinuities of two media is achieved through precise consideration of six boundary conditions at these interfaces.

However, owing to the unique nature of Rayleigh waves, the amplitude of the wave mode transition from surface wave to body wave is also contingent upon the angle formed by the interface of the medium. Simply addressing the reflection and refraction of Rayleigh waves does not fully satisfy the requisite boundary conditions. Consequently, whether it is a reflected or transmitted Rayleigh wave, the mathematical challenges are more intricate, and the analytical methods are more demanding.

In 1961, Lapwood [19] delved into the reflection and transmission of Rayleigh waves at specific angles, presenting theoretical conclusions. Building on these findings, Maurice James [20] conducted an in-depth study on the reflection and transmission of Rayleigh waves. The foundational concept in Maurice James' work is that the residual difference between the corresponding stress and displacement on either side of the contact surface should be zero. Given the intricate waveform transformations of Rayleigh waves at interfaces, including their potential transformation into Love waves or Stoneley waves, achieving this residual condition is challenging. Jafar, Zarastvand and Zhou [21] developed an analytical model to determine the sound transmission loss of a doubly curved sandwich shell with various truss core configurations.

Recent studies on building vibrations from train operations highlight key advance-

vibrations in elevated metro depots and over-track buildings, validated by field measurements. He and Tao [24] created a prediction method for urban environments, considering soil–structure interactions. Ma et al. [25] proposed a semi-analytical model for underground train-induced vibrations, validated in the Hefei metro. These studies underscore the importance of predictive modeling and validation in designing effective vibration mitigation strategies for improved urban living conditions.

While the residual cannot feasibly reach zero if Rayleigh waves only assume transmission or reflection forms, Maurice James proposed the selection of suitable transmission and reflection coefficients to minimize the relevant equation. This approach aims to bring the residual as close as possible to the zero condition, acknowledging the complexity of Rayleigh wave transformations and providing a pragmatic solution to enhance the accuracy of theoretical considerations.

$$I_0 = \frac{1}{I_1} \int_0^\infty \left\{ |\sigma_{xx} - \sigma'_{xx}|^2 + |\tau_{xx} - \tau'_{xx}|^2 \right\} dz \\ + \frac{1}{I_2} \int_0^\infty \left\{ |u - u'|^2 + |w - w'|^2 \right\} dz = 0 \qquad (2)$$

In Equation (2), σ_{xx}, σ'_{xx}, τ_{xz}, τ_{xz}', u, u', and w, w' are the normal stresses, shear stresses, vertical displacements and horizontal displacements in the two different media, respectively. The residual I_0 represents the disparity between stress and displacement at the interface, while I_1 and I_2 denote integral expressions of stress and displacement, respectively. These expressions in Equation (3) specifically account for the contributions solely arising from the incident Rayleigh waves at the medium's discontinuity.

$$\begin{aligned} I_1 &= \int_0^\infty \left\{ |\sigma_{xx0}|^2 + |\tau_{xz0}|^2 \right\} dz \\ I_2 &= \int_0^\infty \left\{ |u_0|^2 + |w_0|^2 \right\} dz = 0 \end{aligned} \qquad (3)$$

where σ_{xxo}, τ_{xzo}, u_o, and w_o are the normal stresses, shear stresses, vertical displacements and horizontal displacements at the interface position, respectively.

The illustration in Figure 5 depicts the propagation of Rayleigh waves on the discontinuous surface of a quarter-space medium.

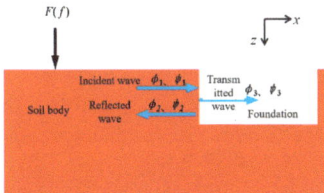

Figure 5. Schematic diagram of Rayleigh wave propagation on the discontinuous surface of a quarter-space medium.

After simplification, the expression for I_0 is obtained:

$$I_0 = (A+J)(1+R)^2 + (B+G)(1-R)^2 + (F+I)T(1-R) \\ + (C+E+H+K)T^2 + (D+L)T(1+R) \qquad (4)$$

In Equation (4), the functions designated by the letters A to L are solely dependent on two media parameters, which can be computed to fixed values. Consequently, within the equation, only R and T remain as the two unknowns.

To minimize the value of I_0, the reflection coefficient R and transmission coefficient T can finally be calculated and determined by $\frac{\partial I_0}{\partial R} = 0$, $\frac{\partial I_0}{\partial C} = 0$.

During the simplification of the equation, the wave circular frequency ω is factored out, and as a result, the widely used transmission coefficient R and reflection coefficient T are only affected by the elastic modulus on both sides of the medium. This is significant because the expression for these coefficients is not limited to simple harmonic waves but can be extended to non-periodic waveforms as well. Any wave, through the Fourier transform, can be decomposed into the superposition of several simple harmonic waves. Each of these simple harmonic waves possesses reflection and transmission coefficients that are independent of the circular frequency. Consequently, the integrals of these simple harmonic waves also have reflection and transmission coefficients that are irrelevant to the circular frequency.

When Rayleigh waves interact with a quarter space, the primary outcomes include the generation of reflected Rayleigh waves (ψ_2, φ_2) and refracted Rayleigh waves (ψ_3, φ_3). Specifically, ψ_1 and φ_1 represent vertically incident waves originating from the positive direction of the x-axis toward the interface between the building foundation and the soil mass (at x = 0). Relevant expressions can be listed from Equations (5) to (7):

$$\psi_1 = A_1 e^{-rz} e^{i(wt-k_r x)}, \varphi_1 = B_1 e^{-sz} e^{i(wt-k_r x)} \tag{5}$$

$$\psi_2 = A_2 e^{-rz} e^{i(wt+k_r x)}, \varphi_2 = B_2 e^{-sz} e^{i(wt+k_r x)} \tag{6}$$

$$\psi_3 = A_3 e^{-r'z} e^{i(wt-k_r' x)}, \varphi_3 = B_3 e^{-s'z} e^{i(wt-k_r' x)} \tag{7}$$

where $A_1 \sim A_3$ and $B_1 \sim B_3$ are constants, while φ_1, φ_2, φ_3 and Ψ_1, Ψ_2, Ψ_3 are the wave potential functions of incident wave, reflected wave and transmitted wave in two different medias, respectively.

According to the boundary conditions, the left-travelling wave $B_i = -b \cdot A_i/(2ik_r s)$ and the right-travelling wave $B_i = b \cdot A_i/(2ik_r s)$, where ($i$ = 1, 2, 3).

The stress and displacement on the left side of the axis x = 0 can be expressed in the following Equations from (8) to (11):

$$u = \frac{\partial \varphi}{\partial x} - \frac{\partial \phi}{\partial z} = ik_r(A_1 - A_2)\left(\frac{2sr}{b} e^{-sz} - e^{-rz}\right) \tag{8}$$

$$w = \frac{\partial \varphi}{\partial z} + \frac{\partial \phi}{\partial x} = -r(A_1 + A_2)\left(-\frac{2k_r^2}{b} e^{-sz} + e^{-rz}\right) \tag{9}$$

$$\begin{aligned}\sigma_{xx} &= \lambda\left(\frac{\partial w}{\partial x} + \frac{\partial u}{\partial z}\right) + \mu \frac{\partial u}{\partial x} \\ &= -\mu(A_1 + A_2)(-be^{-sz} + ae^{-rz})\end{aligned} \tag{10}$$

$$\begin{aligned}\tau_{xz} &= \mu\left(\frac{\partial w}{\partial x} + \frac{\partial u}{\partial z}\right) \\ &= 2\mu i r k_r (A_1 - A_2)(-e^{-sz} + e^{-rz})\end{aligned} \tag{11}$$

where λ and μ are the Lame constants of elastic medium one.

The stress and displacement on the right side of the axis x=0 can be expressed in the following Equations from (12) to (15):

$$u' = ik_r'(A_3 - A_4)\left(\frac{2s'r'}{b} e^{-s'z} - e^{-r'z}\right) \tag{12}$$

$$w' = -r'(A_3 + A_4)\left(-\frac{2k_r'^2}{b} e^{-s'z} + e^{-r'z}\right) \tag{13}$$

$$\sigma_{xx}' = -\mu'(A_3 + A_4)\left(-b'e^{-s'z} + a'e^{-r'z}\right) \tag{14}$$

where the parameters with the superscript symbol ′ represent the parameters of medium two, which have the same meaning as the parameters in medium one.

The residual expression of stress and displacement caused by incident Rayleigh waves on both sides of the interface x = 0 are:

$$\begin{aligned} u_0 &= ik_r A_1 (\tfrac{2sr}{b} e^{-sz} - e^{-rz}) \\ w_0 &= -rA_1(-\tfrac{2k_r^2}{b}e^{-sz} + e^{-rz}) \\ \sigma_{xx0} &= -\mu A_1(-be^{-sz} + ae^{-rz}) \\ \tau_{xz0} &= 2\mu i r k_r A_1(e^{-sz} - e^{-rz}) \end{aligned} \quad (16)$$

Among them,

$$\begin{aligned} s^2 &= k_r^2 - k_\beta^2,\ b^2 = 2r^2 + k_\beta^2 \\ k_\alpha &= \tfrac{w}{c_p},\ k_\beta = \tfrac{w}{c_s},\ k_r = \tfrac{w}{c_R} \end{aligned} \quad (17)$$

The above equations are put into Equations (2)–(8), and in order to minimize I_0, $\frac{\partial I_0}{\partial R} = 0$, $\frac{\partial I_0}{\partial T} = 0$ is defined. The reflection coefficient R and transmission coefficient T can be expressed as:

$$R = \frac{A_2}{A_1} = \frac{X - U}{V - Y} \quad (18)$$

$$T = X + YR \quad (19)$$

Among them,

$$\begin{aligned} X &= \tfrac{-(D+L+F+I)}{(C+E+H+K)},\ Y = \tfrac{-(D-L+F+I)}{2(C+E+H+K)} \\ U &= \tfrac{-(A+J-B-G)}{(F+I-D-L)},\ V = \tfrac{2(A+J+B+G)}{(F+I-D-L)} \end{aligned} \quad (20)$$

The expressions from A to L are deduced in the following Equations from (21) to (34).

$$A = \frac{(c-b)^2}{\alpha_0} \quad (21)$$

$$B = \frac{1}{\alpha_0} \frac{b^2 r}{s} \left[\frac{1}{2r} + \frac{1}{2s} - \frac{2}{r+s} \right] \quad (22)$$

$$C = \frac{1}{\alpha_0} \left(\frac{\mu'}{\mu}\right)^2 \left[\frac{a'^2}{2r'} + \frac{b'^2}{2s'} - \frac{2a'b'}{r+s} \right] \quad (23)$$

$$D = -\frac{2}{\alpha_0} \left(\frac{\mu'}{\mu}\right) \left[\frac{a'a}{r+r'} + \frac{b'b}{s+s'} - \frac{a'b}{r'+s} - \frac{ab'}{r+s'} \right] \quad (24)$$

$$E = \frac{1}{\alpha_0} \left(\frac{\mu'}{\mu}\right)^2 \frac{b^2 r}{s} \left[\frac{1}{2r'} + \frac{1}{2s'} - \frac{2}{r'+s'} \right] \quad (25)$$

$$F = -\frac{8}{\alpha_0} \left(\frac{\mu'}{\mu}\right) k_r' k_r r' r \left[\frac{1}{r+r'} + \frac{1}{s+s'} - \frac{1}{r'+s} - \frac{1}{r+s'} \right] \quad (26)$$

$$G = \frac{k_r^2}{\beta_0} \left[\frac{1}{2r} + \frac{2sr^2}{b^2} - \frac{4sr}{b(r+s)} \right] \quad (27)$$

$$H = \frac{-2k_r' k_r}{\beta_0} \left[\frac{1}{2r'} + \frac{2sr^2}{b'^2} - \frac{4s'r'}{b'(r'+s')} \right] \quad (28)$$

$$I = \frac{-2k_r' k_r}{\beta_0} \left[\frac{1}{r+r'} + \frac{4ss'rr'}{bb'(s+s')} - \frac{2sr}{b(r'+s)} - \frac{2s'r'}{b'(r+s')} \right] \quad (29)$$

$$J = \frac{r^2}{\beta_0}\left[\frac{1}{2r} + \frac{2k_r^4}{b^2 s} - \frac{4k_r^2}{b(r+s)}\right] \tag{30}$$

$$M = \frac{r'^2}{\beta_0}\left[\frac{1}{2r'} + \frac{2k'_r^4}{b'^2 s'} - \frac{4k'_r^2}{b'(r'+s')}\right] \tag{31}$$

$$L = \frac{-2r'r}{\beta_0}\left[\frac{1}{r+r'} + \frac{4k_r'^2 k_r^2}{bb'(s+s')}\right. \\ \left. - \frac{2k_r^2}{b(r'+s)} - \frac{2k_r'^2}{b'(r+s')}\right] \tag{32}$$

$$\alpha_0 = \frac{a^2 s + b^2 r}{2rs} + \frac{s^2 b + b^2 r}{2s^2} - \frac{2abs + 2b^2 r}{s(r+s)} \tag{33}$$

$$\beta_0 = \frac{k_r^2 + r^2}{2r} + \frac{br}{2s^2} - \frac{b}{s} \tag{34}$$

Given the considerable distance between the building foundation and the point of the vibration source, along with the substantial magnitude difference between the soil and the stiffness value of the building foundation, it is assumed that after the incidence of Rayleigh waves on the building foundation, the foundation undergoes rigid motion, resulting in equal vertical displacements everywhere. Therefore, the vertical displacements are denoted as $U_0^{foot} = A_3$ and $U_0^{soil} = A_1$. With these considerations, the relational expression for the coupling loss coefficient (C_l) is as follows:

$$C_l = \frac{U_0^{foot}}{U_0^{soil}} = \frac{A_3}{A_1} = T = X + YR, R = \frac{A_2}{A_1} = \frac{X-U}{V-Y} \tag{35}$$

U_0^{foot} represents the vertical vibration response of the building foundation surface when a building is present on-site, and U_0^{soil} denotes the vertical response value of the free field in the absence of a building foundation, as illustrated in Figure 6.

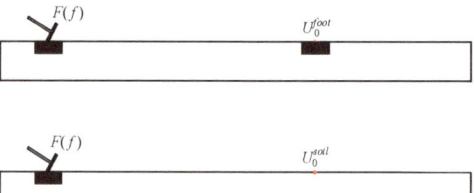

Figure 6. Schematic diagram of the free field and building foundation.

3. Verification of the Derived Results for the Coupling Loss Coefficient

To validate the accuracy of the derived coupling loss coefficient in the soil–structure transfer function, as derived from the Rayleigh wave reflection and transmission theory on the dielectric discontinuity, and to enhance the effectiveness of subsequent calculations of the building vibration responses, a comparative analysis is employed. This method involves comparing the derived results with references in this section. This validation process helps ensure the reliability and correctness of the theoretical framework developed for predicting the coupling loss coefficient in the soil–structure interaction, thereby enhancing confidence in the subsequent calculations related to the building vibration responses.

The data from Colaco et al. [10] are utilized for the soil and building foundation parameters in this study. The selected parameters for the soil medium are as follows:

- Shear wave velocity: $C_{s1} = 150 \text{ m/s}$

- Density: $\rho = 1900$ kg/m^3
- Shear modulus: $G_1 = 42.75$ MPa
- Elastic modulus: $E = 115.425$ MPa

For the building foundation:
- Shear wave velocity: $C_{s2} = 2236.08$ m/s
- Poisson's ratio: $\nu' = 0.2$
- Density: $\rho' = 2500$ kg/m^3
- Shear modulus: $G_2 = 125$ GPa
- Elastic modulus: $E' = 30$ GPa
- Building foundation side length: $B_f = 1.5$ m
- Thickness: $h = 0.6$ m

These parameters will be used for the comparative analysis and validation of the derived coupling loss coefficient in the soil–structure transfer function.

Colaco et al. [26–30] utilized the semi-analytic soil medium model introduced by Bucinskas. In their work, they obtained the soil–structure interaction (SSI) curve based on the Green function in the frequency–wave number domain. This approach effectively captured the relationship between the vibration response of the free field at the building foundation and the vibration response at the center of the building foundation in the presence of a building foundation. In this study, to streamline the calculation process, the soil medium and building parameters are input into a pre-designed table. The coupling loss coefficient is then determined using the Rayleigh wave refraction and transmission theory at various media contact surfaces. Subsequently, the coupling loss coefficient $C_l(f)$ obtained through the reflection and transmission theory of Rayleigh waves on media discontinuities is compared with the $SSI(\omega)$ curve data obtained by. This comparison is conducted through the fitting of the Isqnonlin function, and the results are illustrated in Figure 7. This comparative analysis aims to validate and assess the accuracy of the derived coupling loss coefficient in relation to the SSI curve data obtained.

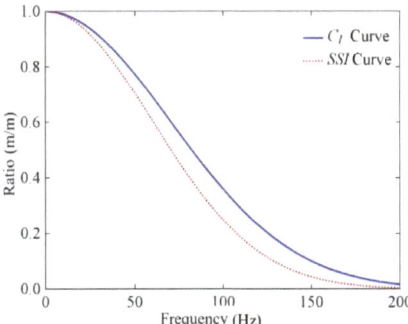

Figure 7. Coupling loss coefficient Cl and SSI (w) curve comparison.

The average ratio between the coupling loss coefficient and the SSI curve, as indicated by the data, is approximately 1.088, and it remains below 1.1. This suggests that the error is below 10%. Moreover, from the observation in Figure 7, it is evident that the coupling loss coefficient and SSI curves are of the same order of magnitude, and their values are closely aligned. The maximum value tends toward 1, and the minimum value tends toward 0. Therefore, the results of the coupling loss coefficient derivation appear reasonable and consistent with the SSI curve data, further supporting the validity of the proposed method.

3.1. Effect of Different Parameters on Coupling Loss Coefficient

The expression for the coupling loss coefficient in the soil–structure transfer function, as derived in Section 2.2 of this paper, is indeed intricate. The expression reveals that alterations in parameters such as the width (B_f), height (h), material properties, and others of the building foundation, as well as variations in the soil mass parameters, can impact the value of the coupling loss coefficient. To investigate the influence of these parameters on the coupling loss coefficient, a control variable method is employed. This method allows for the systematic analysis of how changes in the parameters of the soil mass and the building foundation affect the coupling loss coefficient. This comprehensive analysis aims to discern the laws governing the impact of these parameters on the coupling loss coefficient, providing valuable insights into the behavior of the soil–structure interaction system.

Under standard conditions, the parameters for the soil mass and building foundation are as follows:

For the soil mass:
- Shear wave velocity: C_{s1} = 150 m/s
- Poisson's ratio: ν = 0.35
- Density: ρ = 1900 kg/m^3
- Shear modulus: G_1 = 42.75 MPa
- Elastic modulus: E = 115.425 MPa

For the building foundation:
- Shear wave velocity: C_{s2} = 2236.08 m/s
- Poisson's ratio: ν' = 0.2
- Density: ρ' = 2500 kg/m^3
- Shear modulus: G_2 = 125 GPa
- Elastic modulus: E' = 30 GPa
- Building foundation side length: B_f = 1.5 m
- Thickness: h = 0.6 m

Additionally, the horizontal distance from the Ricker pulse-hammering point to the building foundation is specified as 10 m. Refer to Figures 8 and 9 for a visual representation of the setup.

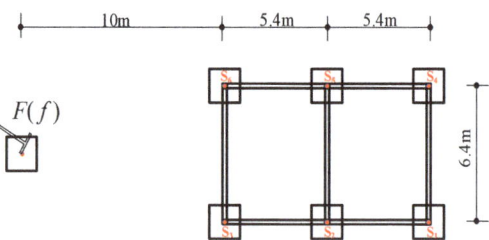

Figure 8. Top view of relationship between the building and the Ricker pulse position.

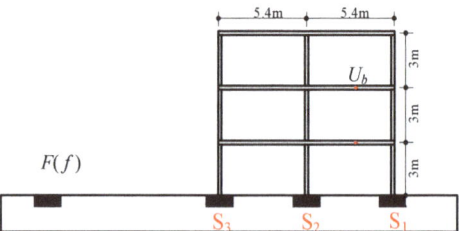

Figure 9. Main view of the relationship with the Ricker pulse position of the building.

The expression of the Ricker pulse is:

$$F(t) = \left[2\left(\frac{\pi(t-t_s)}{T_R}\right)^2 - 1\right] e^{-\left(\frac{\pi(t-t_s)}{T_R}\right)^2} \quad (3-1) \tag{36}$$

where $t_s = 0.1$ s, $T_R = 0.01$ s. The time domain and frequency domain of the Ricker pulse are shown in Figures 10 and 11.

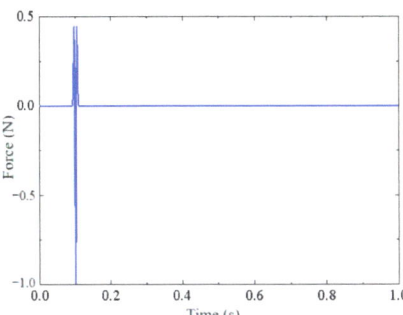

Figure 10. Ricker pulse time domain curve.

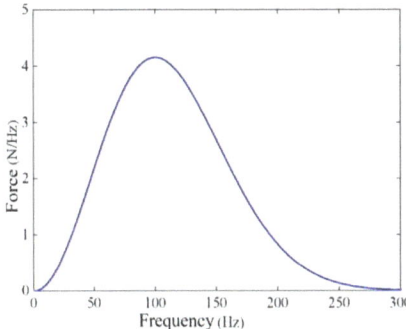

Figure 11. Ricker pulse frequency domain curve.

3.2. Angle Effect on Coupling Loss Coefficient

In the analysis, the horizontal projection point O of the hammering point is kept at a fixed distance $L_{OA} = 10$m from the center A of the building foundation. The damping distance of the Rayleigh wave in front of the building foundation varies with the change in the included angle θ, impacting the dynamic interaction. The analytical diagram considering the angle effect on the coupling loss coefficient is depicted in Figure 12. Additionally, Figure 13 illustrates the influence of different angles on the coupling loss coefficient. These figures provide visual representations of how the included angle θ affects the coupling loss coefficient and help in understanding the dynamic interaction under varying conditions.

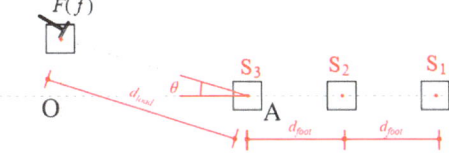

Figure 12. Analytical diagram considering the angle effect on the coupling loss coefficient.

From Figure 13, when the incidence angle θ is within the range of 0 to 45 degrees, assuming no change in the distance L_{OA} (10m) between the horizontal projection point O and the building foundation center A, the propagation distance of the wave in front of the building foundation gradually increases with the rise of the included angle θ. Consequently, the wave damping leads to a gradual reduction in the soil–foundation interaction. As a result, the coupling loss coefficient will progressively increase. Conversely, when the incidence angle θ is in the range of 45 to 90 degrees, the change in the rule is the opposite, leading to different trends in the coupling loss coefficient.

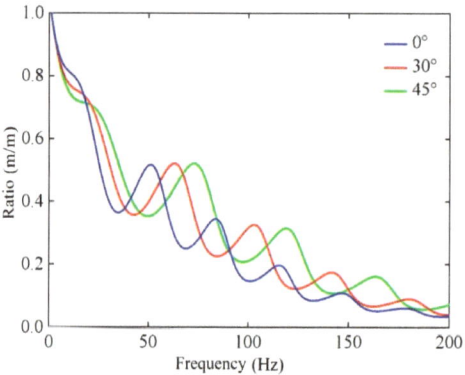

Figure 13. Analytical diagram of the different angles' effect on the coupling loss coefficient.

3.3. Effect of Building Foundation Parameters on Coupling Loss Coefficient

3.3.1. Effect of Building Foundation Size

The variation in the building foundation parameters induces changes in the stiffness of the foundation, subsequently impacting the coupling loss coefficient in the soil–structure interaction. Different building foundation sizes are considered, represented by $Q_1 = 1.5 \text{ m} \times 1.5 \text{ m} \times 0.6 \text{ m}$, $Q_2 = 2.0 \text{ m} \times 2.0 \text{ m} \times 0.8 \text{ m}$, and $Q_3 = 2.5 \text{ m} \times 2.5 \text{ m} \times 1.0 \text{ m}$. By applying the expression of the coupling loss coefficient derived in Section 2.2, the coupling loss coefficient in the soil–structure transfer function of foundation S_3 is observed to change with the building foundation size under the influence of different sizes Q, as illustrated in Figure 14. This analysis provides insights into how alterations in the building foundation size can affect the soil–structure interaction dynamics.

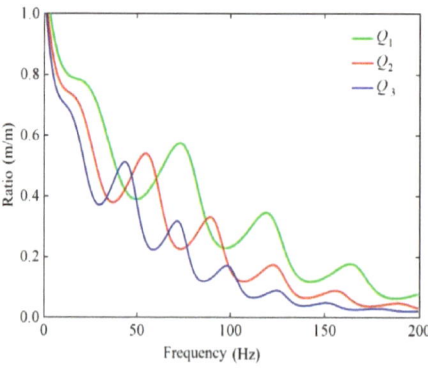

Figure 14. Different Q–coupling loss coefficient curve of building foundation S_3.

building foundation and the free field soil diminishes, resulting in an increased interaction force between the soil and the building. The increase in size corresponds to an increase in the stiffness of the building foundation, and consequently, a larger disparity in the effective stiffness between the soil mass and the foundation. This intensifies the interaction between the soil mass surrounding the foundation and the building foundation, making the interaction more pronounced.

3.3.2. Effect of Building Foundation Density

Under three working conditions, where the density of the building foundation is $\rho'1 = 2400$ kg/m^3, $\rho'2 = 2500$ kg/m^3, and $\rho'3 = 2600$ kg/m^3, with the other parameters unchanged, the calculation equation for the coupling loss coefficient is applied. The resulting variations in the coupling loss coefficient of the soil–structure transfer function with different building foundation densities are illustrated in Figure 15. This analysis provides insights into how changes in the building foundation density affect the coupling loss coefficient and, consequently, the soil–structure interaction dynamics.

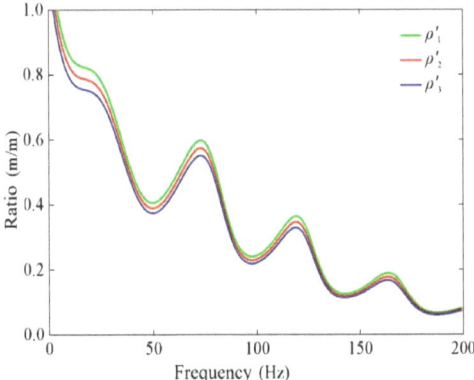

Figure 15. Influence of different building foundation densities on the coupling loss coefficient.

3.3.3. Effect of Elastic Modulus of Building Foundation

Under three different working conditions, where the elastic modulus of the building foundation is selected as $E'_1 = 28.0$ GPa, $E'_2 = 30.0$ GPa, and $E'_3 = 32.5$ GPa, with the other parameters unchanged, the calculation equation for the coupling loss coefficient is applied. The resulting variations in the coupling loss coefficient of the soil–structure transfer function with different building foundation elastic moduli are illustrated in Figure 16. This analysis provides insights into how changes in the elastic modulus of the building foundation affect the coupling loss coefficient and, consequently, the soil–structure interaction dynamics.

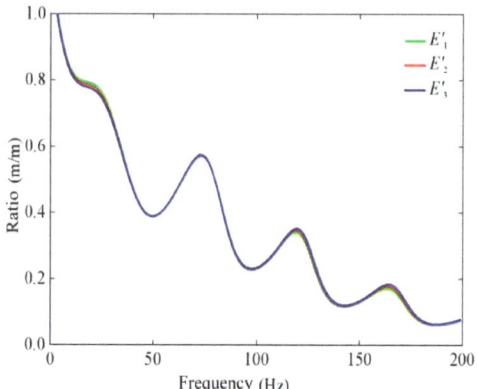

Figure 16. Influence of different elastic moduli of the building foundation on the coupling loss coefficient.

When the elastic modulus of the building foundation (E') changes from 28.0 GPa to 32.5 GPa, the coupling loss coefficient gradually decreases in the frequency band from 10 to 30 Hz. This suggests that the displacement ratio between the building foundation and the free field soil mass decreases gradually within this specific frequency range. However, there is no significant effect on the coupling loss coefficient when there are changes in the elastic modulus of the building foundation in other frequency bands. This observation indicates that the impact of the elastic modulus on the coupling loss coefficient is frequency-dependent and more pronounced within the specified frequency range.

3.4. Effect of Soil Parameter Variation on Coupling Loss Coefficient

3.4.1. Effect of Soil Mass Elastic Modulus E on Coupling Loss Coefficient

To explore the impact of changes in the elastic modulus of the soil mass on the coupling loss coefficient in the soil—structure transfer function, different values for the soil elastic modulus (E) were considered: $E_1 = 100$ MPa, $E_2 = 150$ MPa, $E_3 = 200$ MPa, and $E_4 = 250$ MPa. By utilizing the expression of the coupling loss coefficient introduced in Section 2.2, the resulting variations in the coupling loss coefficient of the soil–structure transfer function with different soil mass elastic moduli are illustrated in Figure 17. This analysis sheds light on how alterations in the soil elastic modulus influence the coupling loss coefficient and, consequently, the soil–structure interaction dynamics.

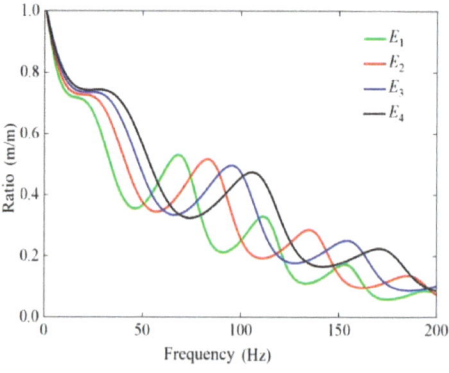

Figure 17. Influence of different elastic moduli of the soil on the coupling loss coefficient.

the displacement ratio between the building foundation and the soil in the free field increases gradually. With the increase in the soil elastic modulus (E), the stiffness of the soil mass becomes larger, resulting in a smaller stiffness ratio between the soil and the building foundation. Consequently, the interaction between the soil mass and the building foundation becomes less pronounced, leading to an increase in the coupling loss coefficient.

3.4.2. Effect of Soil Mass Density on Coupling Loss Coefficient

For three different working conditions, where the soil density is selected as $\rho_1 = 1900$ kg/m³, $\rho_2 = 2000$ kg/m³, and $\rho_3 = 2100$ kg/m³, with the other parameters unchanged, the calculation equation for the coupling loss coefficient is applied. The resulting variations in the coupling loss coefficient of the soil–structure transfer function with different soil mass densities are illustrated in Figure 18. This analysis provides insights into how changes in the soil density affect the coupling loss coefficient and, consequently, the soil–structure interaction dynamics.

From Figure 18, it is evident that as the density ρ of the soil mass changes from 1900 kg/m³ to 2100 kg/m³, the coupling loss coefficient gradually increases. This indicates a gradual increase in the displacement ratio between the building foundation and the free field soil mass. The increase in the coupling loss coefficient is attributed to the increasing density of the soil mass. As the soil density increases, the contact area between the soil particles per unit volume also increases, enhancing the overall stiffness of the soil mass. This, in turn, results in a gradual increase in the stiffness ratio between the soil and the concrete foundation, making the interaction between the soil mass and the concrete foundation less pronounced.

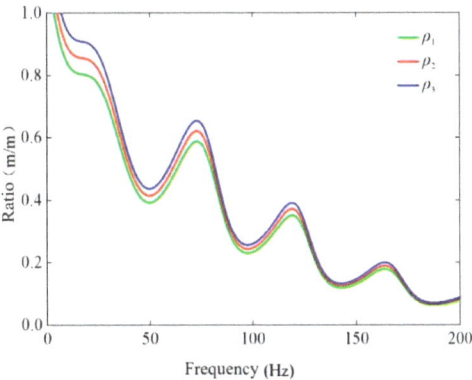

Figure 18. Influence of different soil densities on the coupling loss coefficient.

3.4.3. Effect of Soil Poisson's Ratio on Coupling Loss Coefficient

For three different working conditions, where the Poisson's ratio of the soil is selected as $\nu_1 = 0.2$, $\nu_2 = 0.3$, and $\nu_3 = 0.4$, with the other parameters unchanged, the calculation equation for the coupling loss coefficient is applied. The resulting variations in the coupling loss coefficient of the soil–structure transfer function with different Poisson's ratios of the soil mass are illustrated in Figure 19. This analysis provides insights into how changes in the Poisson's ratio of the soil affect the coupling loss coefficient and, consequently, the soil–structure interaction dynamics.

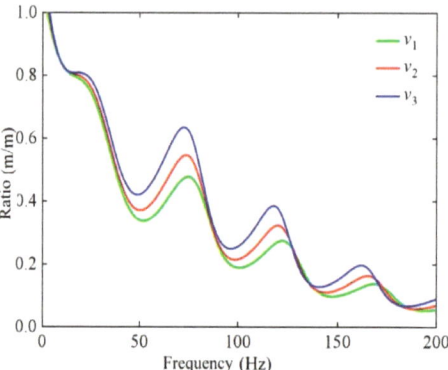

Figure 19. Influence of different Poisson's ratios of the soil on the coupling loss coefficient.

From Figure 19, it is apparent that as the Poisson's ratio (ν) of the soil mass changes from 0.2 to 0.4, the coupling loss coefficient in the soil–structure transfer function increases gradually. This implies a gradual increase in the displacement ratio between the building foundation and the soil mass. Moreover, the interaction force between the soil mass and the building foundation decreases gradually. The observed trends suggest that changes in the Poisson's ratio of the soil affect the soil–structure interaction dynamics, influencing the coupling loss coefficient accordingly.

4. Case Study: Building Foundation Response under Ricker Pulse

The vertical displacement expression of the building foundation is:

$$U_0^{foot}(f) = U_0^{soil}(f) \cdot C_l(f) \tag{37}$$

The expression for the coupling loss coefficient $C_l(f)$ has been derived in Section 2.2. Consequently, the vertical free field displacement of the building foundation under the action of a Ricker pulse can be denoted as $U_0{}^{soil}$ (as shown in Figure 20). The vertical displacement $U_0{}^{foot}$ of the building foundation under the action of a Ricker pulse can then be determined using this expression.

Figure 20. Schematic diagram of the vertical displacement of the free field at S3 of the building foundation under the action of an ER pulse.

4.1. Solution of Vertical Displacement Response of Free Field under Ricker Pulse

In Equation (38), the expression for the vertical displacement of the free field particle at the building foundation under a simple harmonic load $F(t) = Pe^{i\omega t}$ is as follows:

$$U_0^{soil} = \frac{P}{\mu}\bar{\xi}\sqrt{\frac{1}{r\lambda_r}}\cos\left(\omega t - k_r r - \frac{\pi}{4}\right) \tag{38}$$

The data of the case in Section 3.1 are input into the above equation, and the free field displacement $U_0{}^{soil}$ curve at S_3 of the building foundation under a Ricker pulse is obtained as follows.

From Figure 21, it can be observed that the frequency corresponding to the peak vertical displacement is approximately 120 Hz. This aligns with the peak of the Ricker pulse excitation frequency spectrum curve.

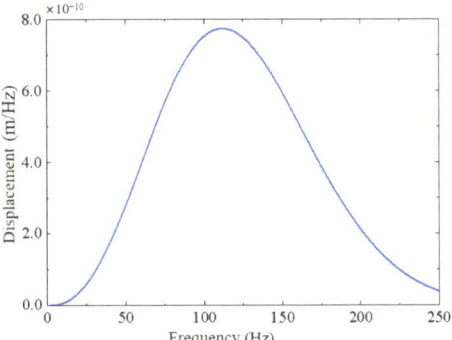

Figure 21. Vertical displacement curve of the free field at S3 of the building foundation under the action of an ER pulse.

4.2. Solution for the Coupling Loss Coefficient of Building Foundation

According to the expression C_l of the coupling loss coefficient in Section 2.2, the coupling loss coefficient curves of S_1, S_2, S_3 of the building foundation can be drawn in Figure 22.

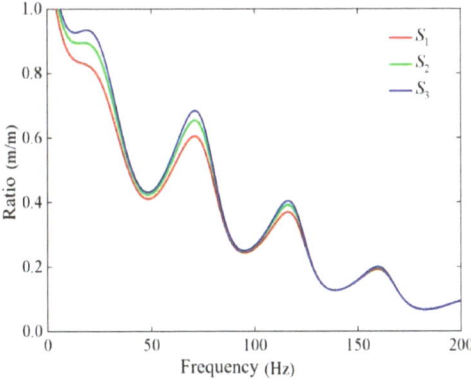

Figure 22. Curve of the coupling loss coefficient of the building foundation.

Recognizing the proximity of the coupling loss coefficient curves for different foundations, a Gaussian curve is contemplated for fitting to derive the coupling loss coefficient for the entire building. The expression is:

$$G(\gamma_R) = \exp^{-\frac{(\gamma_R)^2}{2c^2}}, \gamma_R = \frac{\omega B_f}{4\pi C_s} \qquad (39)$$

The variable c serves as a real constant, representing the standard deviation. The determination of the standard deviation c value is achieved through fitting the coupling loss coefficient curves of the building foundations S_1, S_2, and S_3 using the lsqnonlin function in MATLAB. The resulting coupling loss coefficient curves, denoted as C_l, fitted to the building foundations S_1, S_2, and S_3, are depicted in Figure 23.

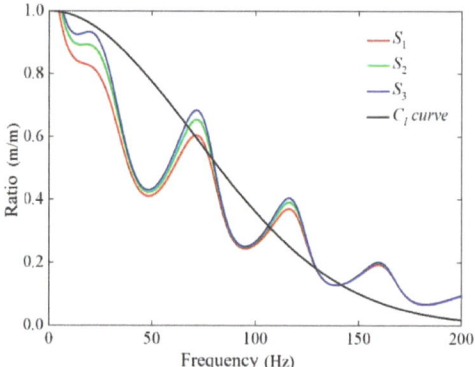

Figure 23. Fitted coupling loss coefficient curves.

4.3. Solution of Vertical Displacement of Building Foundation

Considering that the vibration response of the building foundation is given by $U_0^{foot}(f) = U_0^{soil}(f) \cdot C_1(f)$, the vertical displacement curve of the free field at the building foundation S3 under Ricker pulse action (Figure 21) is multiplied by the fitting coupling loss coefficient, as illustrated in Figure 24. Consequently, the vertical displacement values $U_0^{foot}(f)$ of the building foundation S_3 under a Ricker pulse are obtained, as depicted in Figure 25.

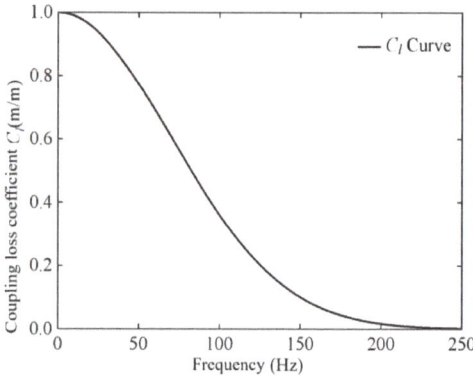

Figure 24. Coupling loss coefficient of the building foundation.

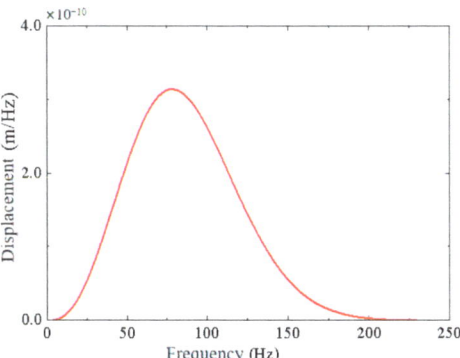

Figure 25. Vertical displacement curve of the S_3 building foundation under a Ricker pulse.

Likewise, the vertical displacement values $U_0^{foot}(f)$ and free field displacement values $U_0^{soil}(f)$ of the building foundation S_2 under a Ricker pulse are illustrated in Figure 26.

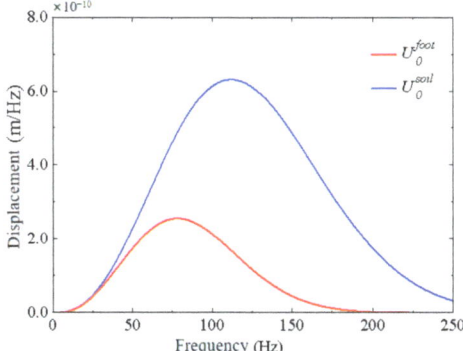

Figure 26. Vertical displacement curve of the building foundation S_2 and vertical displacement curve of the free field under a Ricker pulse.

It is evident from the figure that the frequency corresponding to the peak vibration response of the building foundation is higher than that corresponding to the peak vibration response of the soil mass. This discrepancy can be attributed to the significant differences in the elastic modulus and density between the concrete foundation and the soil mass. Accounting for the coupling loss coefficient helps mitigate the response of the high-frequency vibration component. The vertical displacement value $U_0^{foot}(f)$ and free field displacement value $U_0^{soil}(f)$ of the building foundation Sb under a Ricker pulse are presented in Figure 27.

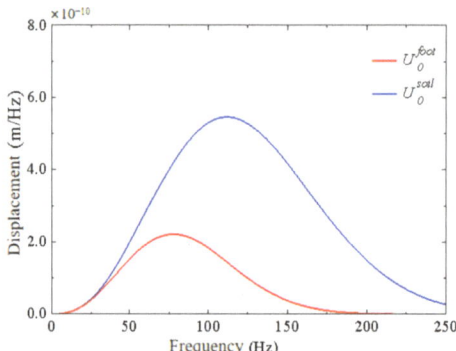

Figure 27. Vertical displacement curve of the building foundation S_1 and vertical displacement curve of the free field under a Ricker pulse.

5. Application Case: Prediction of High-Speed Train-Induced Soil–Structure Vibration

To validate the accuracy of the soil–structure transfer function method in predicting the vibrations induced by high-speed trains on a proposed building, a comparison was performed with actual measurement data. The experimental building is a 12-story reinforced concrete structure with a total height of 36.6 m above ground. The first floor has a height of 3.6 m, while the remaining floors are each 3 m high. The total length is 45 m, comprising 10 rooms, with each bay measuring 4.5 m. The spans are 6 m + 2.4 m + 6 m. The distance between the centerline of the railway track and the building is 12 m, as illustrated in Figure 28. The soil parameters of the experimental site are shown in Table 1.

Figure 28. Test site.

Table 1. Soil parameters at the test site.

Type	Elastic Modulus E/MPa	Poisson	Density (kg/m³)	Thickness (m)
Backfill	29	0.4	1700	1
Sand clay	127	0.33	1880	3
Coarse sand	150	0.33	1918	4
Gravel sand	430	0.29	1211	40

The measured values of the maximum vertical vibration acceleration are selected for the analysis. The soil and building parameters at the test site are then substituted into the soil–structure transfer function to calculate the building's vibration response. The vibration response of each floor is analyzed using the vibration acceleration level (VAL) as

The measured values are to be compared with the calculated results from the analytical coupling loss coefficient, as illustrated in Figure 29. It can be observed that the test values of the building vibrations generally show an increasing trend with the number of floors, although it is not consistently monotonic. The results obtained using the soil–structure transfer function method exhibit a monotonic increase and slightly surpass the measured average value. Both numerical trends demonstrate an overall increase with the number of floors, and the values are relatively close, indicating the feasibility of the soil–structure transfer function method in predicting the vibrational response of buildings induced by train activity.

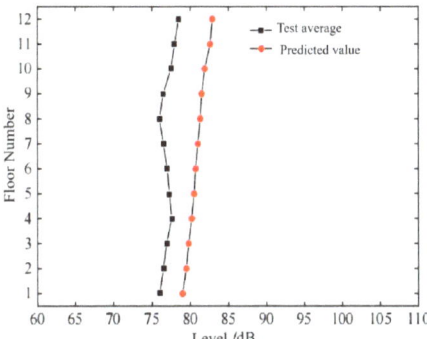

Figure 29. Comparison of the test value of the building's vibration level with the predicted values via the analytical coupling loss coefficient.

6. Conclusions

This paper successfully derives the calculation equation for the transfer function coupling loss coefficient and thoroughly analyzes the factors influencing the coupling loss coefficient, such as the angle, soil mass parameters, and building foundation parameters. Furthermore, the obtained vertical displacement values of the building foundation under a Ricker pulse are compared with references, providing a solid foundation for calculating the building's vibration response.

(1) Advantages of the Coupling Loss Coefficient in the Soil–Structure Transfer Function:

The coupling loss coefficient streamlines the calculation process, eliminating the need for testing and extensive modeling. Theoretical analysis allows for a quick and accurate determination of the displacement transfer relationship between the soil and the building, effectively illustrating their interaction.

(2) Complexity and Parameter Dependencies of the Coupling Loss Coefficient:

The coupling loss coefficient exhibits complexity and dependencies on the building foundation and soil mass parameters. The analysis reveals that increasing the building foundation size results in an increased coupling loss coefficient, indicating a reduction in the displacement ratio between the building foundation and the free field soil mass, leading to decreased interaction force. Moreover, the coupling loss coefficient is influenced by changes in the elastic modulus, density, and Poisson's ratio of the building foundation and soil mass.

(3) Insights from the Coupling Loss Coefficient Curve:

The coupling loss coefficient curve demonstrates values ranging from 0 to 1, signifying that in the presence of a building foundation, Rayleigh wave transmission and reflection occur, mitigating the vibration induced by incident Rayleigh waves.

In summary, the findings of this study offer valuable insights into the coupling loss coefficient, its dependencies, and its role in reducing the vibrations induced by Rayleigh waves. Utilizing the soil–structure transfer function, this research efficiently forecasts the building vibrations induced by train activities. This method, after determining the coupling loss and floor amplification coefficients, integrates with a broad range of free field vibration data. It simplifies and enhances the predictions for structures influenced by various factors, such as construction, seismic events, road traffic, and explosions. By providing a versatile tool, the approach aids in swiftly and accurately anticipating vibrations, especially for prospective buildings near high-speed rail tracks.

Author Contributions: J.Y. came up with the concept and wrote the draft of the manuscript. Z.W. conducted the literature review, replied to the reviewers' comments and revised the final version. X.C. analyzed the data. N.W. checked the computations. N.Z. corrected the model of the simulation and debugged the programming. All authors have read and agreed to the published version of the manuscript.

Funding: The performed research described was financially supported by the Fundamental Research Funds for the Central Universities (2023YJS043), National Natural Science Foundation of China (52178101), and State Key Laboratory for Track Technology of High-Speed Railway, China (2022YJ175).

Data Availability Statement: The data presented in this article are available on request from the corresponding author.

Conflicts of Interest: The authors declare no conflicts of interest.

References

1. Krylov, V.V. On the Theory of Railway-Induced Ground Vibrations. *J. De Phys. IV* **1994**, *4*, 769–772. [CrossRef]
2. Krylov, V.V. Generation of Ground Vibrations by Superfast Trains. *Appl. Acoust.* **1995**, *44*, 149–164. [CrossRef]
3. Krylov, V.V. Effects of Track Properties on Ground Vibrations Generated by High-Speed Trains. *Acust.-Acta Acust.* **1998**, *84*, 78–90.
4. Kurzweil, L.; Cobb, W.; Dinning, M. *Urban Rail Noise Abatement Program: A Description*; Urban Mass Transportation Administration: Washington, DC, USA, 1980.
5. Hirokazu, T.; Taro, U.; Satoshi, F. Component modal synthesis method for 3D seismic response analysis of subground-foundation-superstructure systems. *World Earthq. Eng.* **1988**, *4*, 43–47.
6. Yue, J. Measurement and Numerical Simulation Analysis of building vibration caused by subway traffic. *J. Petrol. Geophys.* **2020**, *41*, 2756–2764.
7. Lei, F. *Research on Prediction Method of Environmental Vibration Caused by High-Speed Train Based on Random Forest Algorithm*; Beijing Jiaotong University: Beijing, China, 2020.
8. Colaço, A.; Costa, P.A.; Amado-Mendes, P.; Calçada, R. Vibrations induced by railway traffic in buildings: Experimental validation of a sub-structuring methodology based on 2.5D FEM-MFS and 3D FEM. *Eng. Struct.* **2021**, *240*, 112381. [CrossRef]
9. Colaço, A.; Costa, P.A.; Castanheira-Pinto, A.; Amado-Mendes, P.; Calçada, R. Experimental Validation of a simplified soil-structure interaction approach for the prediction of vibrations in buildings due to railway traffic. *Soil Dyn. Earthq. Eng.* **2021**, *141*, 106499. [CrossRef]
10. Colaço, A.; Barbosa, D.; Costa, P.A. Hybrid soil-structure interaction approach for the assessment of vibrations in buildings due to railway traffic. *Transp. Geotech.* **2022**, *32*, 100691. [CrossRef]
11. Sadeghi, J.; Vasheghani, M. Safety of buildings against train induced structure borne noise. *Build. Environ.* **2021**, *197*, 107784. [CrossRef]
12. Farahani, M.V.; Sadeghi, J.; Jahromi, S.G.; Sahebi, M.M. Modal based method to predict subway train-induced vibration in buildings. *Structures* **2023**, *47*, 557–572. [CrossRef]
13. Sadeghi, J.; Vasheghani, M. Improvement of current codes in design of concrete frame buildings: Incorporating train-induced structure borne noise. *J. Build. Eng.* **2022**, *58*, 104955. [CrossRef]
14. Ibrahim, Y.E.; Nabil, M. Finite element analysis of multistory structures subjected to train induced vibrations considering soil-structure in-teraction. *Case Stud. Constr. Mater.* **2021**, *15*, 100592.
15. ISO 14837-1; Mechanical Vibration–Groundborne Noise and Vibration Arising from Rail Systems—Part 1: General Guidance. International Organization for Standardization: Geneva, Switzerland, 2005.
16. Richart, F.E., Jr.; Hall, J.R., Jr.; Woods, R.D. *Vibration of Soil and Foundation*; Prentice-Hall: Saddle River, NJ, USA, 1970.
17. Zhao, R. *Theoretical Research and Numerical Analysis of Effects of Filling Ditch on Environmental Vibration Caused by Running Trains*; Beijing Jiaotong University: Beijing, China, 2017.
18. Du, X.-L. *Theories and Methods of Wave Motion for Engineering*; Science Press: Beijing, China, 2009. (In Chinese)
19. Maurice, J.Y. The reflection and transmission of Rayleigh waves. *Univ. Appl. Phys.* **2012**, *112*, 103520.
20. Lapwood, E.R. The Transmission of a Rayleigh Pulse round a corner. *Geophys. J. R. Astron. Soc.* **1961**, *4*, 174–196. [CrossRef]
21. Asadi Jafari, M.H.; Zarastvand, M.; Zhou, J. Doubly curved truss core composite shell system for broadband diffuse acoustic insulation. *J. Vib. Control* **2023**. [CrossRef]
22. Qiu, Y.; Zou, C.; Hu, J.; Chen, J. Prediction and mitigation of building vibrations caused by train operations on concrete floors.

23. Hu, J.; Zou, C.; Liu, Q.; Li, X.; Tao, Z. Floor vibration predictions based on train-track-building coupling model. *J. Build. Eng.* **2024**, *89*, 109340. [CrossRef]
24. He, L.; Tao, Z. Building Vibration Measurement and Prediction during Train Operations. *Buildings* **2024**, *14*, 142. [CrossRef]
25. Ma, M.; Xu, L.H.; Liu, W.F.; Tan, X. Semi-analytical solution of a coupled tunnel-soil periodic model with a track slab under a moving train load. *Appl. Math. Model.* **2024**, *128*, 588–608. [CrossRef]
26. Bucinskas, P.; Andersen, L.V. Semi-analytical approach to modelling the dynamic behaviour of soil excited by embedded foundations. *Procedia Eng.* **2017**, *199*, 2621–2626. [CrossRef]
27. Bucinskas, P.; Andersen, L.V. Dynamic response of vehicle–bridge–soil system using lumped parameter models for structure–soil interaction. *Comput. Struct.* **2020**, *238*, 106270. [CrossRef]
28. Persson, P.; Andersen, L.; Persson, K.; Bucinskas, P. Effect of structural design on traffic-induced building vibrations. *Procedia Eng.* **2017**, *199*, 2711–2716. [CrossRef]
29. Bucinskas, P.; Ntotsios, E.; Thompson, D.J.; Andersen, L.V. Modelling train-induced vibration of structures using a mixed-frame-of-reference approach. *J. Sound Vib.* **2021**, *491*, 115575. [CrossRef]
30. Bucinskas, P.; Persson, K. Numerical modelling of ground vibration caused by elevated high-speed railway lines considering structure-soil-structure interaction. *Inst. Noise Control Eng.* **2016**, *253*, 7017–7028.

Disclaimer/Publisher's Note: The statements, opinions and data contained in all publications are solely those of the individual author(s) and contributor(s) and not of MDPI and/or the editor(s). MDPI and/or the editor(s) disclaim responsibility for any injury to people or property resulting from any ideas, methods, instructions or products referred to in the content.

Article

Analysis of Train-Induced Vibration Transmission and Distribution Characteristics in Double-Layer Metro Depot

Xinwei Luo [1], Xuan Jiang [2], Qingsong Feng [2,*], Wenlin Hu [3], Qinming Tu [1] and Yanming Chen [2]

1 Guangzhou Metro Design & Research Institute Co., Ltd., Guangzhou 510010, China; luoxinwei@dtsjy.com (X.L.); tuqinming@dtsjy.com (Q.T.)
2 State Key Laboratory of Performance Monitoring Protecting of Rail Transit Infrastructure, East China Jiaotong University, Nanchang 330013, China; 13850579121@163.com (X.J.); ymchen@ecjtu.edu.cn (Y.C.)
3 National Engineering Center for Digital Construction and Evaluation Technology of Urban Rail Transit, China Railway Design Corporation, Tianjin 300308, China; huwenlin@crdc.com
* Correspondence: fqs1978@ecjtu.edu.cn

Citation: Luo, X.; Jiang, X.; Feng, Q.; Hu, W.; Tu, Q.; Chen, Y. Analysis of Train-Induced Vibration Transmission and Distribution Characteristics in Double-Layer Metro Depot. *Buildings* 2024, 14, 1702. https://doi.org/10.3390/buildings14061702

Academic Editor: Shaohong Cheng

Received: 26 April 2024
Revised: 31 May 2024
Accepted: 3 June 2024
Published: 7 June 2024

Copyright: © 2024 by the authors. Licensee MDPI, Basel, Switzerland. This article is an open access article distributed under the terms and conditions of the Creative Commons Attribution (CC BY) license (https://creativecommons.org/licenses/by/4.0/).

Abstract: When urban subway trains run in the depot, they can cause vibration and noise, which affects the safety and reliability of the structure under the track, and these transmits to the over-track buildings and often trouble passengers and staff. This paper established a coupling model of a track–metro depot–over-track building based on the structural finite element method and analyzed vibration response and then summarized the vibration transmission and distribution characteristics as the speed changes. The results show that, at train speeds of 20 km/h and 5 km/h, the Z-vibration level difference between the two at the rail is nearly 20 dB, and the vibration can be reduced by 17.9% at most. The difference between the two on the 9 m platform is 6–8 dB and 5–14 dB on the 16 m platform, and the vibration can be reduced by 17.7% at most. The difference between the two in the over-track building is 3–11 dB, and the vibration can be reduced by 13.0% at most. The vibration has the highest energy within a range of 2 m radiating from the center of the line, reaching a maximum of 118.5 dB. The vibration shows a ring-shaped distribution, and the ring-shaped distribution is more pronounced as the train speed increases. In the horizontal direction of the track line, the vibration energy distribution is within a range of −4 m to 11.5 m from the track line. In the longitudinal direction of the track line, the ring-shaped distribution of vibration energy exhibits a periodic pattern. The results provide a reference for the vibration control of the over-track buildings.

Keywords: urban subway; metro depot; over-track building; train-induced vibration; transmission and distribution characteristics

1. Introduction

With the rapid development of urbanization brought about by population gathering in cities, major cities have begun to operate urban rail transit to alleviate the increasing traffic pressure. The top 10 cities with the highest subway mileage in the world are shown in Table 1. As of the end of 2022, a total of 55 cities in mainland China have invested in and operated urban rail transit, with a total length of 10,287.45 km, of which the subway operation line is 8008.17 km, accounting for 77.84%. A total of 489 metro depots and parking lots have been put into operation. The metro depot is the largest area used by the subway system, responsible for the parking and maintenance of subway trains. It can be divided into throat areas, testing lines, parking and inspection garages, and entrance and exit section lines according to the operating area. Developing over-track properties of the metro depots can not only improve land use efficiency and alleviate urban land scarcity but also generate huge commercial value and raise funds for subway operation and maintenance as well as the construction of new lines. Nowadays, the development of over-track properties for metro depots has become a popular trend for major cities to

Table 1. The top 10 cities with the highest subway mileage in the world.

City	Mileage (km)
Shanghai	831
Beijing	783
Guangzhou	621.05
Chengdu	518
Moscow	466.8
Hangzhou	450
Wuhan	435
Chongqing	432.8
Nanjing	427
Shenzhen	419

However, the operation of subway trains can cause environmental vibration, which will transmit to nearby buildings and over-track buildings through the foundation, soil layer, and columns under the track. Not only does it affect the staff in the metro depot, but it also affects the sleep, study, work, and daily life of residents of the building and even affects the safety of the building and the normal use of precision instruments. Zou et al. [1] conducted field measurements of vibration during subway operations at Shenzhen and found that vibration amplification around the natural frequency in the vertical direction of over-track building made the peak values of indoor floor vibration about 16 dB greater than outdoor platform vibration. Then, it is recommended to carefully examine the design of new over-track buildings within 40 m on the platform over the throat area to avoid excessive vertical vibrations and noise. Xia et al. [2] conducted field measurements on the over-track buildings in a certain city and found that, when the train speed was 15–20 km/h, the vibration level was as high as 85 dB. Therefore, the rationality of the design of vibration reduction measures for metro depots is crucial, and the issue of train-induced vibration of the buildings above the depots cannot be ignored.

Many international scholars have conducted extensive research on the train-induced vibration characteristics [3–11]. The research methods for environmental vibration caused by urban rail transit mainly include field measurement and numerical simulation. Chen et al. [12,13] conducted field measurements on the largest underground metro depot in Asia and found that the vibration acceleration level of the top platform in the throat area was about 78 dB, which was 6 dB higher than the nighttime threshold, and the intermediate frequency vibration had a higher vibration level and a smaller attenuation rate. Feng et al. [14–16] conducted field measurements and numerical simulations on different areas and over-track buildings in metro depots, analyzing the differences in vibration attenuation patterns among various areas. However, it must be acknowledged that field measurement methods have certain limitations for vibration prediction. Sanayei et al. [17,18], Zou et al. [19–22], and Tao et al. [23–26] presented impedance-based (wave propagation) model for predicting train-induced floor vibrations in buildings and conducted field measurement to compare the test results with the predicted results of the impedance-based model. These studies indicate that using the impedance-based model to predict train-induced vibration is feasible and has high computational efficiency. Liu et al. [27], Liang et al. [28–30], Zhou et al. [31], and He et al. [32] proposed a deep learning-based approach to identify train-induced vibration segments efficiently for subsequent vibration evaluations. He et al. [33] presented a three-dimensional analytical model that regarded tunnels as cylindrical shells of infinite length, to predict ground vibrations from two parallel tunnels embedded in a full space. He et al. [34] used the potential decomposition, multiple scattering theory and combined it with the transfer matrix method to derive the fundamental solution for the soil-inclusion dynamic interaction in a layered half-space and then used periodic barriers in a layered half space to mitigate railway-induced vibrations. Li et al. [35] proposed a deep learning-based approach to learn the generation, distribution, and dissipation mechanisms of indoor structure-borne noise while also enabling the convenient acquisition of indoor structure-borne noise. Qiu et al. [36] developed a numerical model based on train track coupled dynamic theory and the finite element method to investigate the effectiveness of two mitigation measures implemented in the elevated metro depot. Hu et al. [37] simplified the building floor into a rectangular plate composed of multiple orthogonal structural girders and structural columns in vertical contact with the floor, to obtain the vertical vibrations of the building floor in the time domain.

Existing research is mostly based on conducting field measurements on the metro depot and carrying out numerical simulation for specific buildings or areas and analyzes the vibration transmission and distribution characteristics based on the measurements and simulation results and then takes targeted vibration reduction measures. But there is little research on the impact of different train speeds on the vibration transmission and distribution characteristics of the vibration sources inside the double-layer metro depot and the over-track buildings. The train speed not only affects the vibration characteristics of the vibration source and the over-track building but also has a certain effect on the vibration reduction and noise reduction of the over-track building by controlling the train speed. Based on this, this paper combines the operation zone of a certain metro depot in Guangzhou and then establishes a coupling model of a track–metro depot–over-track building based on the structural finite element method to calculate the vibration response. It further analyzes the vibration response characteristics of the vibration source and over-track building under different train speeds and then summarizes the transmission and distribution characteristics of vibration in a double-layer prefabricated assembly metro depot and provides an useful reference for metro depots to take vibration reduction measures at vibration sources, propagation paths, or sensitive targets during the vibration reduction design stage.

2. Project Profile

This metro depot is a double-layer prefabricated assembly metro depot. It is planned to carry out property development on the cover and adjacent plots. The main structure of the metro depot adopts a double-layer reinforced concrete frame structure. The building plan of the over-track buildings is shown in Figure 1. The main cover platforms of the metro depot include −11.5 m bottom platform, 0 m platform, 9 m platform, and 16 m platform. The project covers an area of 179,500 square meters. Among them, the −11.5 m bottom platform is the negative driving layer of the metro depot, the 0 m platform is the first driving layer of the metro depot, the 9 m platform is the car parking garage and equipment layer, and the 16 m platform is the ground layer of the community.

Figure 1. Building plan of the over-track buildings.

When the train is running on the first floor (the 0 m platform) of the operation zone, vertical vibration analysis is conducted at points such as rails, supporting columns, columns (1.8 m above the 0 m platform), and the ground at a distance of 7.5 m from the rails at the

above the rails are selected to analyze vertical vibration. The layout of analysis points on vibration source and platform is shown in Figure 2. The over-track building has a total of 41 floors, with four households on each floor. The total height of the over-track building is 134 m, with a first-floor height of 6 m, and a standard height of 3.2 m for floors above two. The center of the living room floor, the center of the master bedroom floor, and the center of the secondary bedroom floor are selected to analyze vertical vibration. The layout of analysis points in the over-track building is shown in Figure 3.

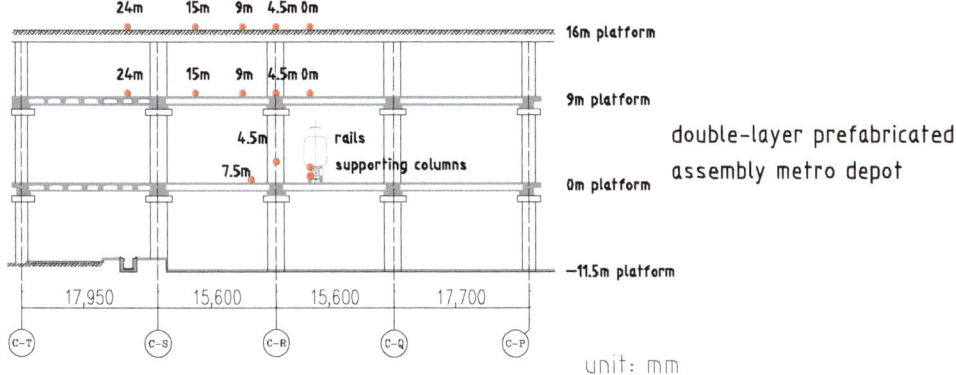

Figure 2. Layout of analysis points on vibration source and platform.

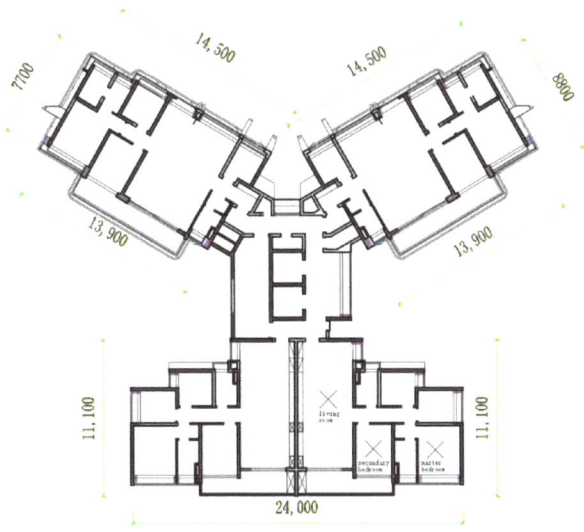

Figure 3. Layout of analysis points in the over-track building.

3. Modeling

3.1. The Vehicle–Track Coupling Model

The coupling model of a track–metro depot–over-track building was established for structural dynamic analysis. The coupling model includes two sub-models. Firstly, the "vehicle–track" coupling model, based on the running train and the track structure in the metro depot, and the multi-body system dynamics simulation software UM 9 is used to calculate the vertical force between wheels and rails. The depot adopts subway 6A model, which is mainly composed of car body, bogie, and wheel pair. The sample of track irregularity is obtained from the short-band processing of the superposition of Sato spectrum in the depot. The "vehicle–track" coupling model is shown in Figure 4.

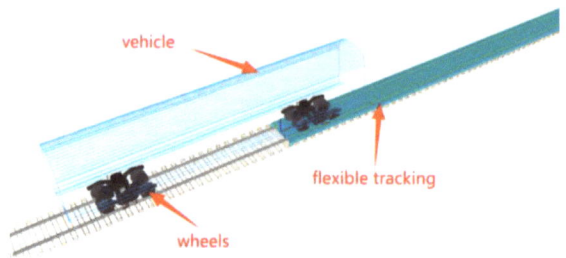

Figure 4. The "vehicle–track" coupling model.

Based on the actual line conditions on site, this article adopts field measured track irregularities of lines with similar working conditions. Due to the fact that the track inspection vehicles (GJ-4 and GJ-5) on the existing lines in China are still unable to accurately detect short-wave irregularities with wavelengths below 1 m, this article uses measured track irregularities in the metro depot in the long-wavelength range (1.5–42 m) and adds Sato spectrum processing in the short-wavelength range to obtain a sample of track irregularity as shown in Figure 5. Using the sample of track irregularity and train motion speed as the input excitation for the "vehicle–track" coupling model, the vertical force between wheels and rails can be calculated.

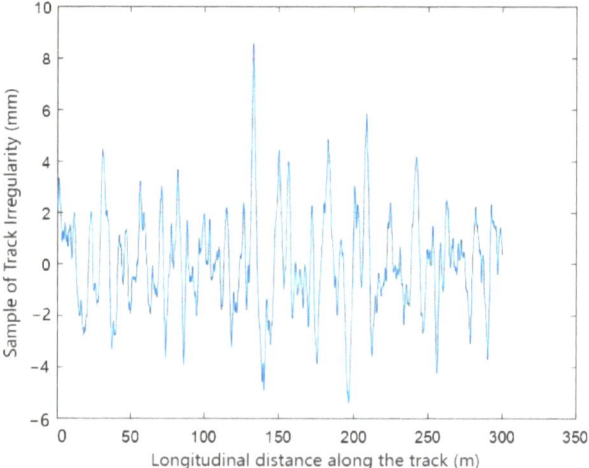

Figure 5. Sample of track irregularity.

3.2. The FEM Model of "Track–Metro Depot–Over-Track Building"

This article considers the actual parameters of the track structure, soil layer, and building and establishes a coupling finite element analysis model of a "track–metro depot–over-track building" using the finite element software, ANSYS 19.0. Using the vertical force between wheels and rails as the input excitation for the "track–metro depot–over-track building" model, the vibration response can be calculated. Among them, solid elements are used for the simulation of soil, and the Mohr–Coulomb (MC) model is chosen as the constitutive model of soil. Beam elements are used for the simulation of beams, columns, and pile foundations, and shell elements are used for the simulation of shear walls and floor

artificial boundary is used to simulate the boundary of the computational domain [38]. The size of the soil is 90 m*30 m*48 m. The building has a total of 41 floors, with a total structural height of 134 m. The first floor of the building is 6 m high, and the standard height for floors above two is 3.2 m. The selection and material parameters of each structural element in the model are shown in Table 2, where the material parameters are set based on the information provided by the design company. The coupling model of "track–metro depot–over-track building" is shown in Figure 6.

Table 2. Selection of structural elements and material parameters.

Name	Poisson's Ratio	Elastic Modulus (MPa)	Density (kg/m³)	Element Selection
Plain fill (the first layer)	0.470	206.63	1760	solid 45
Mucky soil (the second layer)	0.483	85.4	1700	solid 45
Fine sand (the third layer)	0.471	206.4	1950	solid 45
Bearing platform	0.2	3300	2500	solid 45
Pile foundation	0.2	3450	2400	beam 188
Rails	0.3	210,000	7830	beam 188
Supporting columns	0.2	34,500	2400	beam 188
Concrete masonry	0.2	34,500	2400	beam 188
Exterior wall	0.2	33,000	2500	shell 63

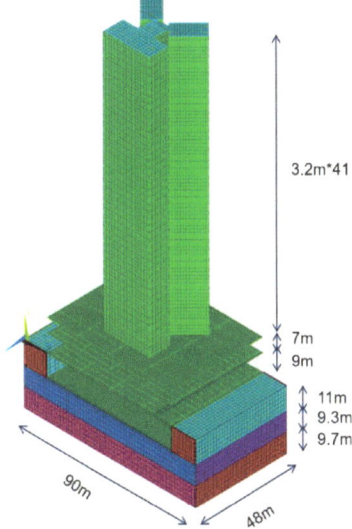

Figure 6. The "track–metro depot–over-track building" coupling model.

3.3. Model Validation

Due to the incomplete construction of the over-track building of this double-layer prefabricated assembly metro depot, in order to verify the rationality and feasibility of the method for establishing the "track–metro depot–over-track building" coupling model, this paper takes a single-layer ground metro depot in Guangzhou as an example to compare the simulation results with the measurement results. The platform above the operation zone is a complete cast-in-place concrete slab that is separated from the throat area and maintenance zone through expansion joints. The pillars in this area are evenly distributed in a chessboard pattern, and the entire platform is divided into small areas. The parameters of the model constructed in this article are basically consistent with the geological exploration soil layer data and building structural parameters provided by the design institute for the metro depot. Therefore, when using ANSYS to establish the single-layer ground metro depot, it is considered that the selection of each structural element and material parameters are consistent with the double-layer prefabricated assembly metro depot model established in this paper, as shown in Table 1. The sample of track irregularity used to calculate the wheel–rail force is the measured sample of track irregularity of the rails in the depot, and the train speed is 2.5 m/s.

The team responsible for measurement uses the SQuadriga III data acquisition instrument for vibration source testing and platform vibration testing and uses trigger sampling for monitoring. PCB-352 vibration accelerometer and PCB-393 vibration accelerometer are used to collect vibration signals. The instrument and accelerometers are shown in Figure 7. Through field measurement, the vibration response of the vibration source in the operation zone was obtained. Measuring points 1 and 2 were placed in the center of the two areas on the platform, with the direction of the measuring line perpendicular to the track line. Measuring point 1 is located directly above the track, and measuring point 2 is located 9 m to the right above the track. The layout of the measuring points is shown in Figure 8. The spectrum analysis of the measured and the simulated vibration frequency at the two measuring points is shown in Figure 9. It can be seen from the figure that the measured and the simulated vibration frequency at the two measuring points is in the range of 20–60 Hz, and the waveforms are similar. The simulation results are in good agreement with the measurement results, indicating that the method for establishing "track–metro depot–over-track building" coupling model is reasonable and feasible.

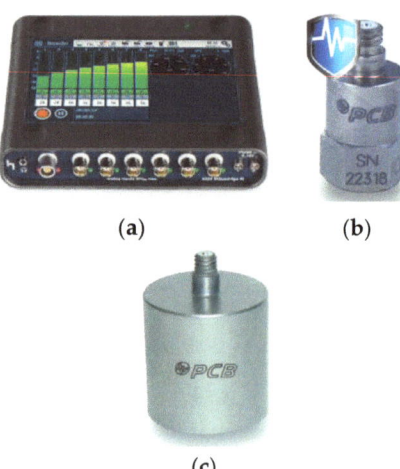

Figure 7. The instrument and accelerometers. (**a**) SQuadriga III data acquisition instrument; (**b**) PCB-352 vibration accelerometer; (**c**) PCB-393 vibration accelerometer.

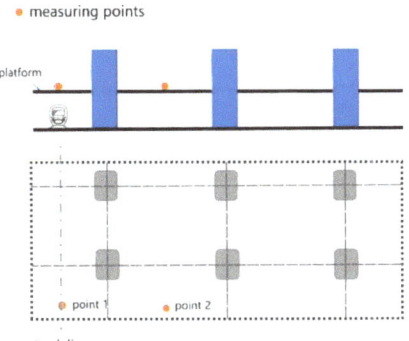

Figure 8. Layout of the measuring points.

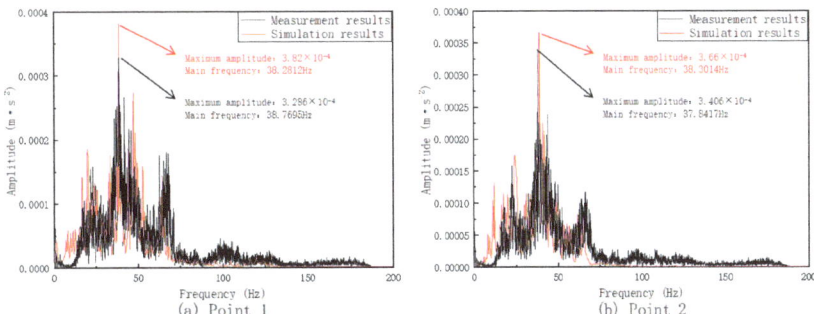

Figure 9. Comparison of measured and simulation calculations.

4. Analysis of Vibration Transmission and Distribution Characteristics

As the speed of the train running in the metro depot is 5–25 km/h, this simulation considers three working conditions: the standard working condition is the train running at the first layer of the operation zone at the speed of 10 km/h (hereinafter referred to as working condition 2), and it sets the comparative working condition of the train running at the speed of 5 km/h (hereinafter referred to as working condition 3) and 20 km/h (hereinafter referred to as working condition 1) in the first layer of the operation zone.

4.1. Analysis of Vibration Response at Vibration Source

The vibration response of each analysis point at the vibration source of the depot under various working conditions is shown in Table 3, and the analysis of vibration frequency domain is shown in Figure 10. It can be seen that, with the decrease in the speed, the vibration response of the vibration source analysis point decreases significantly, and the difference at the rail is the most significant. When the train speeds are 20 km/h and 5 km/h, the difference in Z-vibration level is nearly 20 dB, and the difference in other analysis points is about 7 dB. As the train speed decreases, the peak frequency of vibration decreases significantly. When the train speed is 10 km/h and 5 km/h, the peak frequency of vibration at each analysis point of the vibration source is similar, and the peak frequency of vibration is lower than that of the train speed of 20 km/h. The difference at the rail is the most significant, with a peak vibration frequency of around 160 Hz at a speed of 20 km/h, and around 70 Hz at a speed of 10 km/h and 5 km/h. The higher the train speed, the greater the vibration response of each analysis point. Therefore, the scheme of the trains running at low speeds in the operation zone can be adopted to reduce the vibration response at vibration source, and the vibration at the rail can be reduced by 17.9% at most.

Table 3. The vibration response at the vibration source under various working conditions.

Working Condition	Analysis Point	Peak Acceleration (m/s^2)	Z-Vibration Level (dB)
Train speed of 20 km/h (working condition 1)	Rails	4.01	107.57
	Supporting columns	0.54	91.63
	Columns	0.08	77.82
	The ground at a distance of 7.5 m from the rails	0.12	83.63
Train speed of 10 km/h (working condition 2)	Rails	1.99	97.51
	Supporting columns	0.16	86.51
	Columns	0.03	71.84
	The ground at a distance of 7.5 m from the rails	0.07	81.08
Train speed of 5 km/h (working condition 3)	Rails	2.11	88.36
	Supporting columns	0.14	84.37
	Columns	0.04	69.78
	The ground at a distance of 7.5 m from the rails	0.05	77.31

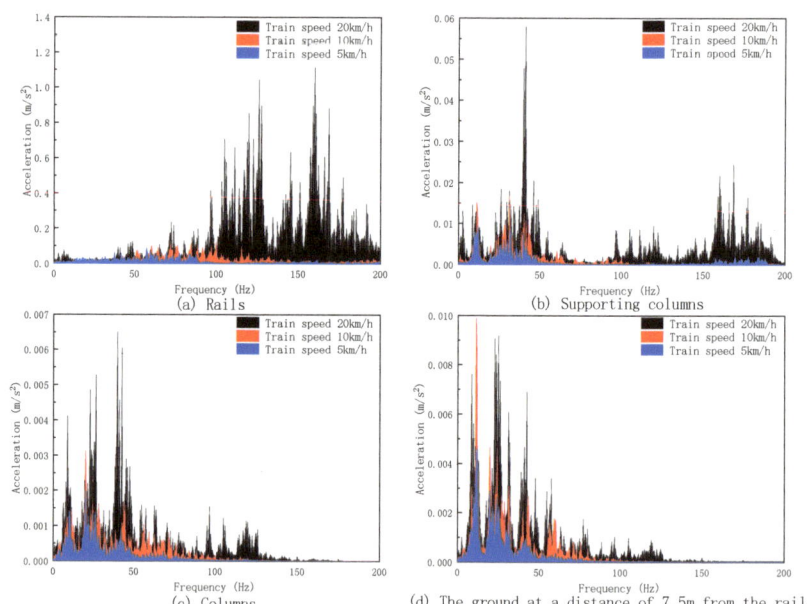

Figure 10. Spectrum of vibration sources at different running speeds.

4.2. Analysis of Vibration Transmission on the Platforms

The plane view of the platform of the operation zone is shown in Figure 11, and the vibration response of each analysis point on the 9 m and 16 m platforms of the operation zone under various working conditions is shown in Figure 12. It can be seen that, with the decrease in the speed, the vibration response of the analysis points on the platforms decreases significantly. Compared with the speed of 20 km/h, the Z-vibration level of each analysis point at the speed of 5 km/h is different by 6–8 dB on the 9 m platform, and the Z-vibration level of each analysis point at the speed of 5 km/h is different by 5–14 dB on the 16 m platform. It can achieve the effect of reducing the vibration response at the platform

platform. When the vibration is transmitted laterally from 0 m to the left above the rails on the platform to 24 m above it, the vibration response generally decreases with increasing distance. The reason for the amplification of vibration at a distance of 9 m and 15 m directly above the rails is that the 9 m and 15 m points are located in the middle of plate 2, while the 4.5 m and 24 m points are located at the edges of plate 1 and plate 3, respectively. The vibration in the middle of plate is generally greater than that at the edge of plate, resulting in the amplification phenomenon.

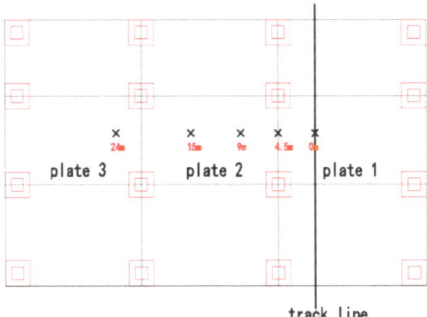

Figure 11. The plane view of the platform of the operation zone.

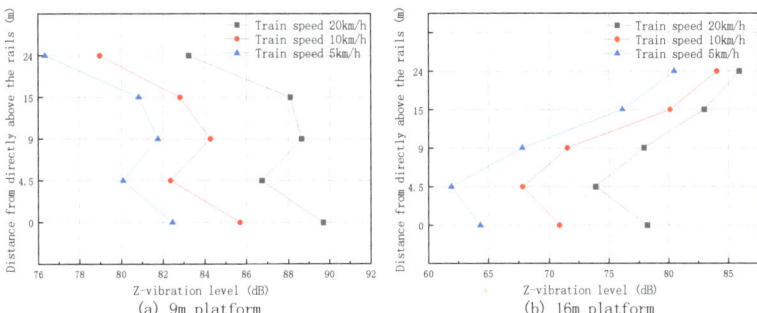

Figure 12. Z-vibration level of platforms at different running speeds.

During the vertical transmission of vibration response from the 9 m platform to the 16 m platform, there is a trend of an initial decrease and then an increase. At the speed of 10 km/h, within a range of 0–15 m to the left above the rails, the Z-vibration level of the analysis point on the 9 m platform are significantly higher than that on the 16 m platform, with a difference of 2–15 dB between the two. Within a range of 15–24 m to the left above the rails, the Z-vibration level of the analysis points on the 9 m platform is smaller than that of the 16 m platform, with a difference of about 5 dB between the two.

The 1/3 octave frequency of each analysis point on the 9 m and 16 m platforms in the operation zone under various working conditions is shown in Figures 13–15. It can be seen that the vibration energy of each analysis point on the platforms attenuates the vast majority in the frequency band above 50 Hz. During the vertical transmission of vibration response from 9 m platform to 16 m platform, the analysis points within the range of 0–15 m to the left above the rails exhibit a certain range of attenuation in the entire frequency band. The analysis points within the range of 15–24 m to the left above the rails have a certain amplification in the frequency bands of 0–20 Hz and 100–200 Hz, and the difference is not significant in the frequency band of 20–100 Hz.

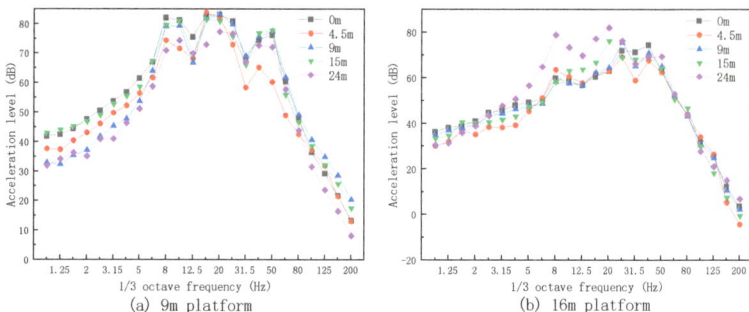

Figure 13. The 1/3 octave frequency of each analysis point on the platforms at a speed of 20 km/h.

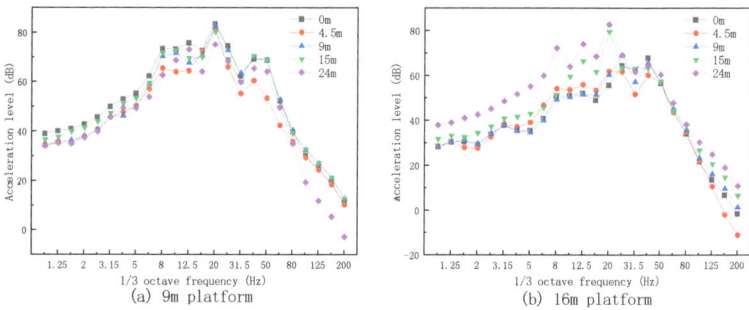

Figure 14. The 1/3 octave frequency of each analysis point on the platforms at a speed of 10 km/h.

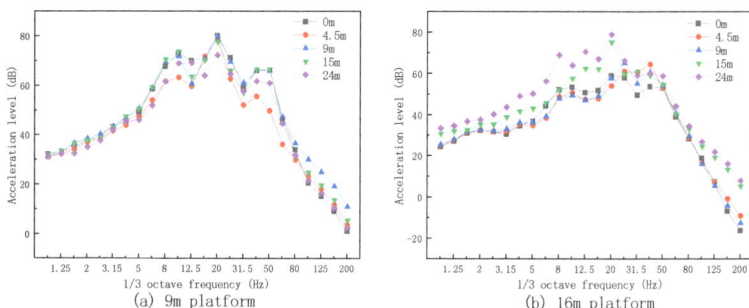

Figure 15. The 1/3 octave frequency of each analysis point on the platforms at a speed of 5 km/h.

4.3. Analysis of Vibration Response in the Over-Track Building

The vibration response of each analysis point in the over-track building under various working conditions is shown in Figure 16. It can be seen that, with the decrease in the speed, the vibration response of the over-track building decreases significantly. Compared with the speed of 5 km/h, the vibration response of each analysis point at the speed of 20 km/h is different by 3–11 dB. According to the standard for limits and measurement methods of vibration in the room of residential building [39], it is found that the Z-vibration level of individual floors at speeds of 10 km/h and 20 km/h exceeds the second-level limit. After reducing the speed, there is no overrun. Therefore, the scheme of the trains running at low speeds in the operation zone can be adopted to reduce the vibration response in the over-track buildings, and the vibration in the over-track buildings can be reduced by 13.0% at most. During the transmission of vibration response from the 2nd floor to the 41st floor,

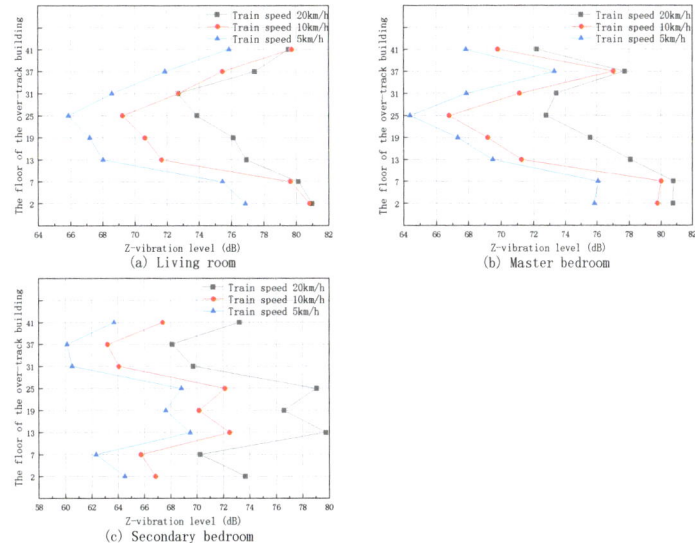

Figure 16. The vibration response in the over-track building at different running speeds.

Taking the living room of the over-track buildings as an example, the 1/3 octave frequency of each floor under various working conditions is shown in Figure 17. It can be seen that the vibration energy of each analysis point in the over-track building attenuates the vast majority in the frequency band above 20 Hz. During the transmission of vibration response from the 2nd floor to the 41st floor, there is a trend of an initial decrease and then an increase. Compared with the speed of 5 km/h, each analysis point has a certain range of attenuation in the full frequency band.

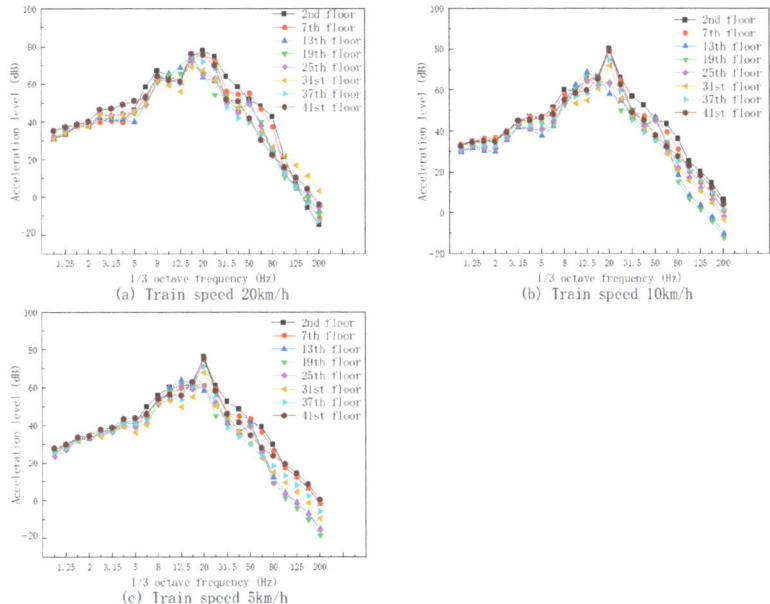

Figure 17. The 1/3 octave frequency of the living room at different running speeds.

4.4. Spatial Distribution of Vibration Energy

To study the distribution law of vibration energy at the vibration source in the operation zone, the plate 1 area on the 0 m platform is selected for analysis. Using the track line as the Z-axis, the distribution of vibration energy at the vibration source under various working conditions is shown in Figure 18. It can be seen that the maximum energy of vibration is within a radius of 2 m centered on the line, showing a ring-shaped distribution, and the ring-shaped distribution is more pronounced as the train speed increases. In the horizontal direction of the track line, the vibration energy distribution is within a range of −4 m to 11.5 m from the track line. The distribution of vibration energy is significantly affected by the boundary effect of the edges, and the vibration at the edge of plate is significantly smaller than that in the middle of the plate. In the longitudinal direction of the track line, the ring-shaped distribution of vibration energy exhibits a periodic pattern and is not affected by the boundary effect of the edges.

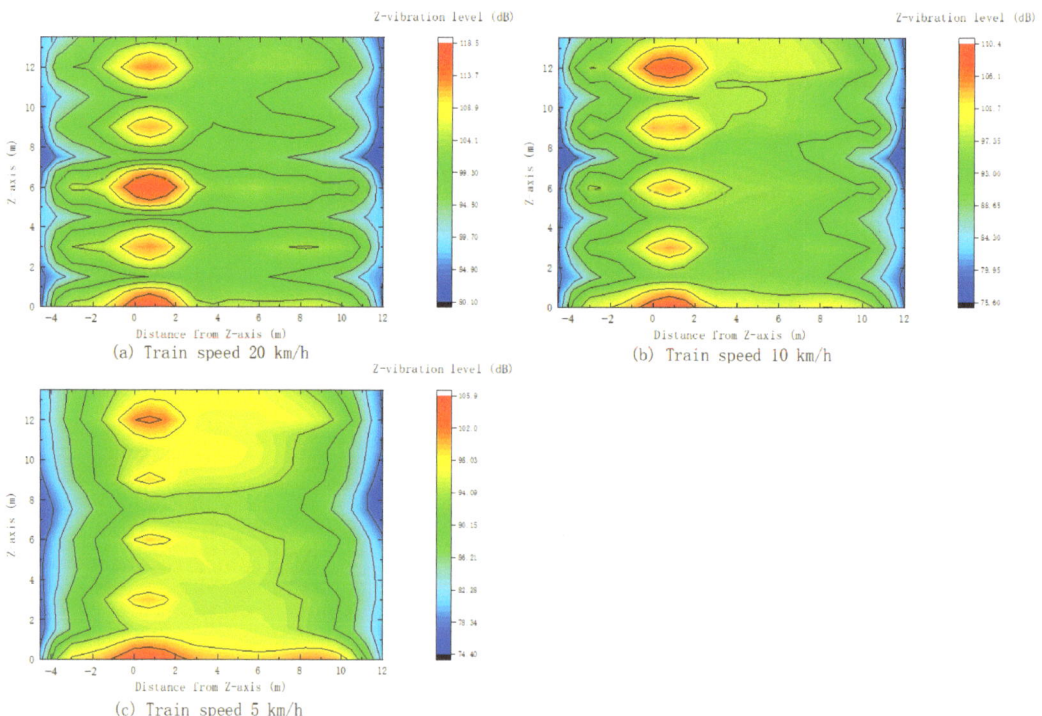

Figure 18. The distribution of vibration energy at the vibration source under various working conditions.

5. Conclusions

This paper established a coupling model of a track–metro depot–over-track building based on the structural finite element method to calculate the vibration response, and it further analyzed the vibration response characteristics of the vibration source and over-track building under different train speeds and then summarized the transmission and distribution characteristics of vibration. It yields the following conclusions:

(1) With the decrease in the speed, the vibration response of the vibration source, platforms, and the over-track building decreases significantly. The Z-vibration level difference at the rail is the most significant. At train speeds of 20 km/h and 5 km/h,

reduced by 17.9% at most. The peak frequency of vibration decreases significantly. The difference between the two on the 9 m platform is 6–8 dB, and the difference between the two on the 16 m platform is 5–14 dB. The vibration on the platforms can be reduced by 17.7% at most. The difference between the two in the over-track building is 3–11 dB, and the vibration can be reduced by 13.0% at most. Therefore, the scheme of the trains running at low speeds in the operation zone can be adopted to reduce the vibration response.

(2) When the vibration is transmitted laterally from 0 m to the left above the rails on the platform to 24 m above it, the vibration response generally decreases with the increasing distance. During the vertical transmission of vibration response from the 9 m platform to the 16 m platform, there is a trend of an initial decrease and then an increase.

(3) The vibration energy of each analysis point on the platforms attenuates the vast majority in the frequency band above 50 Hz. Therefore, when conducting vibration reduction design, engineers should focus on vibrations in the frequency band below 50 Hz. During the vertical transmission of vibration response from the 9 m platform to the 16 m platform, the analysis points within the range of 0–15 m to the left above the rails exhibit a certain range of attenuation in the entire frequency band. The analysis points within the range of 15–24 m to the left above the rails have a certain amplification in the frequency band of 0–20 Hz and 100–200 Hz.

(4) The maximum vibration energy of vibration source in the operation zone is within a radius of 2 m centered on the line, showing a ring-shaped distribution, and the ring-shaped distribution is more pronounced as the train speed increases. In the horizontal direction of the track line, the vibration energy distribution is within a range of −4 m to 11.5 m from the track line. The vibration at the edge of plate is significantly smaller than that in the middle of the plate. In the longitudinal direction of the track line, the ring-shaped distribution of vibration energy exhibits a periodic pattern.

This study explored the structural dynamic response of a double-layer metro depot and an over-track building. An increase in the number of metro depots and over-track buildings makes the vibration reduction design more important. This research provides design references for future double-layer prefabricated assembly metro depot and data for vibration reduction design. In this paper, we found that the vibration energy of each analysis point on the platforms attenuates the vast majority in the frequency band above 50 Hz. And the maximum vibration energy of vibration source in the operation zone is within a radius of 2 m centered on the line. Moreover, this study focused only on the vibration transmission and distribution characteristics, without studying the transmission and distribution characteristics of noise. Future research should extend to include the transmission and distribution characteristics of noise.

Author Contributions: Conceptualization, Q.F.; methodology, X.L.; software, X.J.; validation, X.J. and Y.C.; formal analysis, X.J. and Q.T.; investigation, X.J., Y.C. and W.H.; resources, X.L.; data curation, X.J. and Q.T.; writing—original draft preparation, X.J.; writing—review and editing, X.L. and Q.F.; visualization, W.H.; supervision, Q.F. and W.H.; project administration, X.L.; funding acquisition, Q.F. All authors have read and agreed to the published version of the manuscript.

Funding: This research was funded by National Natural Science Foundation of China (grant number: 52068029).

Data Availability Statement: Data are available upon reasonable request.

Conflicts of Interest: Authors Xinwei Luo and Qinming Tu were employed by the company Guangzhou Metro Design & Research Institute Co., Ltd., Author Wenlin Hu was employed by the company China Railway Design Corporation. The remaining authors declare that the research was conducted in the absence of any commercial or financial relationships that could be construed as a potential conflict of interest.

References

1. Zou, C.; Wang, Y.M.; Moore, J.A.; Sanayei, M. Train-induced field vibration measurements of ground and over-track buildings. *Sci. Total Environ.* **2017**, *575*, 1339–1351. [CrossRef] [PubMed]
2. Xia, H.; Wu, X.; Yu, D.M. Environmental vibration induced by urban rail transit system. *J. Beijing Jiaotong Univ.* **1999**, *23*, 7–13.
3. Triepaischajonsak, N.; Thompson, D.J. A hybrid modelling approach for predicting ground vibration from trains. *J. Sound. Vib.* **2015**, *335*, 147–173. [CrossRef]
4. Gupta, S.; Liu, W.F.; Degrande, G.; Lombaert, G.; Liu, W.N. Prediction of vibrations induced by underground railway traffic in Beijing. *J. Sound. Vib.* **2008**, *310*, 608–630. [CrossRef]
5. Lombaert, G.; Degrande, G.; Vanhauwere, B.; Vandeborght, B.; François, S. The control of ground-borne vibrations from railway traffic by means of continuous floating slabs. *J. Sound. Vib.* **2006**, *297*, 946–961. [CrossRef]
6. Sanayei, M.; Maurya, P.; Moore, J.A. Measurement of building foundation and ground-borne vibrations due to surface trains and subways. *Eng. Struct.* **2013**, *53*, 102–111. [CrossRef]
7. Yang, J.J.; Zhu, S.Y.; Zhai, W.M.; Kouroussis, G.; Wang, Y.; Wang, K.Y.; Lan, K.; Xu, F.Z. Prediction and mitigation of train-induced vibrations of large-scale building constructed on subway tunnel. *Sci. Total Environ.* **2019**, *668*, 485–499. [CrossRef] [PubMed]
8. Feng, Q.S.; Liao, C.M.; Zhang, L. Evaluation of subway vibration influence on human exposure comfort of whole-body vibration. *Noise Vib. Control.* **2021**, *41*, 237–243.
9. Qiu, Y.T.; Zou, C.; Wu, J.H.; Shen, Z.X.; Zhong, Z.X. Building vibration measurements induced by train operation on concrete floor. *Constr. Build. Mater.* **2023**, *394*, 132283. [CrossRef]
10. Kouroussis, G.; Conti, C.; Verlinden, O. Experimental study of ground vibrations induced by Brussels IC/IR trains in their neighborhood. *Mech. Ind.* **2013**, *14*, 99–105. [CrossRef]
11. Liang, R.H.; Ding, D.Y.; Liu, W.F.; Sun, F.Q.; Cheng, Y.L. Experimental study of the source and transmission characteristics of train-induced vibration in the over-track building in a metro depot. *J. Sound. Vib.* **2023**, *29*, 1738–1751. [CrossRef]
12. Chen, Y.M.; Feng, Q.S.; Liu, Q.J.; Jiang, J. Experimental study on the characteristics of train-induced vibration in a new structure of metro depot. *Environ. Sci. Pollut. Res.* **2021**, *28*, 41407–41422. [CrossRef] [PubMed]
13. Chen, Y.M.; Feng, Q.S.; Liu, Q.J.; Liu, W.W.; Luo, X.W. Test and Analysis of Vibration Induced by Train Operation in Sinking Metro Depot Service Shop. *J. Vib. Meas. Diagn.* **2021**, *41*, 532–538+623.
14. Feng, Q.S.; Zhang, Y.L.; Jiang, J.; Wang, Z.Y.; Lei, X.Y.; Zhang, L. Field measurement and evaluation of vibration in different areas of a metro depot. *Earthq. Eng. Eng. Vibr.* **2022**, *21*, 529–542.
15. Feng, Q.S.; Wang, Z.Y.; Liu, Q.M.; Luo, X.W.; Li, J.Y. Vibration characteristics of metro depot upper building under double vibration source excitation. *J. Traffic Transp. Eng.* **2019**, *19*, 59–69.
16. Feng, Q.S.; Wang, Z.Y.; Liu, Q.M.; Luo, X.W.; Luo, K.; Li, J.Y. Comparative analysis of environmental vibration characteristics in different regions of a metro depot. *J. Vib. Shock.* **2020**, *39*, 179–185+200.
17. Sanayei, M.; Kayiparambil, P.A.; Moore, J.A.; Brett, C.R. Measurement and prediction of train-induced vibrations in a full-scale building. *Eng. Struct.* **2014**, *77*, 119–128. [CrossRef]
18. Sanayei, M.; Zhao, N.Y.; Maurya, P.; Moore, J.A.; Zapfe, J.A.; Hines, E.M. Prediction and mitigation of building floor vibrations using a blocking floor. *J. Struct.* **2012**, *138*, 1181–1192. [CrossRef]
19. Zou, C.; Wang, Y.M.; Wang, P.; Guo, J.X. Measurement of ground and nearby building vibration and noise induced by trains in a metro depot. *Sci. Total Environ.* **2015**, *536*, 761–773. [CrossRef] [PubMed]
20. Zou, C.; Wang, Y.M.; Zhang, X.; Tao, Z.Y. Vibration isolation of over-track buildings in a metro depot by using trackside wave barriers. *J. Build.* **2020**, *30*, 101270. [CrossRef]
21. Zou, C.; Moore, J.A.; Sanayei, M.; Wang, Y.M. Impedance model for estimating train-induced building vibrations. *Eng. Struct.* **2018**, *172*, 739–750. [CrossRef]
22. Zou, C.; Moore, J.A.; Sanayei, M.; Wang, Y.M.; Tao, Z.Y. Efficient impedance model for the estimation of train-induced vibrations in over-track buildings. *J. Vib. Control.* **2020**, *27*, 924–942. [CrossRef]
23. Tao, Z.Y.; Wang, Y.M.; Zou, C. Prediction model of vehicle-induced vibration of metro depot superstructure based on impedance and power conservation method. *Zhendong Yu Chongji* **2022**, *41*, 62–67+73.
24. Tao, Z.Y.; Zou, C.; Yang, G.R.; Wang, Y.M. A semi-analytical method for predicting train-induced vibrations considering train-track-soil and soil-pile-building dynamic interactions. *Soil. Dyn. Earthq. Eng.* **2023**, *167*, 107822. [CrossRef]
25. Tao, Z.Y.; Wang, Y.M.; Sanayei, M.; Moore, J.A.; Zou, C. Experimental study of train-induced vibration in over-track buildings in a metro depot. *Eng. Struct.* **2019**, *198*, 109473. [CrossRef]
26. Tao, Z.Y.; Wang, Y.M.; Zou, C.; Li, Q.; Luo, Y. Assessment of ventilation noise impact from metro depot with over-track platform structure on workers and nearby inhabitants. *Environ. Sci. Pollut. Res.* **2019**, *26*, 9203–9218. [CrossRef]
27. Liu, W.F.; Liang, R.H.; Zhang, H.G.; Wu, Z.Z.; Jiang, B.L. Deep learning based identification and uncertainty analysis of metro train induced ground-borne vibration. *Mech. Syst. Signal. Process.* **2023**, *189*, 110062. [CrossRef]
28. Liang, R.H.; Liu, W.F.; Ma, M.; Liu, W.N. An efficient model for predicting the train-induced ground-borne vibration and uncertainty quantification based on Bayesian neural network. *J. Sound. Vib.* **2021**, *495*, 115908. [CrossRef]
29. Liang, R.H.; Liu, W.F.; Li, C.Y.; Li, W.B.; Wu, Z.Z. A novel efficient probabilistic prediction approach for train-induced ground vibrations based on transfer learning. *J. Vib. Control.* **2024**, *30*, 576–587. [CrossRef]

30. Liang, R.H.; Liu, W.F.; Kaewunruen, S.; Zhang, H.G.; Wu, Z.Z. Classification of External Vibration Sources through Data-Driven Models Using Hybrid CNNs and LSTMs. *Struct. Control. Health Monit.* **2023**, *2023*, 1900447. [CrossRef]
31. Zhou, Y.; Pei, Y.L.; Zhou, S.; Zhao, Y.; Hu, J.X.; Yi, W.J. Novel methodology for identifying the weight of moving vehicles on bridges using structural response pattern extraction and deep learning algorithms. *Measurement* **2021**, *168*, 108384. [CrossRef]
32. He, Y.P.; Zhang, Y.; Yao, Y.Y.; He, Y.L.; Sheng, X.Z. Review on the Prediction and Control of Structural Vibration and Noise in Buildings Caused by Rail Transit. *Buildings* **2023**, *13*, 2310. [CrossRef]
33. He, C.; Zhou, S.H.; Guo, P.J.; Di, H.G.; Zhang, X.H. Analytical model for vibration prediction of two parallel tunnels in a full-space. *J. Sound. Vib.* **2018**, *423*, 306–321. [CrossRef]
34. He, C.; Zhou, S.H.; Guo, P.J. Mitigation of railway-induced vibrations by using periodic wave impeding barriers. *Appl. Acoust.* **2022**, *105*, 496–514. [CrossRef]
35. Li, X.M.; Chen, Y.K.; Zou, C.; Wang, H.; Zheng, B.K.; Chen, J.L. Building structure-borne noise measurements and estimation due to train operations in tunnel. *Sci. Total Environ.* **2024**, *926*, 172080. [CrossRef]
36. Qiu, Y.T.; Zou, C.; Hu, J.H.; Chen, J.L. Prediction and mitigation of building vibrations caused by train operations on concrete floors. *Appl. Acoust.* **2024**, *219*, 109941. [CrossRef]
37. Hu, J.H.; Zou, C.; Liu, Q.M.; Li, X.M.; Tao, Z.Y. Floor vibration predictions based on train-track-building coupling model. *J. Build. Eng.* **2024**, *89*, 109340. [CrossRef]
38. Gu, Y.; Liu, J.B.; Du, Y.X. 3D consistent viscous-spring artificial boundary and viscous-spring boundary element. *Eng. Mech.* **2007**, *12*, 31–37.
39. *GB/T 50355-2018*; Standard for Limits and Measurement Methods of Vibration in the Room of Residential Building. Architecture & Building Press: Beijing, China, 2018.

Disclaimer/Publisher's Note: The statements, opinions and data contained in all publications are solely those of the individual author(s) and contributor(s) and not of MDPI and/or the editor(s). MDPI and/or the editor(s) disclaim responsibility for any injury to people or property resulting from any ideas, methods, instructions or products referred to in the content.

Review

Seismic Assessment of Large-Span Spatial Structures Considering Soil–Structure Interaction (SSI): A State-of-the-Art Review

Puyu Zhan, Suduo Xue, Xiongyan Li *, Guojun Sun and Ruisheng Ma

Faculty of Architecture, Civil and Transportation Engineering, Beijing University of Technology, Beijing 100124, China
* Correspondence: xylee@bjut.edu.cn

Abstract: Soil–structure interaction (SSI), which characterizes the dynamic interaction between a structure and its surrounding soil, is of great significance to the seismic assessment of structures. Past research endeavors have undertaken analytical, numerical, and experimental studies to gain a thorough understanding of the influences of SSI on the seismic responses of a wide array of structures, including but not limited to nuclear power plants, frame structures, bridges, and spatial structures. Thereinto, large-span spatial structures generally have much more complex configurations, and the influences of SSI may be more pronounced. To this end, this paper aims to provide a state-of-the-art review of the SSI in the seismic assessment of large-span spatial structures. It begins with the modelling of soil medium, followed by the research progress of SSI in terms of numerical simulations and experiments. Subsequently, the focus shifts towards high-lighting advancements in understanding the seismic responses of large-span spatial structures considering SSI. Finally, some discussions are made on the unresolved problems and the possible topics for future studies.

Keywords: soil–structure interaction; state-of-the-art review; large-span spatial structure; seismic assessment

Citation: Zhan, P.; Xue, S.; Li, X.; Sun, G.; Ma, R. Seismic Assessment of Large-Span Spatial Structures Considering Soil–Structure Interaction (SSI): A State-of-the-Art Review. *Buildings* **2024**, *14*, 1174. https://doi.org/10.3390/buildings14041174

Academic Editor: Fabrizio Gara

Received: 27 February 2024
Revised: 9 April 2024
Accepted: 17 April 2024
Published: 21 April 2024

Copyright: © 2024 by the authors. Licensee MDPI, Basel, Switzerland. This article is an open access article distributed under the terms and conditions of the Creative Commons Attribution (CC BY) license (https://creativecommons.org/licenses/by/4.0/).

1. Introduction

Soil–structure interaction (SSI) has become one of the most concerning research topics in the field of earthquake engineering, especially with the development of complicated structures, i.e., high-rise buildings and large-span spatial structures. The concept of SSI can be traced back to 1936 [1], with the initial emphasis laid on the interaction between the soil and underground structures. In the subsequent decades, extensive investigations delved into SSI, expanding its scope to encompass structures both above and below the ground. Generally speaking, SSI can be categorized into two types, namely the kinematic interaction and inertial interaction. Specifically, the kinematic interaction involves the impact of vibrational feedback from the superstructure on the amplitude and spectral composition of foundation motion. This reduces the acceleration amplitude of the foundation below that of the neighboring free field, simultaneously enhancing the components around the fundamental frequencies of structures. Additionally, the flexibility of the soil foundation contributes to the frequencies and modes of vibration of the superstructure, with softer foundations generally prolonging the structural period. The inertial interaction encompasses the dissipation of energy resulting from the vibrations of the superstructure subjected to inertial forces, attributed to phenomena such as reflection and diffraction [2,3]. A technical report published by the Federal Emergency Management Agency (FEMA) of the U.S. Department of Homeland Security (DHS) [4] distilled both interactions into three primary effects: the flexible foundation effect, kinematic effect, and foundation damping effect.

Structure–soil–structure interaction (SSSI), a specialized case of SSI, has played a

power plant (NPP) complexes through the soil medium for over half a century. It is considered a fundamental dynamic characteristic of NPP reactors [5]. Concurrently, with the ongoing trend of urbanization giving rise to denser complexes to accommodate a growing population, the SSSI effect is gradually becoming integral to the design of large-scale civil structures. This integration is commonly referred to as site–city interaction (SCI). Additionally, the SSSI effect between underground structures and nearby aboveground structures is gaining attention in the ongoing development and utilization of urban underground spaces. This is particularly significant, exemplified by the amplifying effect on low-rise buildings near high-rise structures [6].

When considering the effects of soil and foundations on the seismic response of superstructures, the primary factors can be broadly categorized into four main groups: structural type, foundation form, site effects, and seismic excitation. To date, frame structures and bridge engineering are focal points in SSI research. In both numerical and experimental investigations of SSI, more than half of the addressed superstructures involve multi-layer frame structures and adjacent steel or concrete structures [7]. Sections 3 and 4 will go into more detail on these structures. The dynamic characteristics of the majority of bridge structures are significantly impacted by SSI/SSSI [8,9], as has been demonstrated in subsequent earthquakes [10]; therefore, it has been currently integrated into the bridge design guidelines of many nations and is being used to guide the engineering practice of numerous bridges [11], whereas fewer studies are referring to the SSI effect of large-span spatial structures [12–17]. Large-span spatial structures include shell structures, grid structures, cable-suspended structures, etc., whose vibration subjected to seismic loads is three-dimensional. Since the dynamic characteristics of large-span spatial structures are different from bridges or high-rise buildings, the relatively mature SSI-effect research results for the latter two cannot be directly applied to the design calculation of the former. Meanwhile, when the vertical projection surface of the structure is large or too long and the foundation form adopts an independent pile foundation, it may produce a foundation–soil–foundation interaction (FSFI) effect; therefore, it is necessary to conduct more in-depth studies to investigate the SSI effect of large-span spatial structures. The foundation is a load-bearing structure that connects the superstructure and soil. According to Chinese regulations [18], the foundations are classified as pile foundations, extended foundations, and raft foundations, and the influence of the coefficient of which varies for superstructures according to the SSI/SSSI effect as well [19,20]. One of the key elements altering structural seismic damage is the site effect [21]. Based on the investigation of the 1985 Michoacan, Mexico, earthquake, local site effects were accountable for the uncommonly high ground accelerations [22]. This was because Michoacan is situated on top of a basin with a deep layer of soft soil that causes certain frequencies of ground movements to be dramatically amplified on the surface [23], while the basin has a focusing effect on the propagation of seismic waves within the soil layer [24]. It has been noted that topographic features (such as basins, canyons [25,26], summits, etc.) and soil properties make up the majority of the site conditions. Most of the research on SSI/SSSI is predicated on the assumption of being plain; nevertheless, there are projects constructed on slopes in practice. Moreover, it has been noticed that structures on adjacent slopes are more vulnerable [27], which cannot be disregarded when the ratio of the net distance of the structural foundation from the slope peak to its height is less than 5 [27]. Apart from the slope–foundation–structure system [28,29], the SSSI effects of structures on [30] and near [31] slopes have also been explored, which are known as the topography–soil–structure interaction (TSSI) and topography–structure–soil–structure interaction (TSSSI). Soil properties are a decisive component of SSI, and the numerical modelling techniques for soil determine the ability to realistically predict the propagation of seismic waves [32,33], as will be specified in the next section.

Seismic excitation refers to the entire process of seismic wave propagation from the source upward to the foundation–structure/surface, where the soil traversed by the seismic wave is defined as the free field [34] which is proposed to calculate the vibration response of soil under shear wave excitation, on the premise of assuming that the soil layer is horizontal and homogeneous along the horizontal plane. A function consisting of source effects, travel path effects, and local site effects expresses the free-field ground motion under seismic excitation. Source distance and source count are two categories for the source effect. Based on the distance from the epicentre, earthquakes can be categorized into far-field and near-field ground motion [35], corresponding earthquake groupings are included in the code for the seismic design of buildings [36]. Compared with the former, the latter typically has a longer period and higher amplitude [37], which aggravates the seismic effect of non-flat sites [38]. Meanwhile, the continuous vibration caused by multiple sources probably exacerbates the damage to surface structures [22]. The travel path effect consists of the incoherence effect, wave-passage effect and the attenuation effect [39]. The correlation of seismic waves across separate spaces and times during their propagation is what describes the coherence of seismic waves. When seismic waves travel through uneven soil, the superposition of waves in different spaces on account of reflection and refraction results in the formation of different phases and elimination of interference, which is the incoherence effect. When the structure is so enormous that the time difference between the seismic wave arriving at each point is insurmountable, an occurrence known as the wave-passage effect arises. The attenuation effect is the steady decline in amplitude created by the geometric diffusion of seismic waves in the site's space. Throughout the process of site propagation, seismic waves may shift in space also to time, especially for some large spatial formations [40,41]. The spatial characteristics of ground motion comprise the aforementioned three effects as well as local site effects.

This paper provides a state-of-the-art review of the classification of SSI and the current state of theoretical research. Additionally, it summarizes previous numerical and experimental studies, along with existing research results. The work also delves into the significance of accounting for SSI effects in the seismic design of large-span spatial structures, discussing potential avenues for further studies. It is important to note that while this paper briefly reviews the SSI effect and does not delve into the influence of SSI on structural vibration control, it underscores the importance of considering the SSI effect in the actual design of seismic mitigation or isolation.

2. Modeling of Soil Medium

To simulate soil on SSI, two types of solutions are widely utilized [42], including the direct approach, which is typically realized through the finite element method (FEM) and the substructure approach. The concept of the foundation impedance function is introduced by the boundary substructure approach, whose applications to various sites and foundations are discussed in detail in the technical report (NIST/GCR 12-917-2) [43]. Springs and dampers are used to mimic the stiffness and damping at the soil–foundation interface. The Winkler model and its variants are typically employed in conjunction with the elastic half-space model for deep foundation systems, while the latter model is typically used for shallow foundation systems.

2.1. Winkler Model

Understanding the mechanical properties of the soil surrounding underground construction or foundation is crucial in the exploration of SSI. In the case of pile foundations, a widely used method to simplify soil behavior is the Winkler foundation beam model, which essentially comprises a system of closely spaced discrete spring and dampers that serves as a substitute for the dynamic impedance imposed by the soil on the pile body [44]. The Winkler model, being a one-parameter model, has limitations as it inaccurately assumes that discrete springs cannot replicate continuous soil deformation. In contrast to real-world scenarios, the model predicts only vertical displacement at a point of application. To overcome this limitation, various enhancements have been introduced, incorporating additional parameters such as tension, shear force, and bending moment. This has led to the development of two-parameter and three-parameter Winkler models, including notable variants like the Filonenko-Borodich model, Hetenyi model, Pasternak model, and Kerr model [45].

In extending the elastic foundation formulas, Jemielita [46] synthesized monomial multiparameter formulas into differential formulas for multiple parameters and introduced the n Shear layer-Bending layer-Spring layer (n SBS-layers) concept, where each layer is composed of three sub-layers: the shear layer, bending layer, and spring layer. Expanding on the SBS differential formula, Zhao et al. [47] proposed model types tailored to various superstructures along with their mechanical parameters. Additionally, by integrating the Terzaghi model addressing seepage consolidation, they asserted that the Winkler–Terzaghi model effectively addressed the challenge of soil discontinuity.

2.2. Elastic Half-Space Model

The elastic half-space model is anchored by three translational springs and three rotational springs, with added dampers to account for soil damping and replicating the deformation behavior of the surrounding soil. In light of this, SSI predominantly involves motion interaction and inertia interaction. In instances of high seismic intensity, the impact of inertia interaction takes precedence. Therefore, selecting an appropriate foundation resistance function becomes crucial in accurately characterizing the inertia interaction [48].

Except for the ring augmented by Veletsos and Tang [49], Gazetas [50] provided impedance functions for nearly all foundations, all based on homogeneous soil. Chen [51] introduced the impedance function for layered foundations, employing the Cone model, which essentially substitutes a cone with a specific tensor angle for the semi-infinite body. As earthquakes involve multiple frequencies of seismic waves superimposed, utilizing the parameters suggested by Gazetas [50] directly for calculating dynamic stiffness and damping [52] may not be viable. Instead, more precise algorithms are essential to address the nonlinear behaviors of SSI.

2.3. Finite Element Method (FEM)

The finite element method (FEM) stands out as one of the most effective ways to depict SSI, given its capacity to accurately replicate nonlinear soil behaviors [45]. Therefore, this subsection explicitly details both FEM and its derived methods. To enhance accuracy in FEM, three critical aspects must be considered: the soil constitutive model, artificial boundary conditions, and mechanical properties of contact interfaces. While the previously discussed substructure method also addresses the assessment of Foundation Input Motion (FIM) [43] in the free field, it is crucial to note that FEM, by simulating the variation of soil as seismic waves traverse the site, demands special attention to the input mode of the seismic wave, particularly concerning the artificial boundary. The soil constitutive model is typically categorized into linear–elastic, elastoplastic, and viscoelastic models. The linear–elastic model, based on the generalized Hooke's law, is straightforward but has limitations, such as its inability to describe hysteresis, nonlinearity, and dynamic response deformation accumulation in soil [53]. To address the above limitations, the elastoplastic model, with its primary objective of articulating variation curves for the shear modulus ratio and damping ratio [54], incorporates yield criteria, the flow rule, and hardening rule [55]. Viscoelastic models are divided into two elementary branches: the equivalent linear model and the time–domain hysteretic nonlinear model. To linearize the nonlinear problem, Seed and Idriss [56] initially introduced the equivalent linear model, substituting the equivalent shear modulus and equivalent damping ratio for the original ones. Despite its simplicity and applicative advances, this model is restricted by soil strain and seismic acceleration [57]. For typical constructions, the region where strain exceeds the critical value is usually within a certain fraction of the foundation width on both sides [58]. The shear strain in this part can be adjusted to enhance the calculation accuracy [59]. Masing's nonlinear constitutive model, first proposed in 1926 [60], has undergone modifications to produce various types of Masing models, with notable examples being the Hardin–Drnevich [61] and Ramberg–Osgood [62] models. Subsequently, Martin and Seed [63] established the Davidenkov model, presenting a calculation method following Masing's rules [64]. Based on the double criterion, Pyke [65] proposed the multiple rule, which has been implemented in the Masing model.

When employing FEM, it is essential to implement artificial boundaries that simulate the radiation damping of an infinite foundation. This serves the purpose of transforming an unbounded domain into a bounded one, thereby preventing the reflection of scattered waves at the interception boundary [66]. Global and local artificial boundaries are two categories of artificial boundaries [67]. The former, including the infinite element method (IEM) [68], BEM, etc., offers high levels of precision but is computationally intensive. The other consists of a viscous boundary [69], paraxial boundary [70], Higdon boundary [71], transmission boundary [72–74], viscous-spring artificial boundary [75], and so on. Among these viscoelastic boundaries, which exhibit excellent precision and stability in handling both static and dynamic problems [76], have been shown to consider both the large-scale basin effect and local nonlinear effect [77]. Also, it is one of the artificial boundaries recommended by the seismic design of nuclear power plants in China [75].

Seismic input can be categorized into three methods, namely the displacement input [78], acceleration input [79], and seismic motion input [80]. Since seismic motion input is derived from the viscoelastic artificial boundary, which has a greater fit, it is prioritized while employing the viscoelastic boundary [81]. This method works when the seismic wave propagates in a single homogeneous field [82]; however, it is less effective if the seismic wave traverses two or more soil layers with differing wave impedances. For dealing with this problem, there are generally two methods: the frequency [83] and the time [84] domain method. Although the former is more exact [85], its application in intricate foundations is incalculable. To derive the equivalent input seismic load of the artificial boundary, researchers employ SHAKE91, DEEPSOIL, EERA, etc., learning from the substructure method [86].

The contact interface refers to the soil mass adjacent to the structure with a thickness of 5~10 times the average particle size [87], the mechanical properties of which are intricate, affected by normal stress, structural surface roughness, the particle size of sand, sand type, and homogeneity coefficient of the soil particles [88]. There are three broad approaches to thinking about the contact interface: the shear tests for empirical formulas, soil constitutive model, and mechanical model of contact interface that can be more accurate than the former [89]. For creating an interface model, there are two mainstream methods [90], either utilizing functional equations (Lagrangian or Penalty function) to turn the contact problem into non-contact or establishing a constitutive model of contact element [91], including the zero-thickness Goodman element [91] and Desai thin-layer element [92]. The former element serves when shear failures occur at the contact interfaces, while the other is employed in a situation where shear failure happens to the nearby soil.

As previously stated, for FEM, when the infinite or semi-infinite field is estimated, it is necessary to construct a large large-size mesh structure to simulate the near-field and an artificial boundary is set to absorb the seismic waves from the far-field [93], which may generate a large computational requirement for some large-size models, whereas the boundary element method (BEM) only necessitates the establishment of the mesh of the contact interface between the foundation and the soil. Hence, one of the ways proposed to facilitate the resolution of the dynamic response of the SSI with a complex contact surface is the finite element method–boundary element method (FE–BE coupling method) obtained via introducing BEM based on FEM. The FE–BE coupling method is a method for calculating the superstructures and soil foundation following different methods. Generally, BEM is used to simulate the dynamic behavior of layered soils [94,95], while FEM is employed for superstructures. In some circumstances, only the far-field is calculated using FEM, which is defined as the scaled-boundary finite element method (SBFEM) [96].

The primary objective of BEM is to address Laplace's equation, facilitating the conversion of integral equations within the boundary region of the far-field and the SSI system into boundary integral equations [97]. Once the boundary has been discretized via frequency [98] or time domain [99] methods, the simplified dynamic response equation can be derived [93].

The wave propagation in unbounded fields has been effectively modelled in both the time and frequency domains using the semi-analytical stiffness-based finite element method (SBFEM) [100]. SBFEM stands out due to its exemption from artificial boundary conditions or the fundamental solution requirement, a characteristic not shared by the boundary element method (BEM). This allows seamless coupling with the near-field through the finite element method (FEM) and compensates for the limitations of BEM in mimicking anisotropic soil behavior [101]. SBFEM also generates symmetric dynamic stiffness and unit impulse response matrices, improving computing efficiency [102]. In the extension of the sequential boundary field element (SBFE) to model three-dimensional layered soils, Birk [103] suggested employing a scaling line rather than a proportionate center point. Furthermore, an alternative approach, the indirect boundary element method (IBEM), has been proposed for assessing the seismic response of anisotropic soils [104].

3. Advances in Numerical Simulations on SSI

3.1. Site Effect

The topography and soil properties of a site have discernible impacts on SSI, with the latter playing a pivotal role [105]. When soil nonlinearity is taken into account, the effects of SSI and SSSI extend to a broader range of applications, surpassing the scope covered by the linear elastic model [106].

The propagation of seismic waves within a field varies based on soil properties. Typically, dense soil tends to amplify seismic waves [107], while soft soil may exhibit a similar effect or act as a filter. However, relying solely on soil properties for estimates is inherently inaccurate. The filter effect was exemplified by the softening of soil and the deceleration of waves, as observed in Chen et al.'s investigation into the impact of nonlinearity on SSI [108]. In this study, the arrival time of seismic waves was delayed at the free surface due to the filter effect. It is noteworthy that calculations solely based on soil properties may lead to inaccuracies, especially considering the interlayer reflection coefficient is less than 0 for the weak layer at the bottom, potentially resulting in a transition from forward displacement to reverse displacement. To substantiate the amplification effect of weak soil, Chen et al. [23] conducted an assessment of the seismic response in areas with varying shear velocities (V_S). Their findings indicated that the amplification coefficient of surface acceleration decreased with an increase in the seismic intensity. This trend was attributed to heightened ground motion, which accentuated soil nonlinearity [109]. Notably, not all amplification factors exceeded 1, particularly in cases where the soft soil thickness was less than 4 m. For instance, when the coverage thickness exceeded 6 m, this was verified through both centrifugal simulation [107] and seismic analysis at an actual site [109]. However, the amplification factor was also contingent on the distribution of the weak layer. If a thin, weak soil layer was situated on top of the field, ground acceleration would be exacerbated. In cases where the soft soil was positioned in the middle, its role could vary depending on the depth of burial and thickness. A deeper and thicker layer might act as a seismic isolation layer, while a shallow and thin weak layer could produce the opposite effect [109]. It is crucial to emphasize that, analogous to the phenomenon observed in the 1923 Kanto Earthquake [110], excessively violent ground motion, leading to soil degradation, could result in the dramatic failure of the weak layer, significantly amplifying surface acceleration.

As outlined in [54], a slope devoid of a soft layer demonstrated an amplification effect on vertical ground motion. However, the impact predicted using a nonlinear model, characterized by greater energy consumption, was found to be less significant than that determined using a linear model. The amplification coefficient of ground motion exhibited a decrease correlating with a reduction in the shear wave velocity of the overlying soil, aligning with prior research findings. Notably, it is essential to recognize that the compressive modulus resulting from vertical earthquakes is considerably less than the shear modulus. Despite this, it is crucial to note that the effects of soil nonlinearity on the propagation of P waves in soil are often overlooked. Nevertheless, previous studies have demonstrated that soil exhibits nonlinearity under the influence of strong vertical earthquakes [111]. Shahbazi et al. [112] conducted a comparative analysis of the impact of soft and hard soil on the seismic responses of superstructures, yielding results consistent with those aforementioned. Incorporating SSI considerations, flexible structures built on hard soils were estimated more conservatively. However, it became evident that SSI could not be overlooked when assessing structures characterized by higher stiffness situated on a soft foundation. Furthermore, the dissipative effect of the weak layer proved advantageous in the seismic design for superstructures, particularly when hard soil was positioned above a suitable soft layer capable of supporting the structure without undergoing considerable settlement or liquefaction.

Considering the most unfavorable site conditions, the subsequent review will focus on the numerical studies of the seismic responses of the structures erected on soft soil.

To compare the seismic responses of a structure before and after the involvement of soil, Xu et al. [113] constructed a 15-story concrete building using FLAC 3D considering different shear strengths and ground motions, as illustrated in Figure 1. Within the range of shear strengths considered in this study, the computational results revealed that SSI significantly amplified the maximum base shear, lateral displacement, and inter-story drift. Notably, when the shear strength of the soil was elevated from 65 kPa to 105 kPa, the influence of SSI on the maximum base shear, lateral displacement, and inter-story drift increased by 60%, 100%, and 44%, respectively.

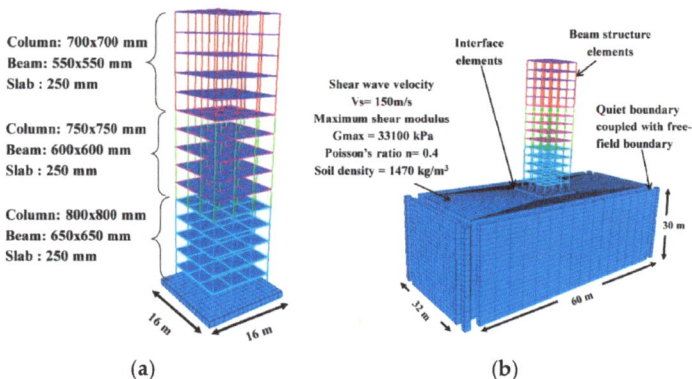

Figure 1. 15-story flexural concrete structure without and with considering SSI [113]: (**a**) fixed-base model; (**b**) SSI model.

Mercado et al. [114] conducted a comparative analysis to assess the impact of linear and nonlinear fields on SSI, as depicted in Figure 2. The results indicated that the SSI effect primarily manifested in an increase in the peak horizontal acceleration of the structure. When compared to the results calculated using the linear model, the nonlinear model exhibited a decrease in the peak acceleration and inter-layer drift of the superstructure, implying that the application of a linear model might be overly conservative. Previous research has also underscored that the error of the equivalent linear model grows with the increment of the site period [59].

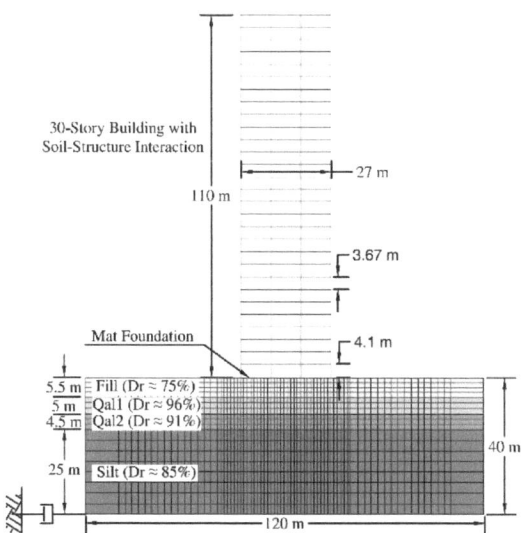

Figure 2. Finite element model of the tall buildings considering SSI [114].

In specific scenarios, particularly when a building is designed in proximity to a slope, the TSSI effect cannot be overlooked. Alitalesh et al. [29], utilizing FLAC 3D, simulated a model as shown in Figure 3 to investigate the influences of soil and slope conditions on both SSI and TSSI. The results indicated that the SSI and TSSI effects were more pronounced when the upper overburden soil layer (h_1) was thinner. Notably, in cases where a soft layer overlays a hard layer, seismic waves were amplified in the upper layer, further affirming the findings observed by Chen et al. [23] in uneven terrains. Additionally, it was revealed that the impact of the slope height on TSSI was more substantial than that of the slope angle. Moreover, when assessing the SSI effect in soils with varying shear wave velocities, it was observed that SSI could be disregarded for flexible superstructures with rigid foundations, aligning with the conclusions drawn by Shahbazi et al. [112].

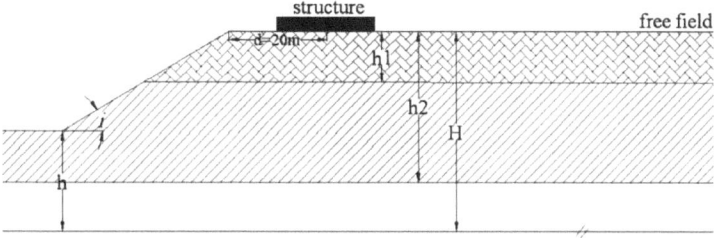

Figure 3. Illustration of TSSI with stratified soil layers [29].

Through adjustments to the geological parameters, slope angle, and slope height, Erfani et al. [115] examined the TSSI and its impact on the seismic response of structures. Their study revealed that slope stiffness decreased with an increase in slope angle, resulting in a simultaneous increase in structural inter-story drift. The effect of the seismic waves reflected by the slope on the structure diminished as the structure moved farther from the slope border. However, although the extra vertical seismic waves generated by the reflection also decreased, the interlayer drift increased, resembling the SSI effect. Notably, when $V_S = 150 \, \text{m/s}$, the influence of TSSI approached that of SSI, illustrating that the contribution of soil properties to SSI was more significant than that of topographic conditions.

Shamsi et al. [116] applied MIDAS GTS/NX to develop the model of SSI, TSSI, and TSSSI, as shown in Figure 4. In this study, a 15-story steel frame was adopted as the benchmark model, and the influences of various parameters were investigated, including the slope height (H), the distance of the structure from the slope (X), the angle of the slope (i), and the distance between the two structures (D). Studies have demonstrated that the effect of X was greater than that of the soil condition, i and H. When numerical simulation results of SSI and TSSI models were examined, it turned out that the slope effect could essentially be omitted when X/H exceeded 2.5~5, and the ratio that referred to the H and soil parameters was more conservative in comparison to earlier research [27]. Where the ratio of inter-story shear and displacement of TSSI and SSI superstructures was more than 1, attributed to the difference in soil shear wave velocity, the precise amplification coefficient still had to be determined based on site conditions. According to the numerical simulation results, TSSSI had more serious damage on the structure immediately adjacent to the slope, triggering the structure to move towards the slope, but this impact is only cause for concern when $D \leq 0.5a$, while the slope effect of the structure far from the slope could be barely noticeable when $D = 0.5a$.

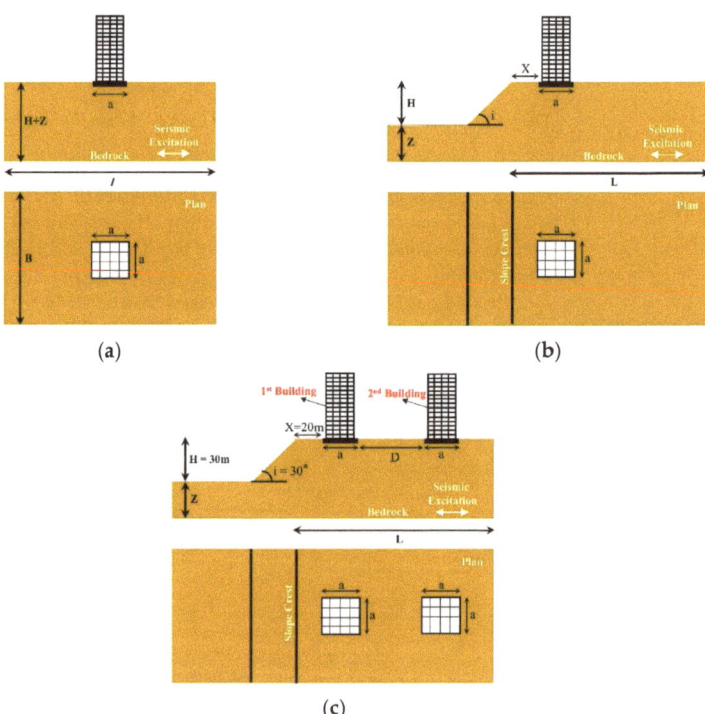

Figure 4. Illustration of SSI under different terrain conditions [116]: (**a**) SSI; (**b**) TSSI; (**c**) TSSSI.

3.2. Seismic Excitation

Seismic acceleration amplitude, vibration frequency, displacement direction, and incident angle constitute the primary factors influencing SSI. In the context of selecting seismic waves, Zhan et al. [117] conducted a study examining the impact of large-scale soft sites subjected to far-field earthquakes. The findings revealed a noticeable prevalence of long-period ground motion in such scenarios. In contrast to artificial waves, the amplification effect on ground motion caused by far-field large earthquakes in large-scale soft sites appeared more pronounced, with the amplification coefficient increasing alongside

Chen et al. [23] based on previously used amplification coefficients, possibly attributed to variations in the spectral composition of seismic waves. An optimized algorithm, as proposed by [118], can effectively determine the site period as follows:

$$T = \sqrt{\sum_{ith}^{N}\left(\frac{4h_{ith}}{v_{ith}}\right)^2 \frac{2H_{ith}}{h_{ith}}} \tag{1}$$

$$H_{ith} = \sum_{n=1}^{i^{th}} h_n - \frac{1}{2}h_{ith} \tag{2}$$

where h_{ith} is the thickness of the ith soil layer, v_{ith} denotes the shear wave velocity of the ith soil layer, H_{ith} presents the midpoint depth of the ith soil layer, and N signifies the total number of soil layers, illustrated in Figure 5. The deep soft soil site with a longer period (1.13~1.28 s) amplified the far-field vibration with a similar peak spectral period, whereas it significantly filtered the ground motion with a larger peak spectral gap (shorter period less than 0.3 s or larger period more than 2.5 s). Subsequent investigations have indicated that distinct seismic waves exhibit comparable effects on a given SSI model when adjusted to the same amplitude and selected based on the response spectrum [59].

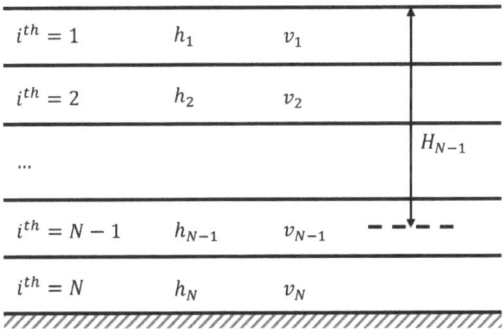

Figure 5. Illustration of the horizontally layered site.

Beyond the amplification of unfavorable stability given the similarity of the soil and seismic wave period, when the frequency of the foundation and superstructure matched that of the seismic wave, it could induce a worse situation, where the structure resonated as a seismic response and reached its peak. Especially for frame construction [119], the destruction of the substructure would be worse than that of the pile foundation. Citing ref. [120], if the seismic wave was an oblique wave incident, the superstructure could resonate at a certain frequency owing to phase fluctuation, which correlated to the stiffness ratio of the foundation and soil or superstructure. When the pile foundation was in a sedimentary clay layer, the foundation stiffness was significantly higher than that of soil, resulting in the vibration of the upper frame by the long-period motion of the ground rather than with the basic period of the foundation. Ergo, the displacement of the pile foundation did not originate from the vibration caused by the upper structure but rather from the soil driving, creating quite low interlayer displacement and damage. The displacement between the superstructure and the foundation was smaller the higher the damping ratio of the superstructure was [112]. Furthermore, the looseness of the backfill layer opposed to the original soil could trigger the surface ground vibration to be further exacerbated [31].

In addition, Zhou et al. [121] investigated the seismic behavior of a concrete chimney under multidimensional earthquakes. It was observed that the structure would be more vulnerable under three-dimensional ground motion than under one-dimensional or two-dimensional ground motion. Meanwhile, the incidence angle would have an impact on the seismic response of the SSI, especially in the case of major earthquakes, which would not be insignificant due to the smaller foundation size, and the most unfavorable input angle is 45° when the PGA is greater than 0.2 g. Dams [122] and frame structures [123] experienced a similar situation, although considering the type of seismic wave, the thickness and qualities of the soil, etc., the unfavorable angle may be larger or less than 45°. Hua [124] reached an analogous conclusion following research of the field under the action of 3D seismic waves that both the thickness of the soil layer and the angle of the incident wave affected the surface amplitude. They would be at maximum when the SH wave was incident perpendicular to the plane in which the P wave and SV wave reside, which was roughly fourfold the incident amplitude when the P wave was incident at an angle of 36° and the SV wave was at an angle of 54°. He further noticed that when the soil layer thickened, the surface amplitude climbed, and then fell, and then returned. Notably, the influence of vertical ground motion on foundation–structure settlement is virtually negligible unless the frequency of the vertical ground vibration approaches the resonance frequency. The nonlinearity introduced in the soil due to vertical motion is also less pronounced compared to that induced by horizontal motion [125].

3.3. Foundation Form

The stiffness of the foundation significantly influences the SSI effect, with the dynamic response at the interface between a flexible foundation and superstructure typically surpassing that of a rigid foundation below [120]. The emphasis on analyzing soil and pile–soil contact in seismic response studies stems from the notable impact of nonlinearities in soil properties and pile–soil interfaces, which often outweigh those associated with superstructure materials [126]. Wu et al. [126] observed that the soil adjacent to the pile tip and pile side exerts varying effects on the seismic response of the superstructure, with the former significantly influencing higher-order modes of the structure and the latter affecting lower-order modes more prominently. Additionally, Zamani and Shamy's study [125] on the soil beneath the foundation revealed that the acceleration amplification factor of the soil beneath a rigid foundation exceeded that of the free field at the same depth, indicating an adverse effect of the foundation on soil motion.

Previous studies on seismic predictions for structures have demonstrated that the inclusion of SSI alters the internal force distribution within the foundation [108]. Consequently, the amplification or reduction factor for the same structure under different foundations varies compared to assumptions made for a rigid foundation. Hokmabadi and Fatahi [19] assessed the sensitivity of foundations to SSI using FLAC 3D. The findings are summarized as follows: (a) the SSI effect effectively reduced the base shear, with the degree of influence being higher for the pile–raft foundation and pile foundation models than for shallow foundations; (b) the maximum sway angles for pile foundations and pile–raft foundations were, on average, 44% and 54% smaller than those for shallow foundations. The sway angles were attributed to the inertial force of the superstructure under seismic forces, causing settlement on one side of the foundation and bulging on the other; (c) the SSI effect exhibited clear amplification of lateral deformation and interlayer displacement, though the amplification was less pronounced for pile foundations and pile–raft foundations compared to shallow foundations. Furthermore, the study revealed that soft soil responded differently to various seismic waves. For instance, under low and medium acceleration, soft soil demonstrated a noticeable amplification impact on seismic waves. However, under excessively large acceleration, the soft soil inhibited the propagation of seismic waves towards the soil surface.

To explore the impact of the foundation type on the seismic response of a mid-rise building considering SSI, Zhang and Far [127,128] created two different foundation types, the end-bearing piled foundation, and classical compensated foundation, as shown in Figure 6. When the shear wave velocity was 150 m/s or 320 m/s, it was demonstrated that the SSI effect magnified the interlayer shear of the piled foundation-supported structure, and yet, for soil with $V_S = 600$ m/s, the SSI effect was not obvious, and amplification factor was even less than 1. When a structure is erected on soft clay

($V_S = 150$ m/s or 320 m/s), the use of a compensated foundation makes it more vulnerable to ground motion. The SSI effect substantially hinders the dynamic response of the structure. Additionally, the interlayer drift experiences significant amplification when SSI is considered for both pile and compensatory foundations. This observation aligns with the findings of Tabatabaiefar et al. [129], namely that SSI had a pronounced effect on the inter-story drift of frame structures.

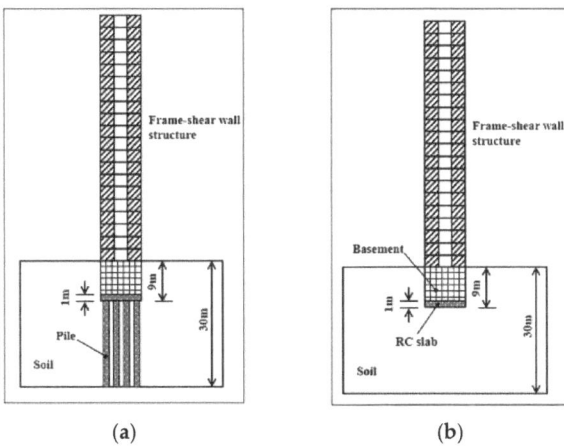

Figure 6. Models of frame–shear wall buildings considering SSI [128]: (**a**) end bearing piled foundation-supported structure; (**b**) classical compensated foundation-supported structure.

3.4. Superstructure

Computational models developed by Monsalve et al. [130], as depicted in Figure 7, reveal that the SSI effect enhances the inter-story drift ratio and acceleration of high-rise buildings. Simultaneously, it extends the period of low-order modes with minimal impact on high-order modes. Furthermore, it is noteworthy that the inclusion of a shear wall can, to some extent, mitigate the increase in inter-story drift while facilitating the expansion of story acceleration.

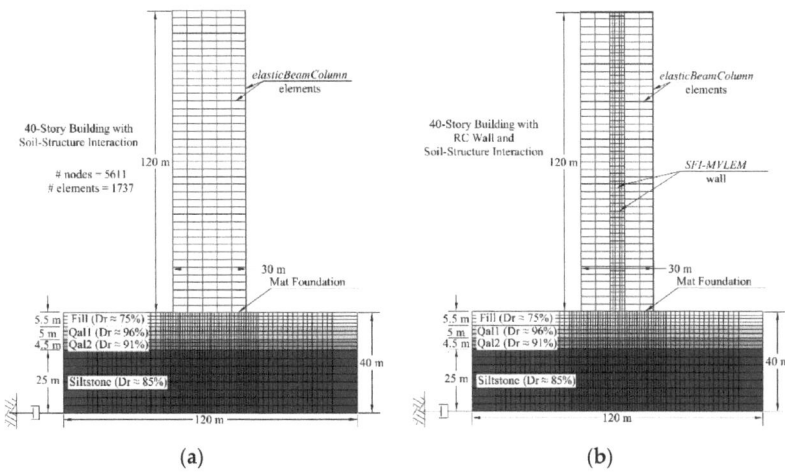

Figure 7. Models of the tall buildings without and with shear walls considering SSI [130]: (**a**) a building model; (**b**) a building model with shear walls.

In the study conducted by Mercado et al. [131], the results underscored the significant influence of the structure's geometry and the shear wave velocity of the soil on the lengthening of structural periods induced by the SSI effect. Notably, structures with larger aspect ratios and foundations in looser soil exhibited a more pronounced extension of periods. These findings align with observations in 1999, where Stewart et al. drew similar conclusions based on an analysis of 77 strong-motion data sets collected from 57 construction sites [132]. They measured the response of the SSI effect on structure in terms of the first-mode period lengthening ratio, $\tilde{T}/\overline{T}$ ($\overline{T}$ denotes the first-mode period without soil modelling, and $\tilde{T}$ denotes the first-mode period with SSI), and their observations revealed that the dimensionless ratio of structure–soil stiffness $1/\sigma$, was the most significant influence factor of $\tilde{T}/\overline{T}$, which could be expressed as

$$1/\sigma = \frac{\overline{h}}{V_S \cdot \overline{T}} \tag{3}$$

where $\overline{h}$ denoted the effective height of the structure. Other parameters, including the structural aspect ratio, foundation type, etc., were negligible.

To investigate the sensitivity of structures with various parameters to distinct soils under the action of four different seismic waves, Zhang and Far [128] established frame structures with a range of stories (20, 30 or 40 floors) and height-width ratios (HWR = 4, 5, 6), as shown in Figure 8. Their findings stated that the rise span had a minor impact on SSI, but the story height could not be omitted. Iida et al.'s survey on steel-framed and reinforced-concrete-framed structures located in Tokyo Bay [119], where resonance occurred due to the proximity of the periods, might help to explain the phenomenon that the latter effect increased nonlinearly and the mid-rise building was more susceptible to the SSI effect than the low-rise and high-rise building.

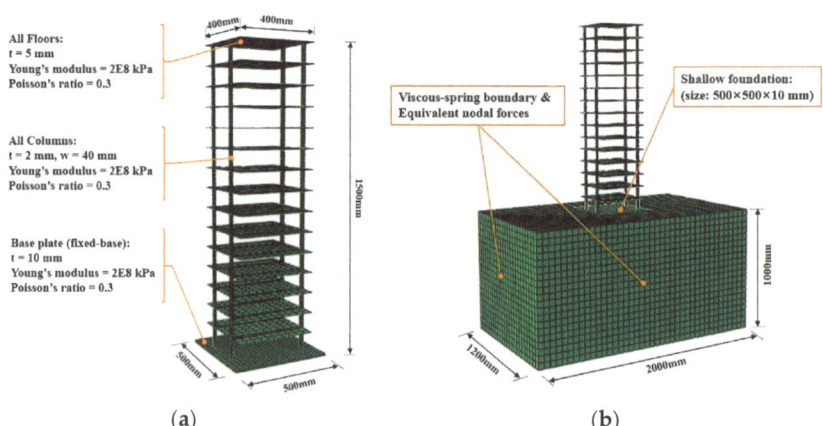

Figure 8. Numerical models with and without considering SSI [128]: (**a**) fixed-base model; (**b**) flexible-base model.

Similar to Alexander et al. [133], Liang et al. [134] found that the SSSI effect might alter the structural response at specific frequencies by 30~50%. Their simulation, utilizing the integral equation boundary element method (IBEM) in two dimensions, indicated that when the limit was surpassed, a building could be treated as an individual entity, with the scaling factor contingent on the distance between structures. The study made it clear that the distance between structures significantly impacted the scope and magnitude of the SSSI effect, persisting even when two flexible foundations were spaced up to 10 times

1985 Michoacan, Mexico, earthquake [135]. Additionally, Vicencio and Alexander's [106] research confirmed that the influence of SSSI could be disregarded in conventional seismic calculations when the distance between structures exceeded almost twice the foundation size. However, the significance of this observation was not definitively tied to soil properties; the lower the shear wave velocity of the soil, the greater the ratio [116].

Another crucial factor influencing the SSSI effect is the height ratio between structures. In Vicencio and Alexander's study [106], SSSI exhibited a significant amplification effect on the lower building, potentially reaching 400%, while causing a suppression effect of nearly 50% on the taller building, especially when the height ratio between the two structures exceeded 1.5 times. Farahania et al. [136] further emphasized this principle, stating that the seismic response of low-rise buildings could be effectively mitigated by increasing the net distance between structures. However, for high-rise buildings, this effect was deemed negligible as the SSI effect was much more intense than the SSSI effect.

Following converting one of the models from the Trombetta et al.'s [137] tests on the SSSI effect in urban neighborhoods (cSSSI model) into a prototype structure with a scaling factor (1:55), as illustrated in Figure 9, Bolisetti and Whittaker [138] carried out a numerical analysis. The simulated structure exhibited that when the combined arrangement was performed in the cSSSI model, SSSI failed to impact the global seismic response of the low-rise and mid-rise structures. In contrast, the SSSI did affect the peak acceleration response of the foundations, particularly in the presence of a deep foundation, which had a restraint effect on the neighboring low-rise building with shallow foundations. In Section 4.3, the experimental results that correlate to this numerical simulation will be enumerated.

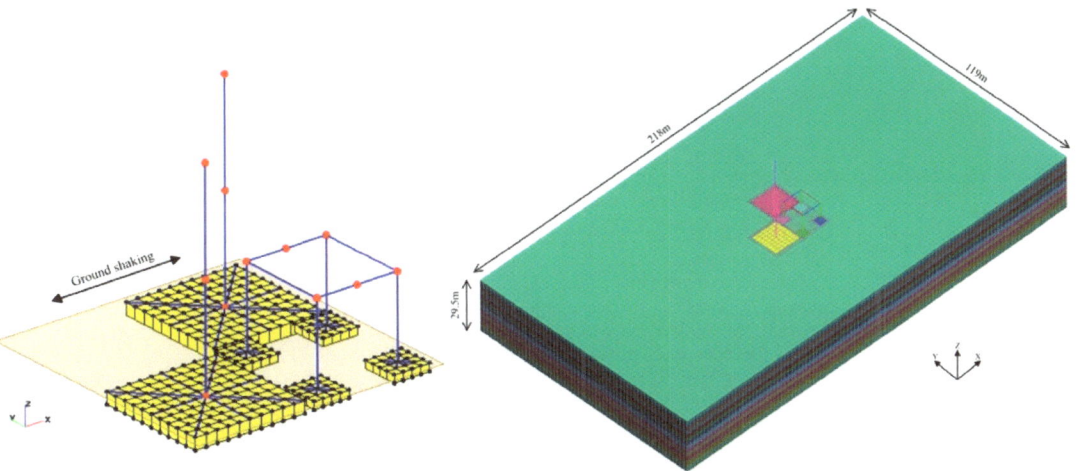

Figure 9. Numerical models of the cSSSI arrangement in SASSI (**left**) and LS-DYNA (**right**) [138].

To assess the influence of SSSI on the extensive clusters of buildings, Vicencio and Alexander [139] introduced a simplified reduced-order model illustrated in Figure 10. The findings indicated that while the SSSI effect on structures located at the corners was not maximized, it did amplify the seismic response of structures parallel to the excitation orientation.

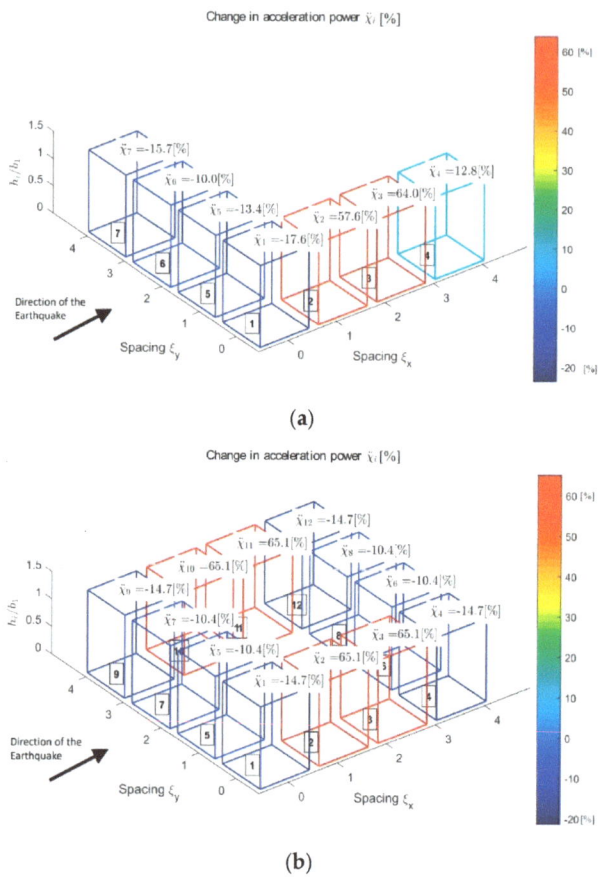

Figure 10. Change in acceleration power due to 3D SSSI [139]: (**a**) L shape arrangement; (**b**) a city block of twelve equispaced identical buildings.

4. Advances in Experimental Studies on SSI

4.1. Site Effect

Rayhani and Naggar [140,141] conducted vibration tests on a rigid structure at a depth of 30 m (1:80 scale) using the C-CORE 5.5-m radius beam centrifuge. That rigid structure was deemed to be a reasonable simulation of a 10-story building. The experimental setup included transducers for settlement measurement and accelerometers, as depicted in Figure 11. Glyben clay was employed during the tests to investigate the influence of soil shear strength on SSI effects. The undrained shear strength of the clay (RG-01) in Figure 12a was 40~60 kPa, with an average shear wave velocity of 73 m/s. The undrained shear strength of the clay (RG-02) in Figure 12b was 40~50 kPa for the top, and 85~95 kPa for the medium, with an average shear wave velocity of 100 m/s. Through a sturdy container, the superstructure was embedded in the clay. The results revealed that the SSI effect amplified the surface peak acceleration significantly, especially in instances of a minor input seismic amplitude. Furthermore, uniform clay (RG-01) amplified structural acceleration less than layered clay (RG-02), but RG-01 amplified near-surface acceleration over RG-02, which might be the consequence of the nonlinearities of the deeper and weaker clay. Also, when the seismic intensity increased, the amplification coefficient of the SSI effect on the amplitude and spectrum of seismic acceleration was minimized allowing researchers to hypothesize a connection between the rise in soil damping under extreme vibration. The acceleration of the free field and clay near the surface under the structure is slightly smaller than the experimental results in the period of 0.2~0.5 s, which might be owed to the rigid boundaries of the model that caused the computed SSI effect to be smaller. The experimental results matched well with those obtained using numerical simulations in FLAC 3D.

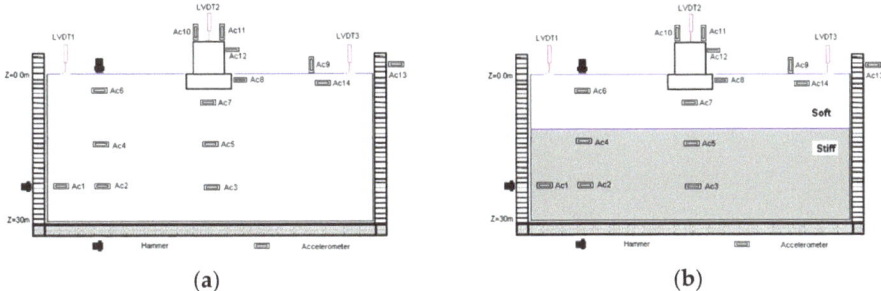

Figure 11. Centrifuge model configuration at prototype scale [140]: (**a**) Model RG-01; (**b**) Model RG-02.

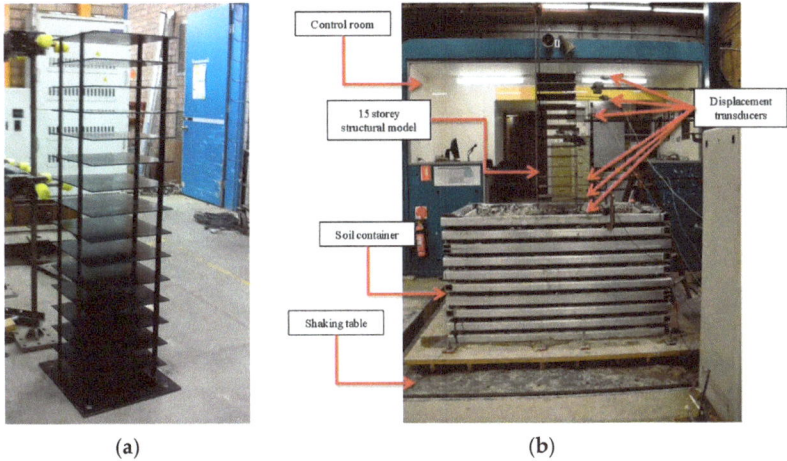

Figure 12. Models of shaking table tests [128]: (**a**) fixed-base model; (**b**) SSI model.

To compare the change in the inter-story shear of the same structure built on different foundations (rigid and soft), Zhang and Far [128] scaled a 15-story frame structure at a ratio of 1:30 and performed shaking table tests, as well as scaled a soil field with $V_S = 200$ m/s at the same ratio. The results revealed that SSI could not be ignored, especially when the shear wave velocity of the soil was lower. To validate the accuracy, the model developed in ABAQUS was compared with the findings obtained from this experiment.

Brennan [28] designed a TSSI centrifuge as in Figure 13, at a 1:50 scale to investigate the slope amplification factor under various conditions. The soil mass affected by TSSI occurred 0.31 H (the slope height was assumed to be H) longitudinally from the soil surface and 0.63 H from the lateral slope, according to the centrifuge test discoveries, with an amplification factor of roughly 1.5~2.0. The range of influence and amplification factors obtained from the experiment were found to be smaller than the numerical results, likely attributed to the errors in the scaled model and the distinct properties of the soil. Nonetheless, this experiment substantiated that the slope height significantly influenced the

TSSI effect, reinforcing the conclusions drawn from the numerical simulations. Additionally, the soil exhibited a filtering or amplification effect on seismic waves at various frequencies.

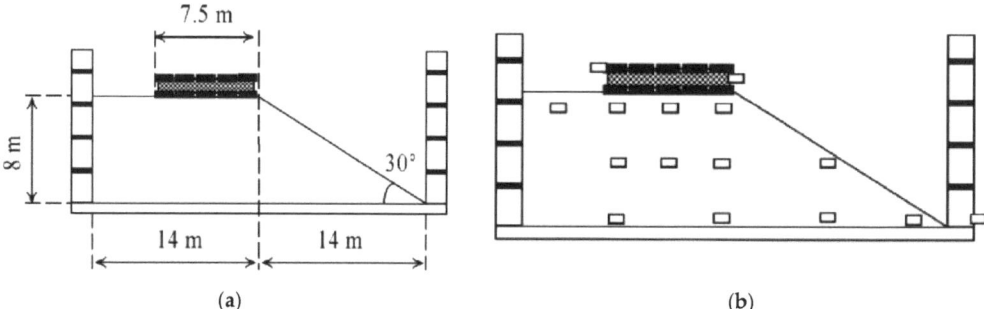

Figure 13. Experimental model of TSSI [28]: (**a**) dimensions; (**b**) accelerometer positions.

4.2. Foundation Form

Through shaking table tests, Shahbazi et al. [112] investigated the response of soil–pile–structure interaction under seismic action using a skid that mimicked the superstructure. The experimental setup consisted of two groups of four piles each, totaling eight helical piles, with distinct lengths and diameters in the two groups, as illustrated in Figure 14. The experiment spanned five days, with the seismic response of the pile group (featuring fixed and pinned connection types) measured during the final two days. The experimental findings indicated comparable seismic response characteristics for both connection types. Assuming a limit of 200 m/s for the soil shear wave velocity, the viscosity of the soil surrounding the piles was found to increase the shear wave velocity. Consequently, the stiffness and intrinsic frequency of the pile group were higher, resulting in a smaller displacement between the structure and piles. However, these effects could be disregarded when the wave velocity exceeded the threshold.

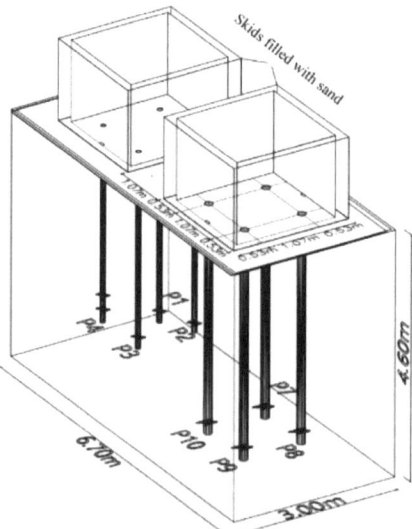

Figure 14. Schematic 3D view of pile groups and skids on shake table [112].

As depicted in Figure 15, Wang et al. [142] conducted a shaking table test on a six-story concrete frame at a 1:30 scale to assess the influence of the subsurface structure on SSI with varying foundation types, ranging from independent to box foundations. According to the experimental results, the SSI effects with the two different foundations caused a variation of up to 20% in the acceleration response of the soil. Due to the robust confinement provided by the box foundation to the superstructure, the overall acceleration response of the box foundation-supported structure was smaller than that of the independent foundation. However, the top layer of the box foundation exhibited a more pronounced amplification effect on the seismic response, characterized by a noticeable whiplash effect. Additionally, the researchers noted that the foundation type had a limited impact on the horizontal spectral characteristics of the structure but a more substantial influence on the acceleration amplitude and vertical spectrum.

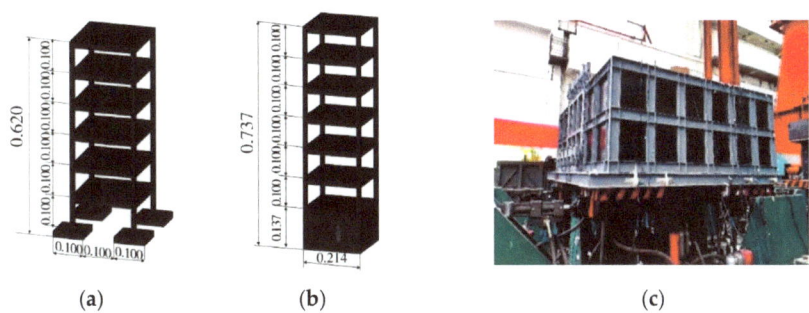

Figure 15. Superstructure diagram and test site [142]: (**a**) frame structure with independent foundation (unit: m); (**b**) frame structure with integral box foundation (unit: m); (**c**) soil container.

4.3. Superstructure

To assess the influence of soil–foundation–structure interaction (SFSI) and SSSI on the seismic response of structures, Trombetta et al. [137,143,144] conducted centrifuge experiments, as depicted in Figure 16. In Figure 16A, test-1 and test-2 were executed by adjusting the separation between two structures: a low-rise frame structure with a shallow embedded foundation and a mid-rise frame structure with a substantial basement. The tests affirmed that the SFSI effect indeed extended the period and increased damping. Moreover, the structure with a basement, capable of inhibiting permanent deformation, was found to be more susceptible to the SFSI effect. Test 2, on the other hand, demonstrated that the presence of deep footing influenced the moment–rotation behavior of the neighboring shallow foundations. As the constraints on shallow foundations increased, the system damping decreased, and the ratio of energy dissipated in superstructure vibration to the total energy dissipated increased. This scenario was less favorable for the seismic response of low-rise buildings. Building on the insights from Test 2, the researchers investigated the impact of SSSI on the seismic response of multiple structures by rearranging buildings. The findings indicated that for small and medium earthquakes, accounting for the SSSI effect was essential. With the intensification of ground motion, the nonlinearity of the SFSI increased gradually, while the SSSI effect diminished gradually, possibly due to a shift in the form of energy release. Additionally, this test confirmed that the vulnerability of structures to the SSSI effect depended on their nature and placement and was unrelated to the strength of the seismic wave.

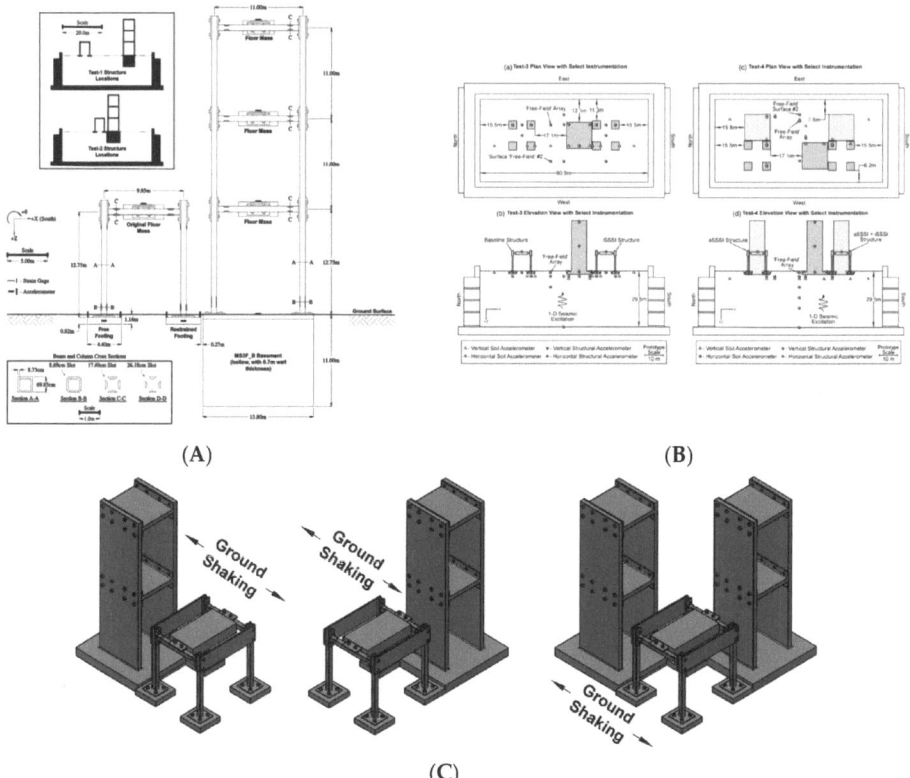

Figure 16. Shaking table test models of SFSI and SSSI: (**A**) setup of Tests 1 and 2 [143]; (**B**) setup of Tests 3 and 4 [144]; (**C**) diagram of iSSSI (Test 3), aSSSI (Tset 4), and cSSSI (Test 4) (from left to right) [137].

Aldaikh et al. [145] conducted a thorough investigation into the SSSI effects on three adjacent buildings through a shaking table test. To minimize the potential errors associated with genuine soil, a block cellular polyurethane foam was utilized, known for its linear and elastic propagation of seismic waves. In Figure 17b, the central building was replaced with an aluminium plate at the center of the foam box, while identical structures were installed on both sides, as depicted in Figure 17c. The test results indicated that the dynamic properties of the adjacent structures had discernible effects on the central building. Specifically, when the surrounding buildings were 10% to 20% higher than those at the center, the SSSI amplification coefficient on the seismic response of the central building could reach up to 170%. Conversely, when the surrounding buildings were 10% to 20% lower, the seismic response of the central building could be reduced by around 30%.

As depicted in Figure 18, Li et al. [146] reduced the scale of two 12-story cast-in-place concrete frame structures to 1:15, placing them on soft ground with a 200 mm separation. The test involved applying two seismic waves along the X and Y directions, and the impact of SSSI on the seismic response of the structures was assessed by varying the initial peak acceleration. The following conclusions were drawn from the test results: (a) the input excitation exceeded the acceleration response at the top of the free-field foundation; (b) as the initial excitation increased, the soil acceleration amplification coefficient, attributed to SSSI, decreased and became similar in both the X and Y

directions; (c) at the pile–soil interface, a disparity in the amplitude of the contact pressure was observed, with greater values at the top or end of the pile and minimal visibility at the pile shaft.

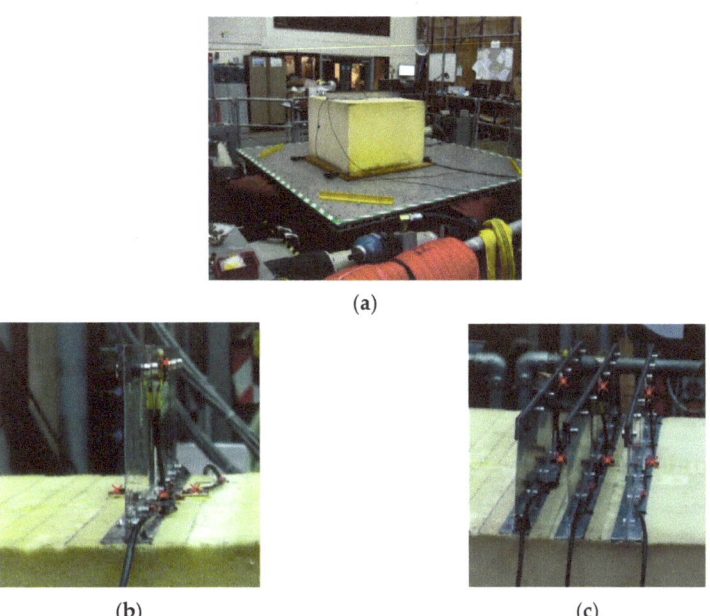

Figure 17. Configurations of experimental models on shaking table [145]: (**a**) block cellular polyurethane foam (soil model); (**b**) central building model; (**c**) parallel buildings model.

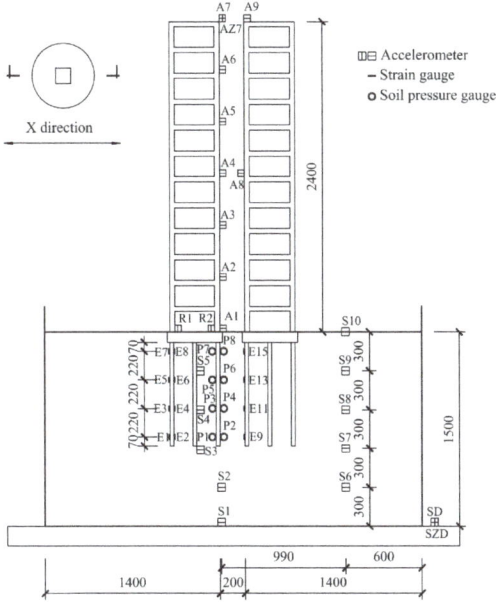

Figure 18. Geometry of model and experimental setup [146].

5. Research Advances on SSI for Large-Span Spatial Structures

In large-span spatial structures, the interaction between upper roofs and lower supporting structures is pivotal. The upper roofs act as a stiffness constraint on the lower structures, while the latter has an amplification effect on the former [147]. It has been demonstrated that when calculating the seismic response of the roof alone, compared to considering the entire large-span space structure with the lower structure in modal analysis, there are distinct differences in the frequencies of the first 10 orders. Simultaneously, nodal accelerations, roof displacements, and internal forces of rods show visible elevations [148]. This indicates that relying solely on the analysis of roofs is unreliable. Existing specifications [36] advocate for the co-calculation of the steel roof and lower supporting structure, emphasizing the need for considering damping ratio values in seismic design for large-span spatial structures. It is important to note that these recommended values were provided without taking into account the influence of different lower support structures and SSI effects.

Simultaneously, in the case of multi-degree-of-freedom (MDOF) structures, every mode contributes to the overall seismic response [149]. In the time history analysis of long-span structures, the most commonly employed method is the mode superposition method, where the key lies in determining the dominant modes [150]. Feng et al. [151] proposed the adoption of the mode contribution ratio to identify the dominant modes of spatial structures under seismic loads, a method that demonstrates greater precision compared to selecting the first 25~30 modes or using the effective mass method. The mentioned studies on SSI make it abundantly clear that SSI influences the modes of the structure, especially when the foundation type is a pile foundation or pile–raft foundation. Consequently, the higher-order modes of the structure exhibit more variation in the presence of SSI, underscoring the significance of considering the soil effects when conducting design or research.

Another crucial aspect of seismic investigation in the realm of large-span spatial structures is the consideration of multidimensional, multipoint, and nonstationary excitation. Typically, seismic response analysis of structures relies on the assumption of uniform excitation, utilizing the consistent input of ground motion. This approach predominantly considers the temporal variability of ground motion while omitting its spatial variation. However, the influence of Multiple Support Excitation (MSE) has garnered increased attention as the structural span expands [152]. Neglecting MSE could potentially lead to seismic design flaws [153]. Although early seismic response studies by Su and Dong [154] on a single-layer latticed shell with a diameter of 100 m suggested that the torsional effect induced by the travelling wave effect of vertical earthquakes could be mitigated by the substantial safety reserve in rods during actual engineering, subsequent seismic response studies on long-span structures and cable structures have revealed the inevitability of spatial variability [41,154]. This variability, observed in existing engineering practices [155–161], can either be amplified or discounted.

Since 2009, the authors' research group has been diligently addressing gaps in our understanding of the seismic response of SSI for large-span spatial structures. This subsection provides an overview of the ongoing work, along with a summary of the corresponding achievements. Utilizing advanced pile–soil models originally tailored for nuclear power plants and high-rise buildings, Luan [12] introduced a three-dimensional uncoupled pile–soil interaction model. This model incorporates the dynamic characteristics of spatial lattice structures subjected to strong seismic effects. Its feasibility was substantiated through a thorough comparison with a fully coupled analysis model. The initial assessment of the SSI effect was conducted using this model in the context of plane truss structures. The assumption of a single-layer site was in accordance with the U.S. code. The results indicated that the seismic responses of the structure on site Class E were notable when compared to those on site Class B, C, and D, and the SSI effect could be ignored. Sun [13] defined the relative stiffness β_1 between the supporting columns and upper roof, and took the site categorization of Chinese code as the soil condition to calculate the β_1 limit

While measuring the structural self-resonance frequency of truss structures erected on site Class I, II, and III, the SSI effect had to be taken into consideration when $\beta_1 > 3$, and if evaluating the internal force response of the members, it had to be involved when $\beta_1 > 6$. For Class I and Class II sites, the SSI effect was hard to disregard when $\beta_1 > 1$, while for Class III sites, the limit value of β_1 was 6 for the numerical calculation of top chord nodes' acceleration. Since then, this pile–soil model has been applied to reticulated shell structures, where the displacements at the top of structures increased with decreasing soil stiffness when the superstructures remained unaltered, although the maximum bending moment of the main ribs might decrease. Similarly, Sun [13] determined the limit value of β_1 that had to be contemplated in terms of the SSI effect utilizing a double-layer Kiewitt dome reticulated shell structure with a span of 60 m as an example. The double-layer shell structure established on site Class II and III had β_1 limits of 0.6 and 0.1, each, whereas the SSI effect was negligible for the structure on site Class I when calculating the internal force of members. Parallel to such, while computing the effect of seismic acceleration of the mesh shells, the structure erected on site Class I could neglect the SSI effect as well. The β_1 limitations for structures built on site Class II and III were 0.2 and 0.1, respectively. Note that the site Class I_0, Class I_1, Class II, and Class III of Chinese code are equivalent to the site Class A, Class B, Class C (D), and Class D of the U.S. code, respectively.

It was demonstrated that the SSI effect played a favorable role in increasing overall damping and extending the structural period. Considering the relatively dense natural frequency of large-span spatial structures, the impact on vibration modes within the reticulated shell structure exceeded that in the truss structure, notably concentrated in the first five orders. Regardless of the SSI effect or the assumption of a rigid foundation, Wang [14] emphasized that the initial vibration mode of a reticulated shell structure tended to be horizontal due to its arched configuration, resulting in increased vertical stiffness and decreased horizontal stiffness. However, the SSI effect expedited the emergence of the vertical vibration mode, particularly manifesting sooner when the soil stiffness was lower. In models with a stiff or rigid base, the vertical vibration mode typically appeared in the 11th order; for medium-soft soil, it was in the 7th order; and for soft soil, it was in the 5th order.

Wang [14] and Wei [15] examined the dynamic stability of single-layer and double-layer reticulated shell structures under step loads, simple harmonic loads, and seismic waves, respectively, aiming to broaden the awareness of the dynamic characteristics of mesh shell structures whilst including SSI. For the single-layer reticulated shell structures, the incorporation of SSI could significantly reduce the seismic responses in terms of both the step and seismic loads, with the reduction ratio exceeding 20%. In other words, the assumption of a rigid foundation fundamentally enhanced the dynamic stability of the superstructure. Concerning simple harmonic loads as the input, the primary determining factor was whether the load frequency fell within the resonance range. The dynamic critical load of the structure was found to be less dependent on whether the SSI effect was considered or not. Additionally, analyzing attentively how the shell structure performed under various seismic effects and on different types of sites, it was discovered that, generally, the structural dynamic instability load tended to show lower values with the smaller the soil stiffness. Yet, the dynamic response of the shell under different sites and seismic waves varied significantly, and this tendency was not strictly linear and had a greater relationship with choices of seismic waves. Moreover, the failure mechanism was connected to the member section and the rise–span ratio. The structure was prone to strength fracture when the member section was smaller, and, in the opposite scenario, the structure was susceptible to buckling failure. Although the site influences the structural failure mechanism, it did not play an irreplaceable role. Wei [15] came to a similar conclusion for their research on the double-layer reticulated shell structure. Involving SSI resulted in a 10% increase in acceleration at the structural apex but a decrease in the lateral displacement. Furthermore, by applying sinusoidal loading, it was found that the plastic ratio of members, the maximum displacement of the structure, and the torsion angle were all amplified collectively when the first period of structural self-resonance and the first-order self-resonance period of the soil coincide. Furthermore, for the sites with different periods, consideration ought to be given as well to the impact of the higher-order vibration mode on the structural response of shells. While the distribution of members' plasticity had been estimated, the area where the maximum displacement appeared differed significantly between the two cases of the structure without and with consideration for the SSI. It indicated that, even though the weak point remained virtually unaltered, the implications of the higher-order vibration modes on the structural impact were unavoidable after adopting SSI. A simplified simulation of SSI was established from Chinese code to attempt to explore the failure mechanism of the double-layer reticulated shell structure under the SSI effect, which contained the sites of Class I0, I1, II, III, and IV, respectively, and concluded that, for structures built on soft ground, extending the period of the structure may approximate the superior period of the input ground vibration, expanding the structural inertia during an earthquake and aggravating its hazard. Furthermore, an analysis based on numerical simulations of existing seismic design provisions indicates that the current seismic codes are suitable for high-rise buildings but may not apply to large-span structures. Instead, the torsional increase coefficients induced by SSI should be carefully considered when identifying the critical nodes in large-span spatial structures. Additionally, a pragmatic reference method for accurately evaluating the seismic performance of double-layer spherical reticulated shell structures was introduced in the form of a failure criterion.

A modified S–R (Sway–Rocking) model was offered by Wang [17] and Liu [16] based on the analysis of previous pile–soil simplified models by this group for the embedded method, the "m" method [162], and the resistance function method. The conclusion was that the resistance function is more accurate compared with other simplified methods for group pile foundations used for large-span spatial structures [12]. Shaking table tests were performed to confirm that the modified model had a better match with the assumed soil field, as illustrated in Figure 19. Liu [16] examined the seismic response of a single-layer cylindrical reticulated shell with various seismic wave incidence angles, foundation types (pile and independent foundations), and the presence or absence of seismic isolation bearings with the modified model, as illustrated in Figures 20–22. He also used shaking table tests to verify the results of the numerical simulation. The outcomes were consistent with our research group's earlier findings, which demonstrated that the SSI effect prolonged the structural period and elevated free-field acceleration by 5% to 30% at the foundation bottom and the soil surface. Simulated evaluation additionally proved that the incident angle profoundly impacted the seismic response of the structure. For example, in the case of P-wave incidence, maximum acceleration in both horizontal and vertical directions could be input proportionately to the ratio of 1:0.65 recommended by the *Code for the seismic design of buildings* [36] when the incidence angle was between 50° and 80°. However, when the incidence angle was between 0° and 50° and between 80° and 90°, the spatial effect of the seismic wave incidence required to be considered was increased. For the S-wave, the seismic response appeared to be enhanced when the input angle ranged between 30° and 60°. The ratio of the peak acceleration in the two horizontal directions at the base of the column fell between 1:0.3 and 1.3. As the intensity of the input seismic wave grew, the mutual coupling between the two horizontals was strengthened and the ratio approached 1:0.85. This range was further narrowed though, when seismic isolation bearings were applied, and, for S-waves, the interval to be pondered was 45°~60°. Based on the S–R modified model, Wang [17] carried out an initial investigation of the seismic response of a suspended dome structure with independent foundations. When the soil was considered, the structural frequency fluctuated similarly to that of the reticulated frames and shells, but its members' and nodes' acceleration responses deviated. The upper reticulated shell node's three displacements increased significantly after the SSI effect was considered, and, with softer soil, the displacements were larger, despite changes being unnoticeable. Meanwhile, the three accelerations of nodes dropped considerably, and the acceleration gradually

of the braces, the maximum tensile stresses of the diagonal members and ring cables, and the initial prestress loss of the ring cables all significantly decreased at the same time as the maximum compressive stresses of the ring and radial members of the reticulated shell increased.

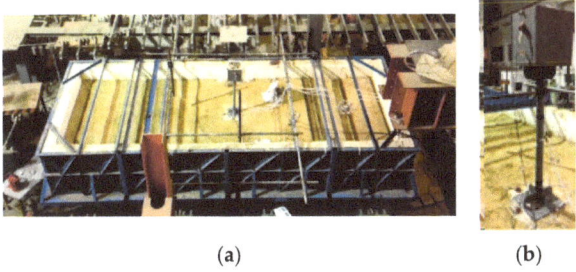

(a)   (b)

Figure 19. Shaking table test of single-column mass structure–soil interaction [17]: (**a**) soil container with experimental model; (**b**) single-column mass structure.

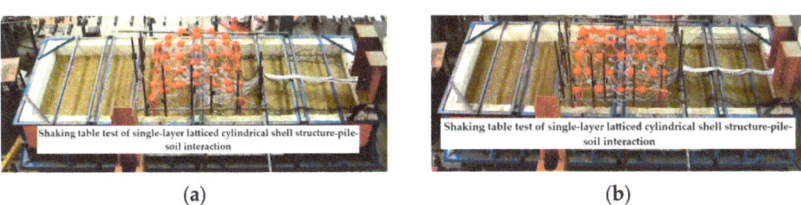

(a)   (b)

Figure 20. Shaking table test of single-layer latticed cylindrical shell structure–pile–soil interaction under vertical incidence of seismic waves [16]: (**a**) vertical arrangement; (**b**) parallel arrangement.

(a)   (b)

Figure 21. Shaking table test of single-layer latticed cylindrical shell structure–independent foundation–soil interaction under vertical and oblique incidence of seismic waves [16]: (**a**) vertical arrangement; (**b**) parallel arrangement.

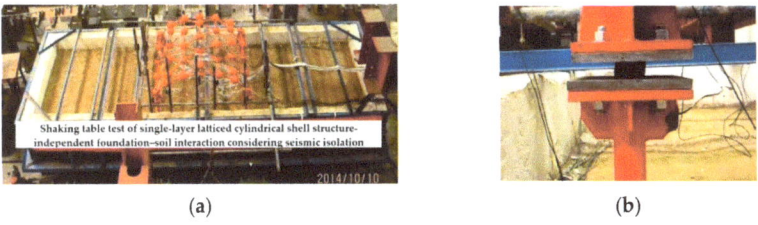

(a)   (b)

Figure 22. Shaking table test of single-layer latticed cylindrical shell structure-independent foundation–soil interaction considering seismic isolation [16]: (**a**) soil container with experimental model; (**b**) layered rubber bearing and its connection.

The author's group is presently engaged in evaluating the SSI effect on suspended dome structures at actual sites, aiming to provide a deeper understanding of the seismic response of prestressed structures. The Chinese code is adopted as a basis for site classification and wave selection to analyze the effects of SSI on the dynamic response of the structure and provide an approximate spectrum of impact. The superstructure depends on an actual construction, the Lanzhou Olympic Sports Center, as illustrated in Figure 23. Earlier studies primarily focused on a simplified model employing a continuous spring unit to replicate the influence of soil on the structure. The soil model was derived from a hypothetical site and lacked refinement, potentially leading to a scenario where the seismic response of the structure deviated from the actual seismic conditions. The two subclasses I0 and I1 of the site Class I, which have a large stiffness and approximate rigid foundation assumption, are not included in the current research scope. Existing results indicate that the sites that need to take the SSI into account primarily are sites Class II, III, and IV. To generate more reliable FEM, the soil models for site Class II, III, and IV in the ongoing study rely on real locations. Seismic waves are selected using two methods: employing standard response spectra and utilizing seismic waves that were measured at the modelled site. It is anticipated that these improvements will enhance the credibility of structural seismic responses at the site, providing empirical coefficients for the SSI effect. These coefficients can be utilized in subsequent endeavors, particularly in the investigations of prestressed structures.

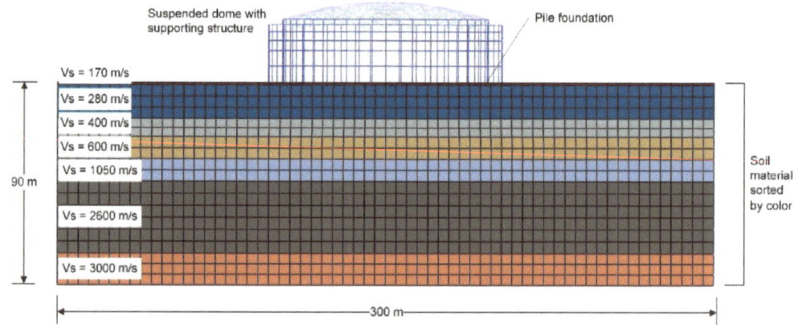

Figure 23. Model of Lanzhou Olympic Sports Center considering SSI.

6. Conclusions and Future Research Directions

In this paper, Section 2 elucidated prevalent solutions for SSI effects, while Sections 3 and 4 comprehensively reviewed site conditions, seismic excitation, the foundation type, and superstructure arrangement's impact on the seismic response of the entire system. These discussions were conducted under the two main categories of numerical and experimental studies for general structures. Section 5 delved into the significance of considering SSI for large-span spatial structures, providing an updated overview of the state of research on SSI effects in this context. In conclusion, this section will summarize key insights drawn from prior investigations and propose research directions for SSI systems involving large-span spatial structures.

6.1. Synthesis of Research Conclusions

(1) Chinese Codes mandate the consideration of the SSI effect for site Class II, III, and IV. For site Class I, SSI can be neglected in the absence of particular topographic conditions, as confirmed by studies on frame structures. However, for specific types of space structures erected on site Class I, the SSI effect cannot be dismissed when the stiffness ratio of the supporting structure to the upper roof surpasses a specified limit. Additionally, factors play decisive role in influencing a building's seismic

(2) While the soil effect generally dominates in most scenarios, it is crucial to note that this is not universally applicable. The impact of a slope on the SSI system is largely contingent on the ratio of the distance between the structure and the slope to the height of the slope. Although values across different papers may vary, there is a common agreement that the slope effect can be disregarded when the ratio exceeds 5. However, a consensus is yet to be reached on the specific amplification factor of the slope effect, and this necessitates examination in light of the specific characteristics of the site.

(3) According to the response spectrum, seismic waves could be selected to reduce the calculation error. Furthermore, a distinct spatial effect is observed, indicating that the seismic responses of structures do not increase linearly with various incidence angles. The determination of the most unfavorable incidence angle under three-dimensional seismic wave input is structure and site-dependent.

(4) Foundations exhibit diverse responses to SSI, with shallow and compensatory foundations being more susceptible compared to pile and pile–raft foundations. Additionally, the SSI effect has the potential to alleviate the base shear force, with pile and pile–raft foundations demonstrating higher discount factors than shallow foundations.

(5) In the case of frame structures, SSI may lead to a reduction in low-order modal frequency and an extension of the structural period. However, owing to their unique structural characteristics, large-span spatial structures should also consider the effects of SSI on high-order modes. Additionally, prior research has established that different superstructures exhibit varying degrees of sensitivity to SSI effects, primarily manifested in the increase or decrease in peak horizontal acceleration, inter-story drift ratio, and lateral displacement of the structure.

(6) Incorporating the nonlinear properties of soil becomes imperative when dealing with long site periods, as the application of a linear soil model introduces greater errors with increasing site periods. The direct method, devoid of the need for the superposition of multiple simplifying assumptions, allows for accurate nonlinear analysis of the SSI system, maintaining optimal accuracy. Unfortunately, its extensive computational demands render it impractical for use in large-scale models within engineering practice.

6.2. Prospective Research Directions

In light of the aforementioned considerations regarding the importance of integrating SSI in the design of large-span spatial structures, and acknowledging the ongoing debate on the applicability of existing research findings from other structures to the study of space structures, coupled with the recognized need for additional numerical and experimental simulations, this section will delve into potential future research directions for large-span spatial structures considering SSI. These directions are based on insights gleaned from previous investigations on other structure types, as well as the ongoing and completed research within the authors' research group.

(1) A more comprehensive and in-depth experimental or theoretical study would be beneficial to explore this aspect of the influence coefficient. *Code for seismic design of buildings* [36] stipulates that, for buildings constructed on sites Class III and IV, when SSI are included, the horizontal seismic shear tends to be diminished. Nonetheless, the group's research [14–17] revealed that, in certain circumstances, the horizontal seismic response was heightened rather than discounted, attributed to the influence of the higher-order vibration modes.

(2) Different numerical simulations have implemented various hypotheses and degrees of simplifying as dynamic interaction spans a wide range of research areas. Hence, the structural and foundation responses offered through several computational models might vary greatly from one another. Most SSI effects generated from two-dimensional modeling, such as dams or frame structures, may not be appropriate for spatial structures since their seismic vibrations are three-dimensional [163]. Meanwhile, there are a variety of spatial structures, along with numerous influencing factors. Certain large projected scale spatial structures (*span* $\geq$ 120 m, *length* $\geq$ 300 m, *cantilever* $\geq$ 40 m) are obliged to

contain multidirectional and multipoint inputs, which involve seismic wave reflection and diffraction. Thus, a more systematic study is necessary for various structural forms.

(3) It is imperative to recognize that diverse subsurface structures influence SSI. Recent articles often employ high-rise buildings and bridges as superstructures to investigate the impact of underground foundation design on the seismic response of SSI systems, with a significant focus on pile foundations and pile–raft foundations. However, the seismic assessment of large-span constructions with different foundation types, particularly in the context of larger underground spaces, is not thoroughly explored. It is advisable to delve deeper, as the presence of a basement imposes greater constraints on the superstructure compared to pile foundations.

(4) Studies conducted thus far on SSI in large-span structures commonly overlook scenarios in which both the superstructure and the soil simultaneously undergo plasticity. The equivalent strain of the soil experiences significant development during rare earthquakes, particularly in sites with soft soil layers, while certain members of the superstructure gradually enter plasticity. To accurately simulate the outcomes of seismic responses and determine whether structures eventually face dynamic instability, strength damage, or support failure, it is crucial to incorporate more sophisticated nonlinear or equivalent linear soil models and robust superstructure models that consider the plastic damage of materials in similar situations.

(5) There remains a discrepancy between experimental results and real ground vibrations, primarily attributed to the predominantly small-scale nature of previous experimental studies on SSI. A critical aspect of this process involves the selection of laboratory soil material. Using genuine soil introduces significant challenges; accurately scaling density and particle size, rearranging internal filler during vibration, and addressing other complexities pose inherent difficulties [145]. Alternatively, using a substitute would hinder the accurate simulation of soil nonlinearity. Furthermore, research on the topographic effects is confined by the limitations of the test site. The author's group conducted a comprehensive examination of the spatial effects of a single cylindrical reticulated shell under oblique incidence of seismic waves through numerical analysis and experiments. [16] Unfortunately, due to the model spans of the sizes of test models being relatively small, the travelling wave effect was not further involved. Furthermore, to ensure the generalizability of test results for practical engineering applications, larger-scale shaking table tests or field testing may be necessary, especially for broader spans.

(6) Further, as highlighted in this introduction, mitigating the impact of SSI is impractical when assessing the vibration control for large-span spatial structures. The vibration-damping effectiveness of the tuned mass damper (TMD) system, founded on the assumption of a rigid foundation, may prove suboptimal on flexible foundations and could potentially yield adverse effects, as evidenced by tests. Hence, further investigations are imperative to scrutinize the influence of SSI, validate their feasibility, and assess their damping effects.

Author Contributions: Conceptualization, X.L. and G.S.; validation, S.X.; writing—original draft preparation, P.Z.; writing—review and editing, P.Z. and R.M.; funding acquisition, X.L. All authors have read and agreed to the published version of the manuscript.

Funding: This research was funded by the National Key Research and Development Program of China (No. 2023YFC3805603).

Data Availability Statement: Not applicable.

Conflicts of Interest: The authors declare no conflicts of interest.

References

1. Kausel, E. Early history of soil-structure interaction. *Soil Dyn. Earthq. Eng.* **2010**, *30*, 822–832. [CrossRef]
2. Wu, S.; Gan, G. Dynamic soil-structure interaction for high-rise buildings. *Dev. Geotech. Eng.* **1998**, *83*, 203–216. [CrossRef]

4. *FEMA 440*; Improvement of Nonlinear static Seismic Analysis Procedures. Applied Technology Council (ATC): Redwood City, CA, USA, 2005.
5. Lou, M.; Wang, H.; Chen, X.; Zhai, Y. Structure-soil-structure interaction: Literature review. *Soil Dyn. Earthq. Eng.* **2011**, *31*, 1724–1731. [CrossRef]
6. Wang, H.; Lou, M.; Chen, X.; Zhai, Y. Structure-soil-structure interaction between underground structure and ground structure. *Soil Dyn. Earthq. Eng.* **2013**, *54*, 31–38. [CrossRef]
7. Abdulaziz, M.A.; Hamood, M.J.; Fattah, M.Y. A review study on Seismic behavior of individual and adjacent structures considering the soil-Structure interaction. *Structures* **2023**, *52*, 348–369. [CrossRef]
8. Naji, M.; Firoozi, A.A.; Firoozi, A.A. A review: Study of integral abutment bridge with consideration of soil-structure interaction. *Lat. Am. J. Solids Struct.* **2020**, *17*, e352. [CrossRef]
9. Jeremić, B.; Jie, G.; Preisig, M.; Tafazzoli, N. Time domain simulation of soil-foundation-structure interaction in non-uniform soils. *Earthq. Eng. Struct. Dyn.* **2009**, *38*, 699–718. [CrossRef]
10. Taslimi, A.; Petrone, F.; Pitarka, A. Characteristics of vertical ground motions and their effect on the seismic response of bridges in the near-field: A state-of-the-art review. *J. Bridge Eng.* **2024**, *29*, 03124001. [CrossRef]
11. Sigdel, L.D.; Al-Qarawi, A.; Leo, C.J.; Liyanapathirana, S.; Hu, P. Geotechnical design practices and soil-structure interaction effects of an integral bridge system: A review. *Appl. Sci.* **2021**, *11*, 7131. [CrossRef]
12. Xue, S.; Luan, X.; Li, X. Effect of soil conditions on the seismic response of spatial structure. *Adv. Mater. Res.* **2011**, *368–373*, 2079–2083. [CrossRef]
13. Sun, Y. Dynamical Performance Analysis of Space Lattice Structure Considering Soil-Foundation-Structure Interaction. Master's Dissertation, Beijing University of Technology, Beijing, China, 2010. (In Chinese)
14. Wang, G. Study on Dynamic Stability of Single-Layer Reticulated Domes Considering Soil-Structure Interaction Subjected to Earthquake Motion. Master's Dissertation, Beijing University of Technology, Beijing, China, 2011. (In Chinese)
15. Wei, X. Failure Mechanism for Double-Layer Latticed Shells Considering Soil-Structure Interaction. Master's Dissertation, Beijing University of Technology, Beijing, China, 2011. (In Chinese)
16. Liu, Y. Seismic Analysis and Experimental Investigation for Soil-Lattice Structure Interaction System. Ph.D. Dissertation, Beijing University of Technology, Beijing, China, 2015. (In Chinese)
17. Wang, G. Numerical Simulation and Experimental Research of Simplified SSI Method of Modified S-R Model. Master's Dissertation, Beijing University of Technology, Beijing, China, 2015. (In Chinese)
18. *GB 55003-2021*; General Code for Foundation Engineering of Building and Municipal Projects. Ministry of Housing and Urban-Rural Development of the People's Republic of China (MOHURD): Beijing, China, 2021. (In Chinese)
19. Hokmabadi, A.S.; Fatahi, B. Influence of foundation type on seismic performance of buildings considering soil-structure interaction. *Int. J. Struct. Stab. Dyn.* **2016**, *16*, 1550043. [CrossRef]
20. Ghazavi, M.; Dehkordi, P.F. Interference influence on behavior of shallow footings con-structed on soils, past studies to future forecast: A state-of-the-art review. *Transp. Geotech.* **2021**, *27*, 100502. [CrossRef]
21. Vergara, J.; Sierra, C.; Sáenz, M.; Jaramillo, J.; Gomez, J. Construction of rational models for topographic effects and size-conditioned-response-spectra. *Soil Dyn. Earthq. Eng.* **2021**, *140*, 106432. [CrossRef]
22. Houston, H.; Kanamori, H. Source characteristics of the 1985 Michoacan, Mexico Earth-quake at periods of 1 to 30 seconds. *Geophys. Res. Lett.* **1986**, *13*, 597–600. [CrossRef]
23. Chen, J.; Chen, X.; Shi, G. Research on seismic response characteristics of sites with deep and soft soils. *J. Disaster Prev. Mitig. Eng.* 2004; *24*, 131–138, (In Chinese) [CrossRef]
24. Jin, D.; Chen, G.; Dong, F. Seismic response of a real basin site considering topography effect and nonlinear characteristic of soil. In Proceedings of the Sixth China-Japan-US Trilateral Symposium on Lifeline Earthquake Engineering, Chengdu, China, 28 May–1 June 2013; pp. 545–552. [CrossRef]
25. Zhao, C.; Valliappan, S. Incident P and SV wave scattering effects under different canyon topographic and geological conditions. *Int. J. Numer. Anal. Methods Geomech.* **1993**, *17*, 73–94. [CrossRef]
26. Zhao, C.; Valliappan, S. Seismic wave scattering effects under different canyon topographic and geological conditions. *Soil Dyn. Earthq. Eng.* **1993**, *12*, 129–143. [CrossRef]
27. Fatahi, B.; Huang, B.; Yeganeh, N.; Terzaghi, S.; Banerjee, S. Three-dimensional simulation of seismic slope-foundation-structure interaction for buildings near shallow slopes. *Int. J. Geomech.* **2020**, *20*, 04019140. [CrossRef]
28. Brennan, A.J.; Madabhushi, S.P.G. Amplification of seismic accelerations at slope crests. *Can. Geotech. J.* **2009**, *46*, 585–594. [CrossRef]
29. Alitalesh, M.; Shahnazari, H.; Baziar, M.H. Parametric study on seismic topography-soil-structure interaction; Topographic effect. *Geotech. Geol. Eng.* **2018**, *36*, 2649–2666. [CrossRef]
30. Raj, D.; Singh, Y.; Kaynia, A.M. Behavior of slopes under multiple adjacent footings and buildings. *Int. J. Geomech.* **2018**, *18*, 04018062. [CrossRef]
31. Madany, M.; Guo, P.J. Structure-soil-structure interaction analysis for lateral seismic earth pressure of deeply buried structure in layered ground. *Int. J. Geomechanics.* **2021**, *21*, 04021217. [CrossRef]
32. Clouteau, D.; Broc, D.; Devésa, G.; Guyonvarh, V.; Massin, P. Calculation methods of structure-soil-structure interaction (3SI) for embedded buildings: Application to NUPEC tests. *Soil Dyn. Earthq. Eng.* **2012**, *32*, 129–142. [CrossRef]

33. Far, H. Advanced computation methods for soil-structure interaction analysis of structures resting on soft soils. *Int. J. Geotech. Eng.* **2019**, *13*, 352–359. [CrossRef]
34. Idriss, I.M.; Seed, H.B. Seismic response of horizontal soil layers. *J. Soil Mech. Found. Div.* **1968**, *94*, 1003–1031. [CrossRef]
35. *GB 17741-2005*; Evaluation of Seismic Safety for Engineering Site. Standardization Administration of the People's Republic of China (SAC): Beijing, China, 2014. (In Chinese)
36. *GB 50011-2010*; Code for Seismic Design of Buildings. Ministry of Housing and Urban-Rural Development of the People's Republic of China (MOHURD): Beijing, China, 2016. (In Chinese)
37. Ni, S.; Li, S.; Chang, Z.; Xie, L. An alternative construction of normalized seismic design spectra for near-fault regions. *Earthq. Eng. Eng. Vib.* **2013**, *12*, 351–362. [CrossRef]
38. Rodriguez-Marek, A.; Song, J. Displacement-Based Probabilistic Seismic Demand Analyses of Earth Slopes in the Near-Fault Region. *Earthq. Spectra* **2016**, *32*, 1141–1163. [CrossRef]
39. Kiureghian, A.D. A coherency model for spatially varying ground motions. *Earthq. Eng. Struct. Dyn.* **1996**, *25*, 99–111. [CrossRef]
40. Harichandran, R.S.; Vanmarcke, E.H. Stochastic variation of earthquake ground motion in space and time. *J. Eng. Mechanics.* **1986**, *112*, 154–174. [CrossRef]
41. Wang, X.; Xue, S.; Cao, Z. Nonstationary response of spatial lattice shells under multiple seismic inputs. In Proceedings of the International Conference on Advances in Building Technology (ABT 2002), Hong Kong, 4–6 December 2002; pp. 595–602.
42. Jaya, K.P.; Prasad, A.M. Embedded foundation in layered soil under dynamic excitations. *Soil Dyn. Earthq. Eng.* **2002**, *22*, 485–498. [CrossRef]
43. Stewart, J.; Crouse, C.B.; Hutchinson, T.C.; Lizundia, B.; Naeim, F.; Ostadan, F. *Soil-Structure Interaction for Building Structures (NIST GCR 12-917-21)*; NIST Pubs: Gaithersburg, MD, USA, 2012.
44. Anoyatis, G.; Lemnitzer, A. Kinematic Winkler modulus for laterally-loaded piles. *Soils Found.* **2017**, *57*, 453–471. [CrossRef]
45. Dhadse, G.D.; Ramtekkar, G.D.; Bhatt, G. Finite element modeling of soil structure interaction system with interface: A review. *Arch. Comput. Methods Eng.* **2021**, *28*, 3415–3432. [CrossRef]
46. Jemielita, G. Governing equations and boundary conditions of a generalized model of elastic foundation. *J. Theor. Appl. Mech.* **1994**, *32*, 887–901.
47. Zhao, X.; Zhu, W.; Li, Y.; Li, M.; Li, X. Review, classification, and extension of classical soil-structure interaction models based on different superstructures and soils. *Thin-Walled Struct.* **2022**, *173*, 108936. [CrossRef]
48. Anand, V.; Kumar, S.R.S. Seismic soil-structure interaction: A state-of-the-art review. *Structures* **2018**, *16*, 317–326. [CrossRef]
49. Veletsos, A.S.; Tang, A.Y. Vertical vibration of ring foundations. *Earthq. Eng. Struct. Dyn.* **1987**, *15*, 1–21. [CrossRef]
50. Gazetas, G. Formulas and charts for impedances of surface and embedded foundations. *J. Geotech. Eng.* **1991**, *117*, 1363–1381. [CrossRef]
51. Chen, W. A dynamical nonlinear effective stress method to evaluate liquefaction of subsoil of building used cone model. *Rock Soil Mech.* **2003**, *24*, 40–44. (In Chinese) [CrossRef]
52. Dutta, S.C.; Roy, R. A critical review on idealization and modeling for interaction among soil-foundation-structure system. *Comput. Struct.* **2002**, *80*, 1579–1594. [CrossRef]
53. Wen, Y.; Yang, G.; Zhong, Z. Multipotential surface elastoplastic constitutive model and its application in the analysis of the phase II cofferdam of the three gorges project. *Geofluids* **2023**, *2023*, 9581827. [CrossRef]
54. Mott, G.; Wang, J. The effects of variable soil damping on soil-structure dynamics. *J. Vib. Control* **2011**, *17*, 365–371. [CrossRef]
55. Chen, W.F.; Saleeb, A.F. *Elasticity and Plasticity*; Yu, T., Wen, X., Liu, Z., Eds.; China Architecture & Building Press: Beijing, China, 2016.
56. Seed, H.B.; Idriss, I.M. *Soil Moduli and Damping Factors for Dynamic Response Analyses*; Technical Report; National Technical Reports Library: Lawrence, KS, USA, 1970. Available online: https://ntrl.ntis.gov/NTRL/dashboard/searchResults/titleDetail/PB197869.xhtml (accessed on 24 July 2023).
57. Bolisetti, C.; Whittaker, A.S.; Coleman, J.L. Linear and nonlinear soil-structure interaction analysis of buildings and safety-related nuclear structures. *Soil Dyn. Earthq. Eng.* **2018**, *107*, 218–233. [CrossRef]
58. Ghandil, M.; Behnamfar, F. The near-field method for dynamic analysis of structures on soft soils including inelastic soil-structure interaction. *Soil Dyn. Earthq. Eng.* **2015**, *75*, 1–17. [CrossRef]
59. Deng, H.; Jin, X.; Gu, M.; Huang, J. Modification of soil dynamic constitutive model in pile-soil-structure interaction analysis. *J. Tongji Univ. (Nat. Sci.)* **2018**, *46*, 1473–1478+1574. (In Chinese) [CrossRef]
60. Masing, G. Eigenspannungen und verfestigung beim messing. In Proceedings of the Second International Congress of Applied Mechanics, Zurich, Switzerland, 12–17 September 1926; pp. 332–335.
61. Hardin, B.O.; Drnevich, V.P. Shear modulus and damping in soils: Design equations and curves. *J. Soil Mech. Found. Div.* **1972**, *98*, 667–692. [CrossRef]
62. Kim, D.S.; Stokoe, K.H. Soil damping computed with Ramberg-Osgood-Masing model. In Proceedings of the 13th International Conference on Soil Mechanics and Foundation Engineering, New Delhi, India, 5–10 January 1994; pp. 211–214.
63. Martin, P.P.; Seed, H.B. *A Computer Program for the Non-Linear Analysis of Vertically Propagating Shear Waves in Horizontally Layered Deposits*; Technical Report; Earthquake Engineering Research Center: Oakland, CA, USA, 1978.
64. Martin, P.P.; Seed, H.B. One-dimensional dynamic ground response analyses. *J. Geotech. Eng. Div.* **1982**, *108*, 935–952. [CrossRef]

66. Liu, J.; Gu, Y.; Du, Y. Consistent viscous-spring artificial boundaries and viscous-spring boundary elements. *Chin. J. Geotech. Eng.* **2006**, *28*, 1070–1075. (In Chinese)
67. Song, E.; Luo, S. A local artificial boundary for transient seepage problems with unsteady boundary conditions in unbounded domains. *Int. J. Numer. Anal. Methods Geomech.* **2017**, *41*, 1108–1124. [CrossRef]
68. Bettess, P. Infinite elements. *Int. J. Numer. Methods Eng.* **1977**, *11*, 53–64. [CrossRef]
69. Lysmer, J.; Kuhlemeyer, R.L. Finite dynamic model for infinite media. *J. Eng. Mech. Div.* **1969**, *95*, 859–877. [CrossRef]
70. Zhang, G.; Zhao, M.; Wang, P.; Du, X.; Zhang, X. Obliquely incident P-SV wave scattering by multiple structures in layered half space using combined zigzag-paraxial boundary condition. *Soil Dyn. Earthq. Eng.* **2021**, *143*, 106662. [CrossRef]
71. Chang, X.; Liu, D.; Gao, F.; Lu, L.; Long, L.; Zhang, J.; Geng, X. A study on lateral transient vibration of large diameter piles considering pile-soil interaction. *Soil Dyn. Earthq. Eng.* **2016**, *90*, 211–220. [CrossRef]
72. Bazyar, M.H.; Song, C. A continued-fraction-based high-order transmitting boundary for wave propagation in unbounded domains of arbitrary geometry. *Int. J. Numer. Methods Eng.* **2008**, *74*, 209–237. [CrossRef]
73. Chen, S.S.; Hsu, W.C. An energy transmitting boundary for semi-infinite structures. *J. Mech.* **2007**, *23*, 159–172. [CrossRef]
74. Liu, T.; Zheng, S.; Tang, X.; Gao, Y. Time-Domain Analysis of Underground Station-Layered Soil Interaction Based on High-Order Doubly Asymptotic Transmitting Boundary. *Comput. Model. Eng. Sci.* **2019**, *120*, 545–560. [CrossRef]
75. Gu, Y.; Bo, J.; Du, Y. 3D consistent viscous-spring artificial boundary and viscous-spring boundary element. *Eng. Mech.* **2007**, *24*, 31–37. (In Chinese)
76. Liu, J.; Li, B. Three-dimensional viscoelastic static and dynamic unified artificial boundary. *Sci. Sin. (Technol.)* **2005**, *9*, 72–86. (In Chinese)
77. Semblat, J.F. Modeling seismic wave propagation and amplification in 1D/2D/3D linear and nonlinear unbounded media. *Int. J. Geomech.* **2011**, *11*, 440–448. [CrossRef]
78. Tsai, H.C. Modal superposition method for dynamic analysis of structures excited by prescribed support displacements. *Comput. Struct.* **1998**, *66*, 675–683. [CrossRef]
79. Tian, Y.; Yang, Q. On time-step in structural seismic response analysis under ground displacement/acceleration. *Earthq. Eng. Eng. Vib.* **2009**, *8*, 341–347. [CrossRef]
80. Liu, J.; Dong, Y. A direct method for analysis of dynamic soil-structure interaction. *China Civ. Eng. J.* **1998**, *31*, 55–64. (In Chinese) [CrossRef]
81. Liu, J.; Bao, X.; Wang, D. The internal substructure method for seismic wave input in 3D dynamic soil-structure interaction analysis. *Soil Dyn. Earthq. Eng.* **2019**, *127*, 105847. [CrossRef]
82. He, J.; Ma, H.; Zhang, B.; Chen, H. Method and realization of seismic motion input of viscous-spring boundary. *J. Hydraul. Eng.* **2010**, *41*, 960–969. (In Chinese) [CrossRef]
83. Kim, M.K.; Lim, Y.M.; Cho, W.Y. Three dimensional dynamic response of surface foundation on layered half-space. *Eng. Struct.* **2001**, *23*, 1427–1436. [CrossRef]
84. Liu, J.; Wang, Y. A 1D time-domain method for in-plane wave motions in a layered half-space. *Acta Mech. Sin.* **2007**, *23*, 673–680. [CrossRef]
85. Zhang, J.; Li, M.; Han, S. Seismic analysis of gravity dam-layered foundation system subjected to earthquakes with arbitrary incident angles. *Int. J. Geomech.* **2022**, *22*, 04021279. [CrossRef]
86. Liu, J.; Tan, H.; Bao, X.; Wang, D.; Li, S. Seismic wave input method for three-dimensional soil-structure dynamic interaction analysis based on the substructure of artificial boundaries. *Earthq. Eng. Eng. Vib.* **2019**, *18*, 747–758. [CrossRef]
87. Saberi, M.; Annan, C.D.; Konrad, J.M. On the mechanics and modeling of interfaces between granular soils and structural materials. *Arch. Civ. Mech. Eng.* **2018**, *18*, 1562–1579. [CrossRef]
88. Uesugi, M.; Kishida, H. Frictional resistance at yield between dry sand and mild steel. *Soils Found.* **1986**, *26*, 139–149. [CrossRef]
89. Zhang, G.; Zhang, J. State of the art: Mechanical behavior of soil-structure interface. *Prog. Nat. Sci.* **2009**, *19*, 1187–1196. [CrossRef]
90. Ma, J.; Wang, D.; Randolph, M.F. A new contact algorithm in the material point method for geotechnical simulations. *Int. J. Numer. Anal. Methods Geomech.* **2014**, *38*, 1197–1210. [CrossRef]
91. Clough, G.W.; Duncan, J.M. Finite element analyses of retaining wall behavior. *J. Soil Mech. Found. Div.* **1971**, *97*, 1657–1673. [CrossRef]
92. Desai, C.S.; Zaman, M.M.; Lightner, J.G.; Siriwardane, H.J. Thin-layer element for interfaces and joints. *Int. J. Numer. Anal. Methods Geomech.* **1984**, *8*, 19–43. [CrossRef]
93. Genes, M.C.; Kocak, S. Dynamic soil-structure interaction analysis of layered unbounded media via a coupled finite element/boundary element/scaled boundary finite element model. *Int. J. Numer. Methods Eng.* **2005**, *62*, 798–823. [CrossRef]
94. Padrón, L.A.; Aznárez, J.J.; Maeso, O. 3-D boundary element-finite element method for the dynamic analysis of piled buildings. *Eng. Anal. Bound. Elem.* **2011**, *35*, 465–477. [CrossRef]
95. Shehata, O.E.; Farid, A.F.; Rashed, Y.F. Practical boundary element method for piled rafts. *Eng. Anal. Bound. Elem.* **2018**, *97*, 67–81. [CrossRef]
96. Wolf, J.P. Response of unbounded soil in scaled boundary finite-element method. *Earthq. Eng. Struct. Dyn.* **2002**, *31*, 15–32. [CrossRef]
97. Tadeu, A.J.B.; Kausel, E.; Vrettos, C. Scattering of waves by subterranean structures via the boundary element method. *Soil Dyn. Earthq. Eng.* **1996**, *15*, 387–397. [CrossRef]

98. Mohammadi, M.; Karabalis, D.L. Dynamic 3-D soil-railway track interaction by BEM-FEM. *Earthq. Eng. Struct. Dyn.* **1995**, *24*, 1177–1193. [CrossRef]
99. Yazdchi, M.; Khalili, N.; Valliappan, S. Dynamic soil-structure interaction analysis via coupled finite-element-boundary-element method. *Soil Dyn. Earthq. Eng.* **1999**, *18*, 499–517. [CrossRef]
100. Chen, X.; Birk, C.; Song, C. Numerical modelling of wave propagation in anisotropic soil using a displacement unit-impulse-response-based formulation of the scaled boundary finite element method. *Soil Dyn. Earthq. Eng.* **2014**, *65*, 243–255. [CrossRef]
101. Lu, S.; Liu, J.; Lin, G. Hamiltonian-based derivation of the high-performance scaled-boundary finite-element method applied to the complex multilayered soil field in time domain. *Int. J. Geomech.* **2016**, *16*, 04015098. [CrossRef]
102. Hassanen, M.; El-Hamalawi, A. Two-dimensional development of the dynamic coupled con-solidatio5 scaled boundary finite-element method for fully saturated soils. *Soil Dyn. Earthq. Eng.* **2007**, *27*, 153–165. [CrossRef]
103. Birk, C.; Behnke, R. A modified scaled boundary finite element method for three-dimensional dynamic soil-structure interaction in layered soil. *Int. J. Numer. Methods Eng.* **2012**, *89*, 371–402. [CrossRef]
104. Liang, J.; Wu, M.; Ba, Z.; Lee, V.W. Surface motion of a layered transversely isotropic half-space with a 3D arbitrary-shaped alluvial valley under qP-, qSV- and SH-waves. *Soil Dyn. Earthq. Eng.* **2021**, *140*, 106388. [CrossRef]
105. Tripe, R.; Kontoe, S.; Wong, T.K.C. Slope topography effects on ground motion in the presence of deep soil layers. *Soil Dyn. Earthq. Eng.* **2013**, *50*, 72–84. [CrossRef]
106. Vicencio, F.; Alexander, N.A. Dynamic interaction between adjacent buildings through non-linear soil during earthquakes. *Soil Dyn. Earthq. Eng.* **2018**, *108*, 130–141. [CrossRef]
107. Lan, J.; Liu, J.; Song, X. Study on the influence of the seafloor soft soil layer on seismic ground motion. *Nat. Hazards Earth Syst. Sci.* **2021**, *21*, 577–585. [CrossRef]
108. Chen, S.; Zhu, X.; Zhao, Y.; Chen, G. Analysis of saturated soil-structure interaction considering soil skeleton nonlinearity. *Earthq. Eng. Eng. Dyn.* **2019**, *39*, 114–127. (In Chinese) [CrossRef]
109. Ding, Y.; Pagliaroli, A.; Lanzo, G. One-dimensional seismic response of two-layer soil deposits with shear wave velocity inversion. In Proceedings of the AIP 2008 Seismic Engineering Conference: Commemorating the 1908 Messina and Reggio Calabria Earth-quake Reggio Calabria, Reggio Calabria, Italy, 8–11 July 2008; Volume 1020, pp. 252–258. [CrossRef]
110. Iida, M. 3D methods for examining soil-building interaction for nonlinear soil behavior based on an input wave field. *Int. J. Geomech.* **2017**, *17*, 04016081. [CrossRef]
111. Iida, M. Three-dimensional nonlinear soil response methods based on a three-component input wave field. *Int. J. Geomech.* **2016**, *16*, 04015026. [CrossRef]
112. Shahbazi, M.; Cerato, A.B.; El Naggar, M.H.; Elgamal, A. Evaluation of seismic soil-structure interaction of full-scale grouped helical piles in dense sand. *Int. J. Geomech.* **2020**, *20*, 04020228. [CrossRef]
113. Xu, R.; Fatahi, B.; Hokmabadi, A.S. Influence of soft soil shear strength on the seismic response of concrete buildings considering soil-structure interaction. In Proceedings of the Advances in Numerical and Experimental Analysis of Transportation Geomaterials and Geosystems for Sustainable Infrastructure (Geo-China 2016), Shandong, China, 25–27 July 2016; pp. 17–24. [CrossRef]
114. Mercado, J.A.; Mackie, K.R.; Arboleda-Monsalve, L.G. Modeling nonlinear-inelastic seismic response of tall buildings with soil-structure interaction. *J. Struct. Eng.* **2021**, *147*, 04021091. [CrossRef]
115. Erfani, A.; Ghanbari, A.; Massumi, A. Seismic behaviour of structures adjacent to slope by considering SSI effects in cemented soil mediums. *Int. J. Geotech. Eng.* **2021**, *15*, 2–14. [CrossRef]
116. Shamsi, M.; Shabani, M.J.; Vakili, A.H. Three-dimensional seismic nonlinear analysis of topography-structure-soi-structure interaction for buildings near slopes. *Int. J. Geomech.* **2022**, *22*, 04021295. [CrossRef]
117. Zhan, J.; Chen, G.; Jin, D. Seismic Response Characteristics of Deep Soft Site with Depth under Far-Field Ground Motion of Great Earthquake. *Adv. Mater. Res.* **2011**, *378–379*, 477–483. [CrossRef]
118. Dobry, R.; Oweis, I.; Urzua, A. Simplified procedures for estimating the fundamental period of a soil profile. *Bull. Seismol. Soc. Am.* **1976**, *66*, 1293–1321. [CrossRef]
119. Iida, M.; Iiba, M.; Kusunoki, K.; Miyamoto, Y.; Isoda, H. Relative seismic risk evaluation of various buildings based on an input wave field. *Int. J. Geomech.* **2017**, *17*, 04017068. [CrossRef]
120. Liang, J.; Jin, L.; Todorovska, M.I.; Trifunac, M.D. Soil–structure interaction for a SDOF oscillator supported by a flexible foundation embedded in a half-space: Closed-form solution for incident plane SH-waves. *Soil Dyn. Earthq. Eng.* **2016**, *90*, 287–298. [CrossRef]
121. Zhou, C.; Tian, M.; Guo, K. Seismic partitioned fragility analysis for high-rise RC chimney considering multidimensional ground motion. *Struct. Des. Tall Spec. Build.* **2018**, *28*, e1568. [CrossRef]
122. Zhang, J.; Li, M.; Han, S.; Deng, G. Estimation of seismic wave incident angle using vibration response data and stacking ensemble algorithm. *Comput. Geotech.* **2021**, *137*, 104255. [CrossRef]
123. Sebastiani, P.E.; Liberatore, L.; Lucchini, A.; Mollaioli, F. A new method to predict the critical incidence angle for buildings under near-fault motions. *Struct. Eng. Mech.* **2018**, *68*, 575–589. [CrossRef]
124. Hua, K.; Fan, L.; Che, J. Characteristics of earthquake ground motion influenced by seismic incident angle. *J. Xi'an Univ. Technol.* **2019**, *35*, 248–255. (In Chinese)
125. Zamani, N.; El Shamy, U. Discrete-element method simulations of the response of soil-foundation-structure systems to multidi-

126. Wu, C.; Yao, Q.; Yan, H.; Li, J. Nonlinear analysis of high-rise building superstructure-pile-soil mass interaction considering contact effect under severe earthquake. *Chin. J. Rock Mech. Eng.* **2011**, *30*, 3224–3233. (In Chinese)
127. Zhang, X.; Far, H. Effects of dynamic soil-structure interaction on seismic behaviour of high-rise buildings. *Bull. Earthq. Eng.* **2022**, *20*, 3443–3467. [CrossRef]
128. Zhang, X.; Far, H. Seismic response of high-rise frame-shear wall buildings under the influence of dynamic soil-structure interaction. *Int. J. Geomech.* **2023**, *23*, 04023141. [CrossRef]
129. Tabatabaiefar, S.H.R.; Fatahi, B.; Samali, B. Seismic behavior of building frames considering dynamic soil-structure interaction. *Int. J. Geomech.* **2013**, *13*, 409–420. [CrossRef]
130. Arboleda-Monsalve, L.G.; Mercado, J.A.; Terzic, V.; Mackie, K.R. Soil-structure interaction effects on seismic performance and earthquake-induced losses in tall buildings. *J. Geotech. Geoenviron. Eng.* **2020**, *146*, 04020028. [CrossRef]
131. Mercado, J.A.; Arboleda-Monsalve, L.G.; Mackie, K. Study of period lengthening effects in soil-structure interaction systems. In Proceedings of the International Foundations Congress and Equipment Expo 2021 (IFCEE 2021), Dallas, TX, USA, 10–14 May 2021; pp. 1–10. [CrossRef]
132. Stewart, J.P.; Seed, R.B.; Fenves, G.L. Seismic Soil-structure interaction in buildings. II: Empirical findings. *J. Geotech. Geoenviron. Eng.* **1999**, *125*, 38–48. [CrossRef]
133. Alexander, N.A.; Ibraim, E.; Aldaikh, H. A simple discrete model for interaction of adjacent buildings during earthquakes. *Comput. Struct.* **2013**, *124*, 1–10. [CrossRef]
134. Liang, J.; Han, B.; Todorovska, M.I.; Trifunac, M.D. 2D dynamic structure-soil-structure interaction for twin buildings in layered half-space I: Incident SH-waves. *Soil Dyn. Earthq. Eng.* **2017**, *102*, 172–194. [CrossRef]
135. Iida, M. Three-dimensional finite-element method for soil-building interaction based on an input wave field. *Int. J. Geomech.* **2013**, *13*, 430–440. [CrossRef]
136. Farahani, D.; Behnamfar, F.; Sayyadpour, H. Effect of pounding on nonlinear seismic response of torsionally coupled steel structures resting on flexible soil. *Eng. Struct.* **2019**, *195*, 243–262. [CrossRef]
137. Trombetta, N.W.; Mason, H.B.; Hutchinson, T.C.; Zupan, J.D.; Bray, J.D.; Kutter, B.L. Nonlinear soil-foundation-structure and structure-soil-structure interaction: Engineering demands. *J. Struct. Eng.* **2015**, *141*, 04014177. [CrossRef]
138. Bolisetti, C.; Whittaker, A.S. Numerical investigations of structure-soil-structure interaction in buildings. *Eng. Struct.* **2020**, *215*, 110709. [CrossRef]
139. Vicencio, F.; Alexander, N.A. Method to evaluate the dynamic structure-soil-structure interaction of 3-D buildings arrangement due to seismic excitation. *Soil Dyn. Earthq. Eng.* **2021**, *141*, 106494. [CrossRef]
140. Rayhani, M.H.T.; El Naggar, M.H. Centrifuge modeling of seismic response of layered soft clay. *Bull. Earthq. Eng.* **2007**, *5*, 571–589. [CrossRef]
141. Rayhani, M.H.T.; El Naggar, H. Physical and numerical modeling of dynamic soil-structure interaction. In Proceedings of the Geotechnical Earthquake Engineering and Soil Dynamics IV, Sacramento, CA, USA, 18–22 May 2008; pp. 1–11. [CrossRef]
142. Wang, G.; Wang, Y.; Sun, F.; Zheng, N. Experimental study on influence of foundation type on seismic response of frame structure and site soil. *World Earthq. Eng.* **2022**, *38*, 96–109. (In Chinese) [CrossRef]
143. Trombetta, N.W.; Mason, H.B.; Chen, Z.; Hutchinson, T.C.; Bray, J.D.; Kutter, B.L. Nonlinear dynamic foundation and frame structure response observed in geotechnical centrifuge experiments. *Soil Dyn. Earthq. Eng.* **2013**, *50*, 117–133. [CrossRef]
144. Trombetta, N.W.; Mason, H.B.; Hutchinson, T.C.; Zupan, J.D.; Bray, J.D.; Kutter, B.L. Nonlinear soil-foundation-structure and structure-soil-structure interaction: Centrifuge test observations. *J. Geotech. Geoenviron. Eng.* **2014**, *140*, 04013057. [CrossRef]
145. Aldaikh, H.; Alexander, N.A.; Ibraim, E.; Oddbjornsson, O. Two dimensional numerical and experimental models for the study of structure-soil-structure interaction involving three buildings. *Comput. Struct.* **2015**, *150*, 79–91. [CrossRef]
146. Li, P.; Liu, S.; Lu, Z.; Yang, J. Numerical analysis of a shaking table test on dynamic structure-soil-structure interaction under earthquake excitations. *Struct. Des. Tall Spec. Build.* **2017**, *26*, e1382. [CrossRef]
147. Fu, X.; Gao, Y.; Yang, X. Whole structure analysis of the large-span spatial structure in Jinan Olympic Stadium. *J. Build. Struct.* **2009**, *30*, 176–182. (In Chinese) [CrossRef]
148. Li, Z.; Wu, Y.; Ding, Y.; Shi, Y.; Zong, L. Amplification effect analysis of substructure on seismic response of large-span spatial hub structure. *J. Disaster Prev. Mitig. Eng.* **2021**, *41*, 823–836. (In Chinese)
149. Kato, S.; Nakazawa, S.; Saito, K. Two-mode based estimation of equivalent seismic loads and static estimation of dynamic response of reticular domes supported by ductile substructures. *J. Int. Assoc. Shell Spat. Struct.* **2006**, *47*, 35–52.
150. Masanao, N.; Yasuhito, S.; Keiji, M.; Toshiyuki, O. An efficient method for selection of vibration modes contributory to wind response on dome-like roofs. *J. Wind Eng. Ind. Aerodyn.* **1998**, *73*, 31–43. [CrossRef]
151. Feng, R.; Zhu, B.; Wang, X. A mode contribution ratio method for seismic analysis of large-span spatial structures. *Int. J. Steel Struct.* **2015**, *15*, 835–852. [CrossRef]
152. Zhao, B.; Wang, Y.; Chen, Z.; Shi, Y. State of seismic response analysis of large-span spatial structures under multi-support excitations. *Res. Steel Struct. Eng.* **2012**, *9*, 188–197. (In Chinese)
153. Bogdanoff, J.L.; Goldberg, J.E.; Schiff, A.J. The effect of ground transmission time on the response of long structures. *Bull. Seismol. Soc. Am.* **1965**, *55*, 627–640. [CrossRef]
154. Su, L.; Dong, S. Seismic analysis of two typical spatial structures under multi-support vertical excitations. *Eng. Mech.* **2007**, *24*, 85–90. (In Chinese)

155. Sun, J.; Zhang, Q. Non-stationary seismic response of long-span space cable structures under multiple support excitation. *World Earthq. Eng.* **2010**, *26*, 156–161. (In Chinese)
156. Liu, F.; Zhen, C.; Xu, Z.; Qian, J.; Zhao, J.; Ke, C.; Wang, C.; Wang, G.; Zhu, Z. Time-history analysis of terminal 3 of the Capital Airport under multi-support and multi-dimension seismic excitation. *J. Build. Struct.* **2006**, *27*, 56–63. (In Chinese) [CrossRef]
157. Shen, S.; Zhang, W.; Zhu, D.; Qian, J.; Pei, Y. Seismic response analysis of two long-span hangars under multiple support excitations. *China Civ. Eng. J.* **2008**, *41*, 17–21. (In Chinese) [CrossRef]
158. Yang, Q.; Liu, W.; Tian, Y. Response analysis of national stadium under specially variable earthquake ground motions. *China Civ. Eng. J.* **2008**, *41*, 35–41. (In Chinese) [CrossRef]
159. Jiang, Y.; Shi, Y.; Wang, Y.; Zhang, Y. Seismic response analysis of large-span gate-type tube truss structures under multi-support excitation. *J. Beijing Jiaotong Univ.* **2009**, *33*, 88–93. (In Chinese)
160. Wang, D.; Zhang, A. Anti-seismic performance experiment about the suspendome structure of badminton gymnasium for 2008 Olympic Games. *Ind. Constr.* **2010**, *40*, 104–108. (In Chinese) [CrossRef]
161. Wang, Y.; Zhao, B.; Ding, D.; Jiang, Y.; Shi, Y.; Chen, Z. The steel structure system comparison of Hefei Xinqiao Airport Terminal based on the seismic response under multi-support excitations. *China Civ. Eng. J.* **2012**, *45*, 72–75. (In Chinese) [CrossRef]
162. JTG 3363-2019; Specifications for Design of Foundation of Highway Bridges and Culverts. Ministry of Transport of the People's Republic of China (MOT): Beijing, China, 2019. (In Chinese)
163. Kato, S.; Iida, M.; Minamibayashi, J. FEM analysis of elasto-plastic buckling loads of single layer latticed cylindrical roofs and estimation of the buckling loads based on the buckling stress concept. In Proceedings of the Asian Pacific Conference on Shell and Spatial Structures (APCS), Beijing, China, 21–25 May 1996; pp. 508–515.

Disclaimer/Publisher's Note: The statements, opinions and data contained in all publications are solely those of the individual author(s) and contributor(s) and not of MDPI and/or the editor(s). MDPI and/or the editor(s) disclaim responsibility for any injury to people or property resulting from any ideas, methods, instructions or products referred to in the content.

Article

Measurements and Evaluation of Road Traffic-Induced Micro-Vibration in a Workshop Equipped with Precision Instruments

Zhijun Zhang [1], Xiaozhen Li [2,*], Xun Zhang [2], Guihong Xu [1] and Anjie Wu [1]

[1] School of Civil Engineering, Guizhou Institute of Technology, Guiyang 550003, China; zhijun0973@163.com (Z.Z.); smileanne@163.com (G.X.); wuanjie163@163.com (A.W.)
[2] State Key Laboratory of Bridge Intelligent and Green Construction, Southwest Jiaotong University, Chengdu 611756, China; zhxunxun@swjtu.edu.cn
* Correspondence: xzhli@swjtu.edu.cn

Abstract: Road traffic transportation has flourished in the process of urbanization due to its advantages, but concurrently it generates harmful environmental vibrations. This vibration issue becomes particularly crucial in production workshops housing precision instruments. However, limited research has been undertaken on this matter. This study aimed to investigate the influence of road traffic-induced vibration on micro-vibrations within a workshop housing precision instruments. A field test was conducted to assess the vibration levels originating from both machinery operation and vehicular traffic. The results indicated that ground-borne vibrations caused by road vehicles decrease with increasing propagation distance, peaking around 10 Hz. Machinery operation vibrations were primarily concentrated above 20 Hz, while vehicular traffic vibrations were more prominent below 20 Hz. Notably, the passage of heavy trucks significantly impacted both ground and workshop vibrations, with vertical vibrations being particularly significant. Within the workshop, the second floor experienced higher vibrations above 20 Hz due to the presence of installed instruments. Importantly, the micro-vibration levels on both floors exceeded the VC-C limit (12.5 μm/s), highlighting the need to account for road traffic and machinery vibrations in workshop design. These data can be utilized to validate numerical models for predicting road traffic-induced vibrations, aiding in vibration assessment during road planning and design.

Keywords: production workshop; environmental vibration; micro-vibration; road traffic; precision instruments

Citation: Zhang, Z.; Li, X.; Zhang, X.; Xu, G.; Wu, A. Measurements and Evaluation of Road Traffic-Induced Micro-Vibration in a Workshop Equipped with Precision Instruments. *Buildings* **2024**, *14*, 1142. https://doi.org/10.3390/buildings14041142

Academic Editor: Honggui Di

Received: 2 March 2024
Revised: 5 April 2024
Accepted: 10 April 2024
Published: 18 April 2024

Copyright: © 2024 by the authors. Licensee MDPI, Basel, Switzerland. This article is an open access article distributed under the terms and conditions of the Creative Commons Attribution (CC BY) license (https://creativecommons.org/licenses/by/4.0/).

1. Introduction

With improvements in the urbanization rate in China, the population is gathering in cities, and the urban road network system has also been upgraded [1,2]. The well-connected traffic network has brought great convenience to people's lives and travel. However, traffic-induced vibration has become a major environmental issue in urban areas, since the distance between the buildings and the traffic lines is decreasing due to space limitations in heavily populated cities [3–7]. The vibration hazard caused by road traffic is thus an urgent social and engineering problem to be solved [8–10].

Research on the vibration and noise problems of buildings, caused by railway traffic, is quite common [11,12]. As for vibration propagation caused by road vehicles running, Watts [13] found that significant vibration responses could be observed within 50 m from the center of the road when heavy vehicles were running on an uneven road. In theoretical investigations, a simplified vibration prediction model was proposed to consider the influence of factors such as vehicle speed, vibration propagation distance, and the maximum height and depth of the uneven pavement on the road [14]. Hunt [15,16] and Lombaert [17] assumed that the effect of running vehicles on the road was a steady-state process. In their studies, the vehicle was simplified into different models, and the influence of different road surfaces on the vibration caused by the vehicle was discussed. An analytical method based on random vibration analysis was proposed to predict the power spectral density of traffic-induced ground-borne vibration by Hao and Ang [18], which showed that the ground-borne vibration was mainly caused by the propagation of the Rayleigh wave. Other numerical models were established to study the vibration response caused by the running vehicle on roads of different roughness [19,20]. In Lak's study [20], the load weight, traffic degree, and road roughness were significant factors that affected the environmental vibration caused by road traffic. However, the pavement structure primarily affected the near-field vibration response in a high-frequency band, and the road–soil transfer function was mainly determined by soil parameters, so the accurate identification of soil dynamic parameters was the key to predicting traffic-induced environmental vibration issues. Road humps and speed cushions were used to control vehicle speeds in specific areas, whereas obvious vibrations were produced when vehicles passed over these speed bumps; in some cases, they could reach perceptible levels in adjacent buildings [21]. A theoretical analysis model was established to study the subject of vehicles passing over speed bumps, and the influences of the vehicle speed, tire stiffness, and the height and width of the unevenness were discussed [22]. Many valuable findings have been obtained from the above theoretical analysis and numerical simulation.

However, any theoretical analysis and numerical simulation cannot consider all objective factors. Prediction uncertainties and inaccuracies resulting from assumptions in modeling cannot be neglected. A site experiment is the most intuitive and accurate means of studying the vibration source excitation and the response of sensitive targets when the vehicle passes, and it is also the most rigorous way of checking and evaluating analytical and numerical solutions. A large number of field experiments have been carried out in the academic and engineering fields to provide accurate model parameters and test data for validating theoretical models [23–29].

To verify a numerical model of environmental vibration, a field experiment was performed by Mhanna et al. [23]. Test vehicles were used in another field test carried out by Taniguchi and Sawada [24]. It was found that the Rayleigh wave played a primary role in the vibration propagation, and its dominant frequency range was located at 10~20 Hz. The field vibration responses induced by different types of vehicles traveling at different speeds were measured, and it was observed that the vibration responses mainly dominated in 10~40 Hz and the vibration frequency induced by buses was concentrated in the relatively narrow frequency band of 10~12.5 Hz [25]. A detailed field measurement of the environmental vibration caused by road traffic in Montreal was conducted, and it was observed that the dominant frequency band of vibration caused by trucks was concentrated between 10 and 20 Hz [26]. Another field test was executed to study the vibration response of a free field beside a two-way subway line and road traffic, and the different characteristics of vibration induced by the subway trains and road vehicles were discussed [27]. To find out the influence of road traffic on ancient buildings, an experimental study on the vibration of ancient buildings caused by road traffic vehicles was performed [28]. The vibration response caused by the vehicle passing through different types of deceleration devices was researched using experiments [29]. Obvious frequency characteristics of road traffic-induced environmental vibration were discovered in the above-mentioned studies.

In contrast, most experimental studies have focused on vibration problems encountered with intensive vibration levels, and relatively little attention has been paid to the impact of micro-vibration. With the development of the social industrialization process, the problem of micro-vibration has gradually been exposed [30–33]. For workshops with precision instruments, micro-vibration may affect the normal operation and working accuracy of precision instruments [34]. An experimental study was conducted to evaluate the influence of heavy-vehicle-induced vibration on precision instruments, which found that the amplitude of heavy vehicles is about 3~5 times that of small cars [35]. Another study focused on the attenuation of ground-borne vibrations by applying isolation measures for sensitive areas with precision instruments [36,37].

In this study, a random sampling analysis was conducted to assess traffic flow conditions on the expressway. Then, a micro-vibration field test was carried out for the production workshop of microelectronics enterprises adjacent to an urban expressway in Xiamen, China. Special attention is paid to the site vibration on the ground and the micro-vibration inside the production workshop due to road traffic. This study differs from previous works primarily in two aspects: firstly, we conducted a vibration test in a workshop operating continuously, with data collected every hour capturing both production and traffic-induced vibrations; secondly, we strategically placed measurement points along the traffic–factory path to cross-verify the traffic-induced vibration characteristics in the workshop, enhancing the reliability of the analysis. To obtain a better understanding of such a complex dynamic system, the measurement data were analyzed both in the time and frequency domains, and the test results provide references for similar research.

2. Field Test Site and Experiment Scheme

2.1. Field Test Site

The microelectronics enterprise is located off the northern side of an urban expressway in Xiamen, China, as shown in Figure 1a, where the No. 3 production workshop is in operation. It is a high-tech workshop for the production of new-type displays and is equipped with precision instruments for the production of low-temperature polysilicon (LTPS) and color filters (CFs). The distance from the edge of the urban expressway to the production workshop is approximately 82 m. There are six lanes in both directions making up the main carriageway of the urban expressway. Additionally, on the auxiliary carriageway, there are four lanes in both directions. The design speed of the expressway is 80 km/h. To the southwest of the No. 3 production workshop, the nearest distance from the workshop to the elevated rail transit line is about 65 m (see Figure 1a).

The overall length and width of the No. 3 production workshop is 507.9 m × 252.6 m, with a building height of about 26 m. There is no production equipment in the dust-free room on the first floor of the workshop, which is used as a ventilation space. The core area in the workshop is on the second floor. There is plenty of fully automated production equipment on the second floor, which operates continuously over 24 h. Therefore, the vibration sources for precision instruments in the workshop mainly include (1) vibration generated by the operation of machinery and equipment in the production workshop; (2) vibration of air conditioning and ventilation equipment in the dust-free room; (3) instantaneous shock and vibration caused by personnel walking; (4) vibration from the nearby construction site; and (5) vibration caused by vehicles running on the road in the factory area and urban roads outside the factory area.

2.2. Measurement Scheme

Since the environmental vibration caused by road traffic is low-frequency vibration, the vibration attenuation rate is slow during the process of propagation. Micro-vibration is one of the key issues to be considered and solved in the production process of the display screen. At the same time, the elevated rail transit line under planning may also have a significant adverse impact on the normal production activities in the No. 3 production workshop after it is put into operation. At present, the primary task is to find out the micro-vibration level of the No. 3 workshop and the significant influential factors of the existing vibration sources. Therefore, three test sections were arranged in different areas of the No. 3 workshop. Meanwhile, to find out the attenuation characteristics of vibration on the propagation path, test section 4 # is set on the transmission route between test section 3 # and the expressway. The four test sections are all within the green dotted circle in Figure 1a, which is illustrated in Figure 1b in detail.

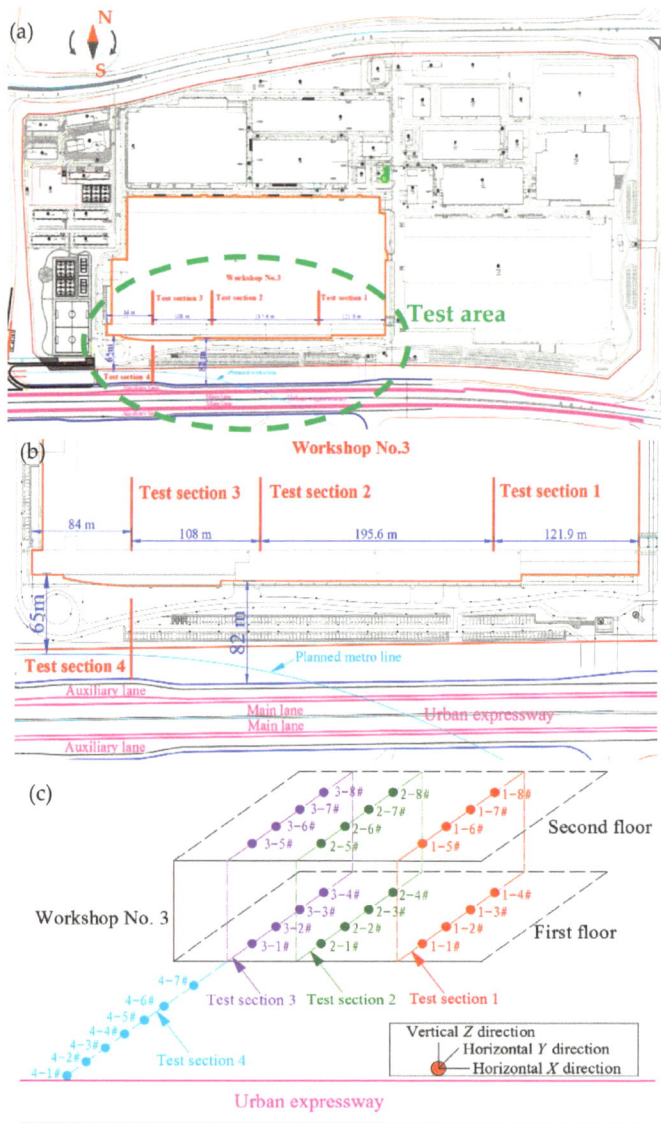

Figure 1. Test scheme for the microelectronics enterprise beside the urban expressway: (**a**) plan of the microelectronics enterprise; (**b**) locations of the four test sections; and (**c**) illustration of measuring points.

The layout of test points for the four test sections is shown in Figure 1c. Test sections 1 #, 2 #, and 3 # were located in the workshop. Each test section contained eight measuring points, of which measuring points 1 # to 4 # were arranged on the first floor, measuring points 5 # to 8 # were installed on the second floor, and the intervals between the measuring points were 20 m. Test section 4 # was located on the transmission path between the urban expressway and the workshop. Seven measuring points of test section 4 # were arranged on

the measuring points 4-6 # and 4-7 # is 15 m. Each measuring point of the four test sections was equipped with three sensors to record the vibration speed in the X, Y, and Z directions, as shown in Figure 1c.

Based on the Technical Specifications for Environmental Vibration Monitoring (HJ 918-2017) [38], the monitoring of the regional environmental vibration was divided into two periods: daytime and nighttime monitoring. Daytime refers to the period between 6:00 a.m. and 10:00 p.m. in one day, and nighttime refers to the period from 10:00 p.m. to 6:00 a.m. on the next day. The time interval between the two tests was one hour. According to the above-mentioned specifications, micro-vibration monitoring in workshop No. 3 was carried out on test sections 1 #, 2 #, and 3 # for 24 h, respectively, and the monitoring was performed each hour with a duration of 10 min. Measurement was conducted for test section 4 # in the period between 7:30 p.m. and 8:30 p.m. when a large number of vehicles were running on the urban expressway. In total, 35 groups of data were collected, and the duration of each data collection was 30 s.

For test sections 1#, 2#, and 3 # in the No. 3 workshop, sensors were installed on the first and second floors of the building. The dust-free room on the first floor has no production equipment, and the sensors were directly installed on the first floor slab, as shown in Figure 2. The core production area was on the second floor, and the production equipment was placed on the waffle board. When installing the sensors on the second floor slab, the waffle board needed to be removed first, and the sensors were installed on the floor slab under the waffle board. The second floor was equipped with ventilation holes, and the sensors were installed on the floor beside the ventilation hole, as shown in Figure 3. In Figure 3, test sections 1 # and 2 # on the second floor were located in the yellow light area; the surrounding production equipment was dense, and test section 3 # was located in the white light area, which had the least equipment of the three test sections.

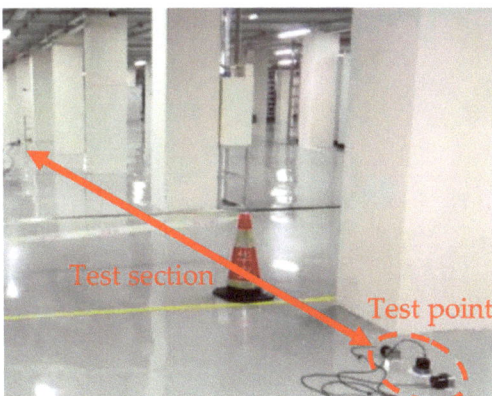

Figure 2. Installation of sensors on the first floor slab.

The measuring points of test section 4 # were placed on the ground between the urban expressway and the workshop. The surface of the green belt is covered with weeds and low trees. The installation procedure for the sensors is shown in Figure 4. Firstly, the surface floating soil needed to be removed, and a square foundation pit was excavated to place a square steel plate with an L-shaped angle steel and four small holes at the four corners. Then, four steel drills were inserted into the soil through the holes to fix the plate. Finally, sensors were mounted on the steel plate in the X, Y, and Z directions.

In the experiment, Type 891-II low-frequency accelerometers were employed, boasting a measurement range of 0 to 4 g and an effective frequency range spanning from 0.5 Hz to 80 Hz. The data acquisition system utilized for the measurements was the INV3060S, a product of the Beijing Oriental Institute of Vibration and Noise Technology. This advanced system supports up to 16 channels and offers a sampling frequency range extending from

6.25 Hz to 102.4k Hz. The sampling frequency was 1024 Hz. Random sampling was conducted beside the road. About three to five heavy trucks passed by per minute in one direction. Regarding trucks, they typically had multiple axles, ranging from double to five axles, with axle weights varying between several tons to ten tons. The traffic condition on the expressway is shown in Figure 5.

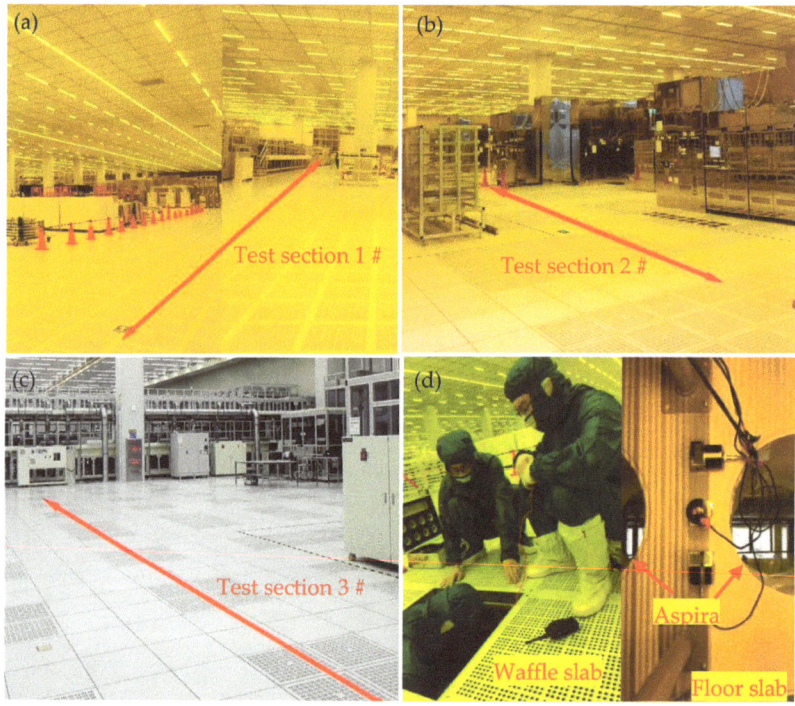

Figure 3. Placement of test points on the second floor: (**a**) test section 1 #; (**b**) test section 2 #; (**c**) test section 3 #; and (**d**) installation of sensors on the second floor slab.

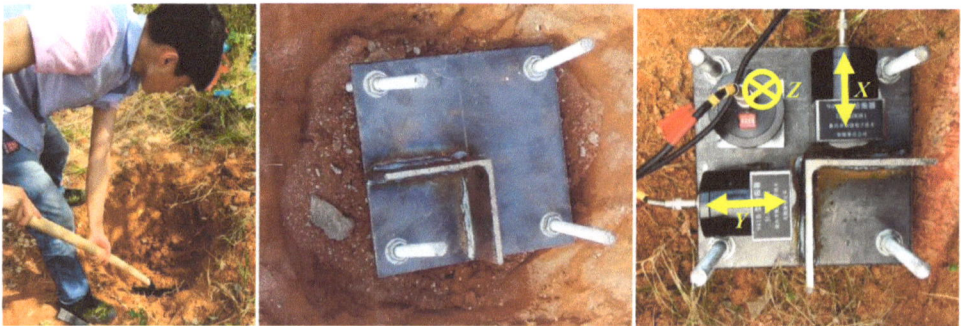

Figure 4. Vehicles on the urban expressway beside the workshop.

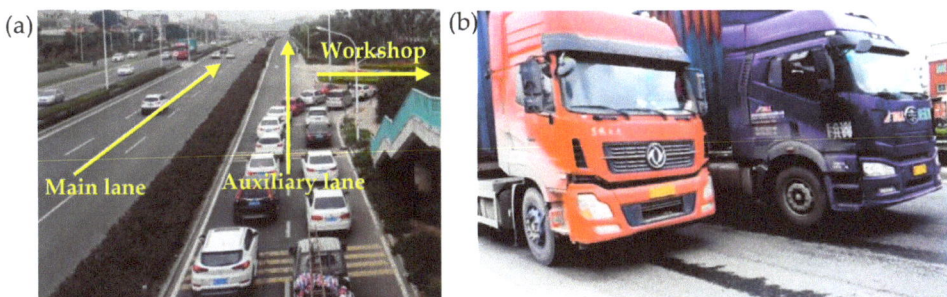

Figure 5. Vehicles on the urban expressway beside the workshop: (**a**) traffic flow on the expressway and (**b**) heavy trucks on the road.

3. Vibration on the Transmission Route

Test section 4 # was set on the site of the transmission route between the road and the workshop. Before the field test, a random sampling study of the expressway's traffic flow condition was undertaken. From 7:30 to 8:30 p.m., three to five heavy trucks went in one direction every minute, while other vehicles, including various types of cars, passed by continuously. To examine the effects of the actual random traffic flow, the duration of each test was 30 s. Generally, during the test period, the excitation of heavy trucks and cars could be recorded. The characteristics of the vibration responses were analyzed in time and frequency domains. The heavy vehicle shown in Figure 5b represents trucks.

3.1. Time History Analysis of Ground Vibration

The background vibration responses on the ground were quite minimal when there were no moving vehicles on the road. With the excitation of road traffic vehicles, the vibration responses on the ground increased to some extent, and while the heavy vehicles passed by, the ground-borne vibration responses were magnified significantly. The vibration responses of the measuring points decreased gradually with the increase in the distance from the road. Figure 6 shows the time histories of velocity responses in the vertical, X, and Y directions at locations 4-1 #, 4-5 #, and 4-7 #.

The comparison of time histories in the three directions in Figure 6 shows that the vertical vibration response was the largest, and the vibration response in the Y direction was the smallest. In Figure 6, both the background vibration and traffic-induced vibration can be observed in response to the three directions. The vibration responses of the ground measurement points in the periods of 7 s to 23 s and 27 s to 30 s were generated by the passage of road vehicles (except for heavy trucks) and were greater than those of background vibration. The vibration response from 23 s to 27 s increased significantly due to the passage of heavy trucks. Additionally, the closer to the road, the more the vibration response increased. From Figure 6a, the peak response caused by heavy vehicles passing by can be observed at locations 4-1 # and 4-5 #, corresponding to 23.77 s and 24.48 s, respectively, from which the time lag phenomenon can be observed.

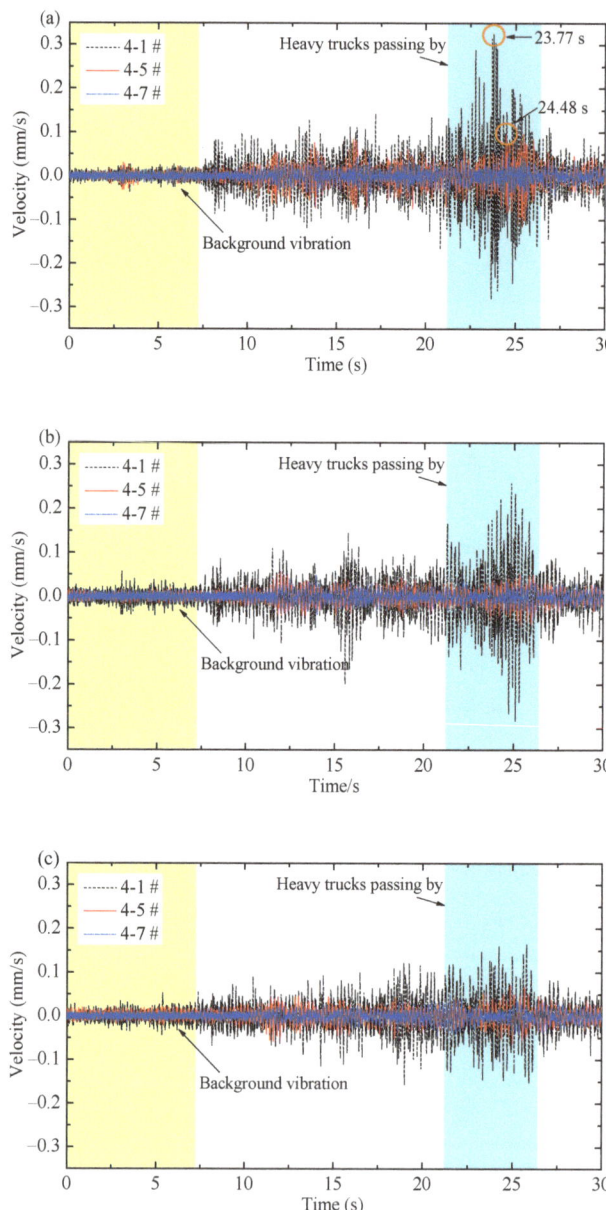

Figure 6. Time histories of vibration velocity at test section 4 # in the (a) vertical direction; (b) X direction; and (c) Y direction.

3.2. Fourier Spectrum Analysis of Ground Vibration

The dominant frequency band of ground-borne vibration in the vertical, horizontal, X, and Y directions caused by road traffic was concentrated within 20 Hz. The peak frequency of the vertical vibration appeared at 12 Hz, and the peaks of the horizontal vibration responses in the X and Y directions appeared around 6 Hz and 12 Hz. During the propagation of

distance from the road. The fast Fourier transform (FFT) spectra of vibration velocity in the vertical, horizontal, X, and Y directions at locations 4-1 #, 4-3 #, 4-5 #, and 4-7 # are given in Figure 7.

It can be seen from Figure 7 that the vertical vibration amplitude was greater than that in the horizontal X and Y directions, and the vibration amplitude in the horizontal Y direction showed the smallest response, which is consistent with the conclusion of the analysis from Figure 6. The vibration responses in the three directions decreased gradually with the increase in the propagation distance.

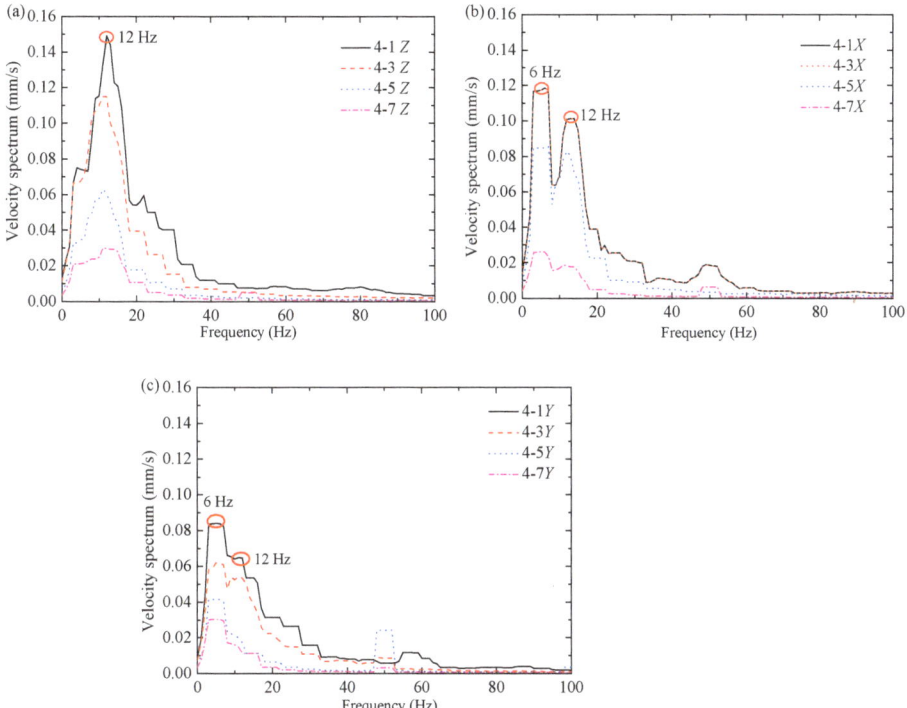

Figure 7. Spectra of vibration velocities at test section 4 # in the (**a**) vertical direction; (**b**) X direction; and (**c**) Y direction.

3.3. 1/3 Octave Spectrum of Ground Vibration

Based on the description in Section 2.2, 35 groups of test data were recorded for test section 4 # during the period of 7:30 p.m. to 8:30 p.m., when there were a large number of vehicles passing through the urban expressway. The duration of each test was 30 s. According to the 35 groups of test data, a 1/3 octave analysis was carried out. Due to the differences in the test data, a 95% confidence interval was employed for 1/3 octave spectra. The 95% confidence limits of vertical vibration responses at measuring points 4-1, 4-3, 4-5, and 4-7 # are shown in Figure 8. It can be seen from Figure 8 that the dominant frequency band of the ground-borne vibration response was within 20 Hz, and the peak frequency of the vibrations at 4-1 # and 4-3 # was located at 10~20 Hz. With the increase in propagation distance, the peak frequency of the vibration responses at 4-5 # and 4-7 # shifted to 10 Hz, and the vibration amplitude decreased, which was especially evident in the dominant frequency band.

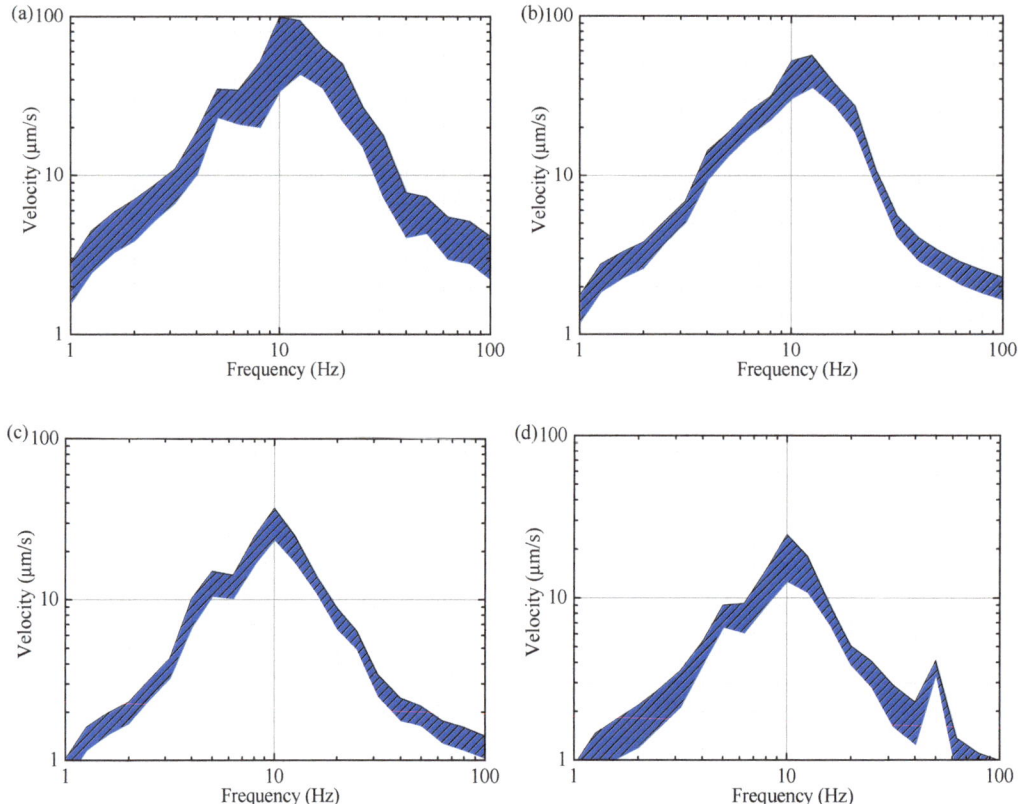

Figure 8. The 1/3 octave spectra of vertical velocity at the following test points: (**a**) 4-1 #; (**b**) 4-3 #; (**c**) 4-5 #; and (**d**) 4-7.

3.4. Vibration Level of Ground Vibration

To further investigate the ground-borne vibration induced by road traffic, the vibration was evaluated by use of the speed vibration level L_V in decibels, expressed as

$$L_V = 20\lg\left(\frac{V_{rms}}{V_{ref}}\right), \qquad (1)$$

where L_V is the velocity level in decibels, V_{rms} is the effective value (root mean square value) of vibration velocity in m/s, and V_{ref} is reference velocity, of which the value is 2.54×10^{-8} m/s, according to the regulations of the Department of Railway Transportation in the U.S. [39].

For 35 groups of test data, the vertical velocity vibration level L_V was calculated using Equation (1). Due to the differences that exist in the test data, a 95% confidence interval was employed, as shown in Figure 9. In Figure 9, the total vibration levels at all test points on the ground are exhibited. The vertical velocity vibration level decreased gradually with the increase in distance from the road, indicating that the vibration energy gradually decreased during the vibration propagation, which follows the vibration propagation rule in Ref. [40].

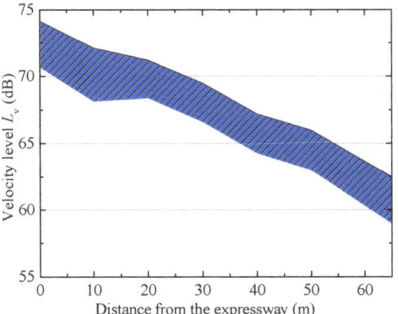

Figure 9. Vertical velocity levels of test points from 4-1 # to 4-7 # at test section 4.

4. Vibration in the Workshop

The production workshop is located near the urban expressway. The road traffic-induced micro-vibration may affect the normal operation of the instruments and equipment in the workshop. Therefore, it was necessary to conduct a detailed study on the influence of road traffic on the workshop.

4.1. Time History Analysis of Workshop Vibration

The road traffic-induced vibration in the workshop was attenuated during the vibration propagation from the first floor to the second floor. The vibration of the second floor was significantly affected by the operation of the production equipment. The time histories of the vertical vibration velocity at locations 2-4 # and 2-8 # are shown in Figure 10. Test point 2-4 # is on the first floor, and 2-8 # is on the second floor.

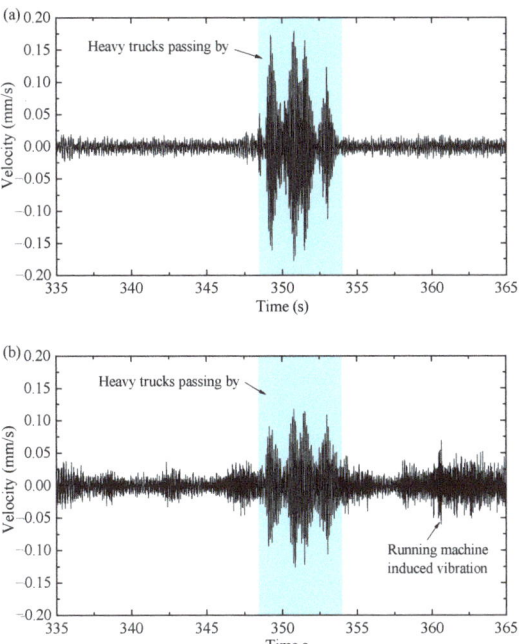

Figure 10. Time histories of vertical vibration at test points 2-4 # and 2-8 #: (**a**) first floor and (**b**) second floor.

It can be observed from Figure 10 that the response increased significantly during the period of 348 s to 354 s when heavy trucks passed by. However, the response amplitude of the second floor was smaller than that of the first floor in this period, which indicates that the traffic-induced vibration was attenuated with the vibration propagation from the first floor to the second floor. Compared to the responses in the other periods (335 s to 348 s and 354 s to 365 s), it can be observed that the vibration amplitude in Figure 10b is greater than that in Figure 10a. Since the production equipment on the second floor operated continuously for 24 h, the most likely reason for the increase in vibration on the second floor slab in the periods of 335 s to 348 s and 354 s to 365 s is the contribution of vibration due to the operation of production machinery and equipment.

4.2. Fourier Spectrum Analysis of Workshop Vibration

The peak frequency of workshop vibrations in the vertical direction caused by the road traffic was concentrated around 10 Hz. The vibration in this frequency range was attenuated with the vibration propagation from the first floor to the second floor in the workshop. The vibration of the second floor slab was affected by the excitation induced by the production equipment, which resulted in localized peaks in the frequency band above 20 Hz. FFT spectra of vibration velocity in the vertical direction at all locations of test section 2 # are shown in Figure 11.

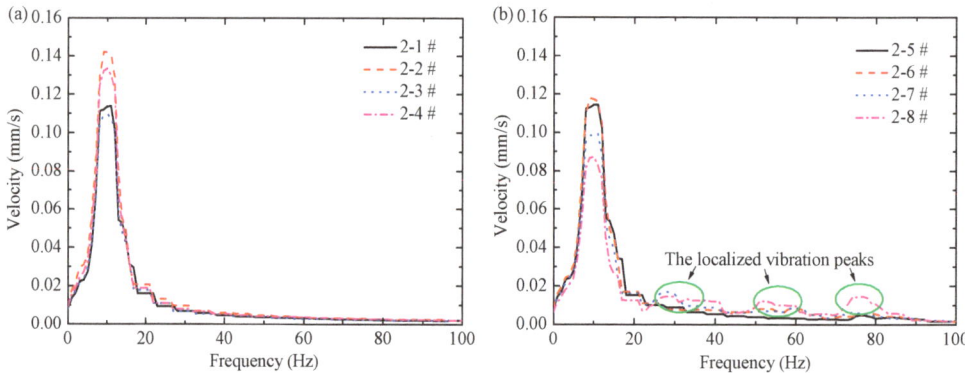

Figure 11. Spectra of vertical vibration at test section 2 # on the (**a**) first floor and (**b**) second floor.

For the vertical vibration response of the workshop in Figure 11, the FFT spectrum waveforms of the vibration response at the test locations on the first and second floors were similar, and the vibration peaks appeared around 10 Hz. The vibration amplitude of the second floor slab was smaller than that of the first floor in frequencies around 10 Hz, suggesting that the dominant frequency range of road traffic-induced vibration was around 10 Hz, which agrees with Refs. [24–26]. The traffic-induced vibration decreased when it was transmitted from the first floor to the second floor. In addition, the localized vibration peaks of the second floor can be observed at 30 Hz, 50~60 Hz, and 70~80 Hz in Figure 11b. The amplitudes of the localized vibration peaks increased compared with Figure 11a, and the most likely reason for this phenomenon is the vibration effects generated by the operation of the equipment on the second floor.

4.3. 1/3 Octave Spectrum of Workshop Vibration

During the normal operation of sensitive equipment, the vibration at the placement position of the equipment was quite severe. There is a special vibration criterion for the evaluation of micro-vibration which takes a 1/3 octave spectrum of the effective value of velocity as the evaluation index [41]. Seven vibration levels of vibration criterion (VC) limit value was reduced step by step (VC-G had the most rigorous standard and the smallest vibration limit) [41]. Table 1 shows the vibration limit from VC-A to VC-G in VC standard [41]. A detailed description can be found in the work of Amick [42]. According to the precision requirements of the production equipment in this study, the vibration in the workshop needed to be kept within the VC-C limit.

On the second floor slab, the area of test section 2 # was densely populated with production machinery, the vibration response of which was analyzed in detail to explore the vibration characteristics in the workshop. The 1/3 octave spectra of the vibration velocity at each measurement point in the vertical, horizontal X, and Y directions on the first and second floors are illustrated in Figures 12–14, respectively. The test results have obvious discreteness. The reasons for this are two. On one hand, the vibrations of different measurement points caused by the operation of equipment were different. On the other hand, random traffic flow on the expressway contributed to significant differences in vibration responses.

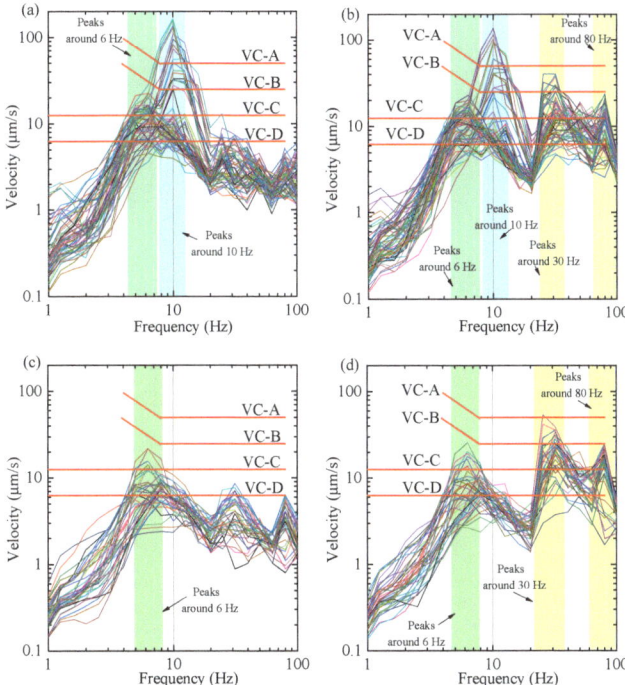

Figure 12. The 1/3 octave spectra of vibration velocities in the vertical direction (**a**) on the first floor in the daytime; (**b**) on the second floor in the daytime; (**c**) on the first floor at night; and (**d**) on the second floor at night.

Table 1. Vibration limits from VC-A to VC-G.

Vibration Level	Frequency Range (Hz)	Vibration Limit
VC-A	4~8	The acceleration does not exceed 260 µg.
	8-80	The velocity does not exceed 50 µm/s.
VC-B	4~8	The acceleration does not exceed 130 µg.
	8~80	The velocity does not exceed 25 µm/s.

Table 1. Cont.

Vibration Level	Frequency Range (Hz)	Vibration Limit
VC-C	1~80	The velocity does not exceed 12.5 μm/s.
VC-D	1~80	The velocity does not exceed 6.25 μm/s.
VC-E	1~80	The velocity does not exceed 3.12 μm/s.
VC-F	1~80	The velocity does not exceed 1.56 μm/s.
VC-G	1~80	The velocity does not exceed 0.78 μm/s.

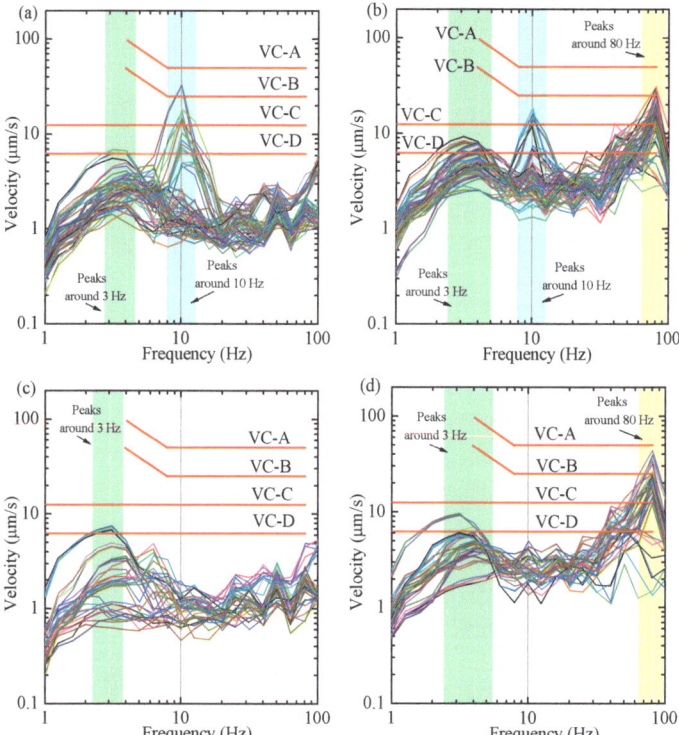

Figure 13. The 1/3 octave spectra of vibration velocity in the X direction (**a**) on the first floor in the daytime; (**b**) on the second floor in the daytime; (**c**) on the first floor at night; and (**d**) on the second floor at night.

Figure 12a,b show the 1/3 octave spectra of vertical vibration at the measuring points on the first and second floors in the daytime (6:00 a.m. to 10:00 p.m), and the nighttime spectra (10:00 p.m. to 6:00 a.m. on the next day) are presented in Figure 12c,d. In the daytime, the vertical vibration response of the measuring points on the first and second floors far exceeded the VC-C limit in a wide frequency range. At night, some test conditions of the vertical vibration response on the first floor exceeded the VC-C limit of around 6 Hz, while the vertical vibration response on the second floor in the peak frequency band of around 30 Hz and 80 Hz also exceeded the VC-C limit.

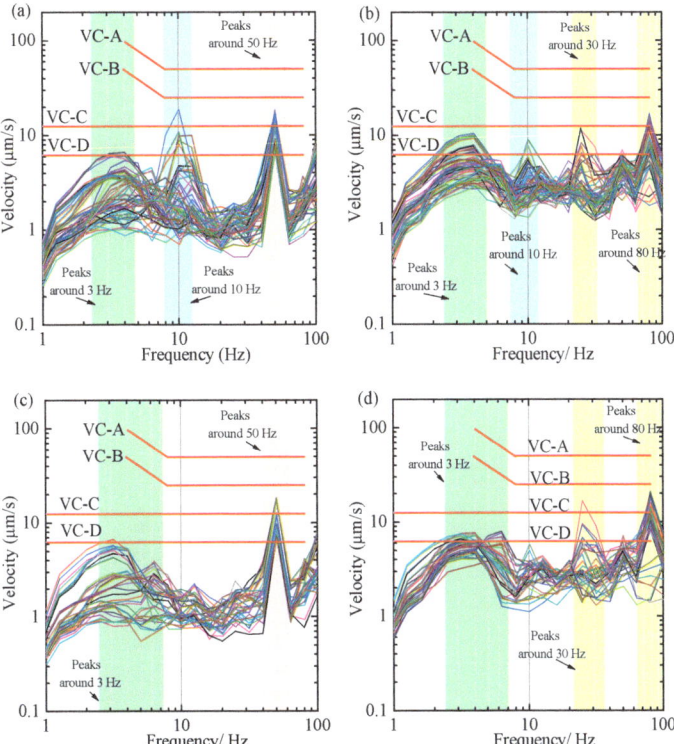

Figure 14. The 1/3 octave spectra of vibration velocity in the Y direction (**a**) on the first floor in the daytime; (**b**) on the second floor in the daytime; (**c**) on the first floor at night; and (**d**) on the second floor at night.

Considering the effects of the road vehicles, part of the test results shows response peaks around 6 Hz in Figure 12a–d, and a larger response peak appears around 10 Hz in Figure 12a,b. The reason is primarily due to the large traffic flow and rapid passage of heavy vehicles during the daytime, resulting in a peak response near 10 Hz. While there was relatively less traffic flow and few heavy vehicles at night, no response peaks around 10 Hz were observed. In the frequency ranges around 30 Hz and 80 Hz, obvious vibration response peaks can be observed in Figure 12b,d, and response amplitudes are significantly higher than those in Figure 12a,c. This is likely due to the 24 h uninterrupted operation of the production facilities on the second floor. So, for the daytime and nighttime, no significant differences in vibration responses around 30 Hz and 80 Hz can be observed in Figure 12b,d. It can be seen from Figure 12d that the vibration generated by the operation of the production facilities at night even exceeded the contribution of road vehicles to the vibration on the second floor slab.

The 1/3 octave spectra of vibration in the X direction at the measuring points on the first and second floors during the daytime are illustrated in Figure 13a,b, and the nighttime spectra are presented in Figure 13c,d. The X-direction vibration response of test section 2 # in the production workshop was smaller than that in the vertical direction. In the daytime, only a few test data on the first and second floors exceeded the VC-C limit in the peak frequency band. At night, the vibration responses on the first floor were very small, and all of them failed to reach the VC-C limit. Additionally, X-direction vibration responses on the second floor were only greater than the VC-C limit in the peak frequency band around 80 Hz.

The 1/3 octave spectra of vibration in the Y direction at the measuring points on the first and second floors during the daytime are indicated in Figure 14a,b, and the nighttime spectra are presented in Figure 14c,d. The Y-direction vibration response of test section 2 # was similar to the vibration in the X direction and much smaller than that in the vertical direction. In the daytime, very few test data on the first floor exceeded the VC-C limit in the peak frequency ranges around 10 and 50 Hz, and vibrations on the second floor only exceeded the VC-C limit at the peak frequency of 80 Hz. At night, only individual data on the first floor exceeded the VC-C limit at the peak frequency of 50 Hz, while partial vibration responses on the second floor exceeded the VC-C limit at 30 and 80 Hz.

4.4. Vibration Level of Workshop Vibration

To study the vibration level of the first- and second floor slabs in the workshop at different moments, Equation (1) is used to obtain the velocity vibration level of the responses in the workshop. The mean and standard deviation of the vibration levels at the four measurement points on the same floor are then found. Taking test section 1 # as an example, the mean and standard deviation of the vibration response on the first floor are calculated according to the responses of the measuring points 1-1 # to 1-4 #, and the mean value and standard deviation of the responses on the second floor are obtained according to the responses of measuring points 1-5 # to 1-8 #. The mean and standard deviation of vibration responses on the first- and second floor slabs at the three test sections in the workshop are shown in Figure 15.

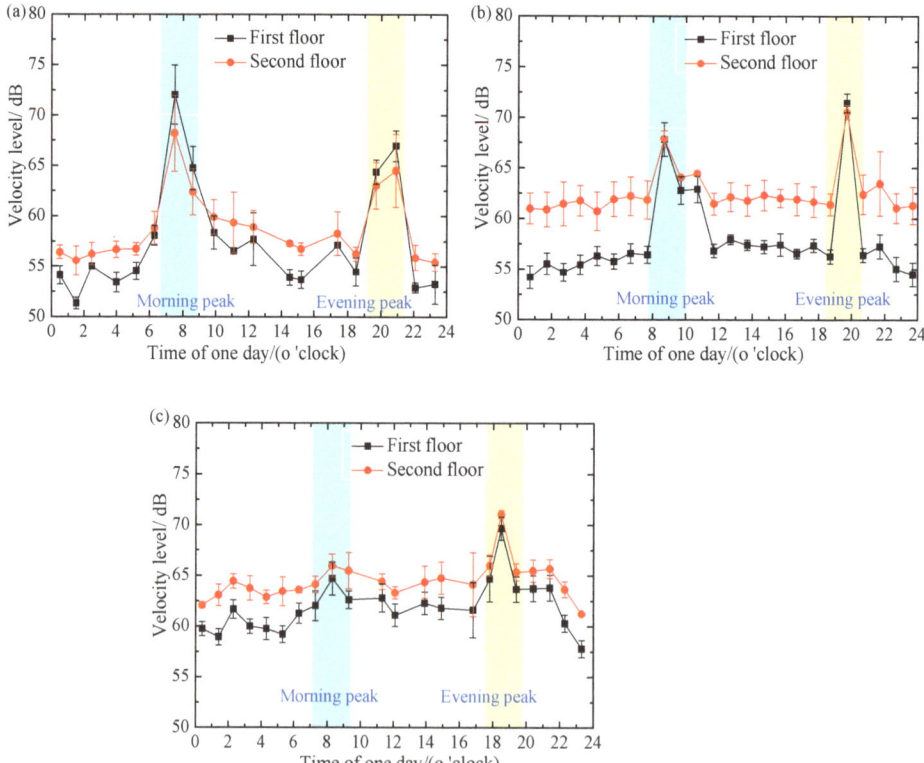

Figure 15. Vertical vibration levels of the workshop during one day at (**a**) test section 1 #; (**b**) test section 2 #; and (**c**) test section 3 #.

As shown in Figure 15, the square and circular dots represent the means of the vibration responses, and the vertical short line represents the standard deviation of the vibration responses. It can be seen from Figure 15 that the mean value and standard deviation of the vibration levels in the workshop have obvious variations at different moments, mainly due to the random road traffic flow on the expressway. However, it can be observed that the mean values of the vibration level magnify with the increased traffic flow in the morning and evening peak periods. During the three days when the three test sections were measured, respectively, the morning peak periods were uniform, which appeared in the period of 8:00 a.m. to 9:00 a.m., while the evening peaks primarily appeared in the period of 7:00 p.m. to 9:00 p.m.

Comparing the vibration characteristics of the first and second floor slabs at the three test sections, it can be found that except for the morning and evening peak periods, the velocity vibration level of the second floor slab is greater than that of the first floor slab, which is due to the vibration excitation caused by the operation of production facilities on the second floor. However, during the morning and evening peak periods, due to the increase in the road traffic flow, the contribution of vibration induced by road traffic increases, and the differences in vibration levels between the first and second floors decrease. It can even be observed in Figure 15a that the vibration level on the first floor slab exceeds that on the second floor slab due to the dominant increase in road traffic flow during the morning and evening peak periods. Therefore, it is necessary to take measures to control the vibration response caused by running vehicles, such as limiting the load capacity of heavy vehicles, controlling the vehicle speed, improving the pavement grade, and reducing the pavement's roughness.

5. Conclusions

The rapid construction of urban roads improves people's travel, yet it may cause environmental vibration issues. In the current study, we conducted field measurements to investigate the effect of road traffic-induced vibration on the micro-vibration of a workshop equipped with precision instruments. The test equipment was divided into two groups. One group was set on the ground to measure the vibration response on the propagation path. The other group was arranged in the workshop to collect the velocity signals on the first and second floors. The test results were analyzed in the time and frequency domains. Relevant data and findings can not only guide the design of vibration reduction schemes for road traffic environmental vibration but also validate potential numerical model predictions of traffic-induced micro-vibration. Our main conclusions are listed as follows:

The traffic-induced ground-borne vibration on the transmission route attenuates with the distance from the road and shows peak responses around 10 Hz. The passage of heavy vehicles causes remarkable vibration response both on the ground and in the workshop, among which the vertical vibration is the most significant.

The workshop is subject to vibration excitation from the operation of production equipment (above 20 Hz) in addition to vibration stimulation from vehicles (below 20 Hz). In the daytime, the vertical vibration responses on the first and second floors in the workshop far exceed the VC-C limit, but only a few test conditions of the vibration responses in the horizontal X and Y directions exceed the VC-C limit in the peak frequency range. At night, only individual test conditions of the vibration responses on the first and second floors exceed the VC-C limit in the peak frequencies.

During the periods of morning and evening peaks with heavy road traffic flow, the vertical vibration responses on the first and second floors in the workshop increased significantly. However, the second floor vibrates more than the first floor above 20 Hz, which can be attributed to the stimulation of instruments installed there.

Due to the heavy traffic flow in the daytime, the vertical vibration response of the floor slab in the workshop can easily exceed the VC-C limit. In addition, the vibration generated by the operation of the production equipment in the workshop cannot be ignored, which, in turn, has an obvious impact on the normal running of the production equipment.

Therefore, it is necessary to explore vibration reduction measures for the production equipment. Moreover, due to the planning of a metro line, in the future, there will be an elevated metro line located 65 m away from the workshop, which, after it is completed and put into operation, will have adverse impacts on the workshop. It is necessary to further study the impacts of vibration induced by the subway line. Apart from this effect, we did not delve into specific details such as the axle weight, vehicle speed, or model information of the passing vehicles. Future research should aim to quantify these factors and conduct a more nuanced analysis of the building's vibration response to specific sources of vehicle-induced vibration.

Author Contributions: Conceptualization, Z.Z. and X.L.; methodology, X.L.; investigation, Z.Z. and X.Z.; writing—original draft preparation, Z.Z. and G.X.; writing—review and editing, Z.Z., X.L. and X.Z.; visualization, A.W. All authors have read and agreed to the published version of the manuscript.

Funding: This research was funded by the National Natural Science Foundation of China (Grant Nos. 52008123 and 52368059) and the Startup Project for High-Level Talents of Guizhou Institute of Technology (No. XJGC20190651).

Data Availability Statement: The original contributions presented in the study are included in the article, further inquiries can be directed to the corresponding author.

Acknowledgments: Thanks are extended to the anonymous reviewers whose suggestions improved this manuscript.

Conflicts of Interest: The authors declare no conflicts of interest.

References

1. Khajehdezfuly, A.; Shiraz, A.A.; Sadeghi, J. Assessment of vibrations caused by simultaneous passage of road and railway vehicles. *Appl. Acoust.* **2023**, *211*, 109510. [CrossRef]
2. Li, Z.; Cao, Y.; Ma, M.; Xiang, Q. Prediction of ground-borne vibration from random traffic flow and road roughness: Theoretical model and experimental validation. *Eng. Struct.* **2023**, *285*, 116060. [CrossRef]
3. Czech, K.R. The impact of the type and technical condition of road surface on the level of traffic-generated vibrations propagated to the environment. *Procedia Eng.* **2016**, *143*, 1358–1367. [CrossRef]
4. Hao, H.; Ang, T.C.; Shen, J. Building vibration to traffic-induced ground motion. *Build. Environ.* **2001**, *36*, 321–336. [CrossRef]
5. Guo, M.; Ni, M.Y.; Shyu, R.; Ji, J.S.; Huang, J. Automated simulation for household road traffic noise exposure: Application and field evaluation in a high-density city. *Comput. Environ. Urban Syst.* **2023**, *104*, 102000. [CrossRef]
6. Zhang, Z.; Li, X.; Zhang, X.; Fan, J.; Xu, G. Semi-analytical simulation for ground-borne vibration caused by rail traffic on viaducts: Vibration-isolating effects of multi-layered elastic supports. *J. Sound Vib.* **2022**, *516*, 116540. [CrossRef]
7. He, C.; Zhou, S.; Guo, P. An efficient three-dimensional method for the prediction of building vibrations from underground railway networks. *Soil Dyn. Earthq. Eng.* **2020**, *139*, 106269. [CrossRef]
8. Ertugrul, O.L.; Ozkan, M.Y.; Ulgen, D. Attenuation of Traffic Induced Ground Borne Vibrations Due to Heavy Vehicles. In Proceedings of the Fifth International Conference on Recent Advances in Geotechnical Earthquake Engineering and Soil Dynamics and Symposium in Honor of Professor I. M. Idriss, San Diego, CA, USA, 24–29 May 2010.
9. Hajek, J.J.; Blaney, C.T.; Hein, D.K. Mitigation of Highway Traffic-Induced Vibration. In Proceedings of the 2006 Annual Conference of the Transportation Association of Canada, Charlottetown, PEI, Canada, 17–20 September 2006.
10. Hu, Z.; Tian, L.; Zou, C.; Wu, J. Train-induced vibration attenuation measurements and prediction from ground soil to building column. *Environ. Sci. Pollut. Res.* **2023**, *30*, 39076–39092. [CrossRef] [PubMed]
11. Qiu, Y.; Zou, C.; Hu, J.; Chen, J. Prediction and mitigation of building vibrations caused by train operations on concrete floors. *Appl. Acoust.* **2024**, *219*, 109941. [CrossRef]
12. Li, X.; Chen, Y.; Zou, C.; Wang, H.; Zheng, B.; Chen, J. Building structure-borne noise measurements and estimation due to train operations in tunnel. *Sci. Total Environ.* **2024**, *926*, 172080. [CrossRef]
13. Watts, G.R. *Traffic-Induced Ground-Borne Vibrations in Dwellings*; Transport and Road Research Laboratory: Crowthorne, UK, 1987.
14. Watts, G.R. The generation and propagation of vibration in various soils produced by the dynamic loading of road pavements. *J. Sound Vib.* **1992**, *156*, 191–206. [CrossRef]
15. Hunt, H. Modelling of road vehicles for calculation of traffic-induced ground vibration as a random process. *J. Sound Vib.* **1991**, *144*, 41–51. [CrossRef]
16. Hunt, H. Stochastic modelling of traffic-induced ground vibration. *J. Sound Vib.* **1991**, *144*, 53–70. [CrossRef]
17. Lombaert, G.; DeGrande, G. The experimental validation of a numerical model for the prediction of the vibrations in the free field
18. Hao, H.; Ang, T.C. Analytical

19. Agostinacchio, M.; Ciampa, D.; Olita, S. The vibrations induced by surface irregularities in road pavements—A Matlab® approach. *Eur. Transp. Res. Rev.* **2014**, *6*, 267–275. [CrossRef]
20. Lak, M.A.; Degrande, G.; Lombaert, G. The effect of road unevenness on the dynamic vehicle response and ground-borne vibrations due to road traffic. *Soil Dyn. Earthq. Eng.* **2011**, *31*, 1357–1377. [CrossRef]
21. Watts, G.R.; Krylov, V.V. Ground-borne vibration generated by vehicles crossing road humps and speed control cushions. *Appl. Acoust.* **2000**, *59*, 221–236. [CrossRef]
22. Ducarne, L.; Ainalis, D.; Kouroussis, G. Assessing the ground vibrations produced by a heavy vehicle traversing a traffic obstacle. *Sci. Total Environ.* **2018**, *612*, 1568–1576. [CrossRef]
23. Mhanna, M.; Sadek, M.; Shahrour, I. Numerical modeling of traffic-induced ground vibration. *Comput. Geotech.* **2012**, *39*, 116–123. [CrossRef]
24. Taniguchi, E.; Sawada, K. Attenuation with distance of traffic-induced vibrations. *J. Jpn. Soc. Soil Mech. Found. Eng.* **1979**, *19*, 15–28. [CrossRef] [PubMed]
25. Al-Hunaidi, M.O.; Rainer, J.H. Remedial measures for traffic-induced vibrations at a residential site. Part 1: Field tests. *Can. Acoust.* **1991**, *19*, 3–13.
26. Al-Hunaidi, M.O.; Rainer, J.H.; Tremblay, M. Control of traffic-induced vibration in buildings using vehicle suspension systems. *Soil Dyn. Earthq. Eng.* **1996**, *15*, 245–254. [CrossRef]
27. Qu, S.; Yang, J.; Zhu, S.; Zhai, W.; Kouroussis, G.; Zhang, Q. Experimental study on ground vibration induced by double-line subway trains and road traffic. *Transp. Geotech.* **2021**, *29*, 100564. [CrossRef]
28. Crispino, M.; D'Apuzzo, M. Measurement and prediction of traffic-induced vibrations in a heritage building. *J. Sound Vib.* **2001**, *246*, 319–335. [CrossRef]
29. Astrauskas, T.; Janueviius, T.; Grubliauskas, R. Vehicle speed influence on ground-borne vibrations caused by road transport when passing vertical traffic calming measures. *Promet-Traffic Transp.* **2020**, *32*, 247–253. [CrossRef]
30. Ding, D.Y.; Liu, W.N.; Gupta, S.; Lombaert, G.; Degrande, G. Prediction of vibrations from underground trains on Beijing metro line 15. *J. Cent. South Univ. Technol.* **2010**, *17*, 1109–1118. [CrossRef]
31. Ding, D.Y.; Gupta, S.; Liu, W.N.; Lombaert, G.; Degrande, G. Prediction of vibrations induced by trains on line 8 of Beijing metro. *J. Zhejiang Univ. Sci. A* **2010**, *11*, 280–293. [CrossRef]
32. Qu, S.; Yang, J.; Zhu, S.; Zhai, W.; Kouroussis, G. A hybrid methodology for predicting train-induced vibration on sensitive equipment in far-field buildings. *Transp. Geotech.* **2021**, *31*, 100682. [CrossRef]
33. He, C.; Zhou, S.; Di, H.; Guo, P.; Xiao, J. Analytical method for calculation of ground vibration from a tunnel embedded in a multi-layered half-space. *Comput. Geotech.* **2018**, *99*, 149–164. [CrossRef]
34. Wang, Y.; Zhang, H.; Ji, Y.; Xiong, M.; Shen, Z. Micro-vibration prediction and evaluation analysis of the structural foundation of a precision instrument plant across the reservoir. In Proceedings of the 8th International Conference on Hydraulic and Civil Engineering: Deep Space Intelligent Development and Utilization Forum (ICHCE), Xi'an, China, 25–27 November 2022; pp. 1287–1291.
35. Chen, X.; Jiang, J.; Hu, Y.; Sheng, T.; Tang, K. Experimental study on the influence of precision instruments caused by heavy vehicles vibration. *J. Phys. Conf. Ser.* **2021**, *2044*, 12113. [CrossRef]
36. Zhang, Y.; Wang, X.; Lu, C.; Liu, K.; Song, B. Influence of road traffic vibration on micro-dynamic response of precision instrument vibration isolation platforms. *Eng. Proc.* **2023**, *36*, 48. [CrossRef]
37. Ulgen, D.; Ertugrul, O.L.; Ozkan, M.Y. Measurement of ground borne vibrations for foundation design and vibration isolation of a high-precision instrument. *Measurement* **2016**, *93*, 385–396. [CrossRef]
38. HJ918-2017; Technical Specifications for Environmental Vibration Monitoring. Ministry of Environmental Protection of the People's Republic of China: Beijing, China, 2017.
39. Hanson, C.E.; Towers, D.A.; Meister, L.D. *Transit Noise and Vibration Impact Assessment (FTA-VA-90-1003-06)*; Department of Transportation Federal Transit Administration (FTA): Washington, DC, USA, 2006.
40. Li, H.; He, C.; Gong, Q.; Zhou, S.; Li, X.; Zou, C. TLM-CFSPML for 3D dynamic responses of a layered transversely isotropic half-space. *Comput. Geotech.* **2024**, *168*, 106131. [CrossRef]
41. Gordon, C. Generic criteria for vibration-sensitive equipment. In Proceedings of the SPIE—The International Society for Optical Engineering, San Jose, CA, USA, 4–6 November 1991; Volume 1619, pp. 71–85.
42. Amick, H.; Gendreau, M.; Busch, T.; Gordon, C. Evolving criteria for research facilities: I-Vibration. In Proceedings of the SPIE Conference 5933: Buildings for Nanoscale Research and Beyond, San Diego, CA, USA, 31 July–4 August 2005.

Disclaimer/Publisher's Note: The statements, opinions and data contained in all publications are solely those of the individual author(s) and contributor(s) and not of MDPI and/or the editor(s). MDPI and/or the editor(s) disclaim responsibility for any injury to people or property resulting from any ideas, methods, instructions or products referred to in the content.

Article

Research on the Dynamic Response of the Multi-Line Elevated Station with "Integral Station-Bridge System"

Xiangrong Guo * and Shipeng Wang

School of Civil Engineering, Central South University, Changsha 410075, China; wsp1095299975@163.com
* Correspondence: csuguoxr@csu.edu.cn

Abstract: Elevated stations serve as critical hubs in urban rail transit engineering. The structure of multi-line "building-bridge integrated" elevated stations is unique, with intricate force transfer paths and challenges to clarify dynamic coupling from train vibrations, necessitating the study of such stations' train-induced dynamic responses. This paper presents a case study of a typical "building-bridge integrated" elevated station, utilizing the self-developed finite element software GSAP-V2024 to establish a simulation model of a coupled train–track–station system. It analyzed the station's dynamic response under various single-track operating conditions and the pattern of the vibration response as the speed changes. Additionally, the study examined lateral vibration response changes in the station under double, quadruple, and sextuple train operations at the same speed. Findings reveal that the station's vertical responses generally increase with speed, significantly outpacing lateral responses. Under single-track operations, dynamic responses vary across different types of track-bearing floors and frame structures with different spans. With an increase in the number of operating train lines, the station's vertical response grows, with lateral responses being neutralized in the mid-span of the triple-span frame structure and amplified at the edges. These results provide a reference for the structural design of multi-line "building-bridge integrated" elevated stations.

Keywords: building–bridge integration; train-induced vibration; multi-line railway; dynamic response; elevated station

Citation: Guo, X.; Wang, S. Research on the Dynamic Response of the Multi-Line Elevated Station with "Integral Station-Bridge System". *Buildings* **2024**, *14*, 758. https://doi.org/10.3390/buildings14030758

Academic Editor: Fabrizio Gara

Received: 14 February 2024
Revised: 3 March 2024
Accepted: 8 March 2024
Published: 11 March 2024

Copyright: © 2024 by the authors. Licensee MDPI, Basel, Switzerland. This article is an open access article distributed under the terms and conditions of the Creative Commons Attribution (CC BY) license (https://

1. Introduction

As urban areas expand, challenges like rising population density, growing transportation demand, and worsening environmental pollution escalate. The urban rail transit, a pivotal solution, facilitates connectivity across city clusters. Elevated stations, vital in these projects, significantly contribute to passenger distribution and transport connectivity, and accelerate development along railway lines [1], garnering more focus from the construction sector.

Elevated stations are classified into three principal types according to the connection of track beams to station buildings: "building-bridge integrated", "building-bridge separated", and "building-bridge combined" structures. In the building–bridge integrated design, tracks are laid directly on the track-bearing floor, utilizing a unified frame structure. This design ensures a simple grid of frame columns, reduced overall height, and enhanced structural rigidity. The building–bridge separated design features independent bridge structures from station buildings, offering a straightforward structural system with clear force transfer paths. The building–bridge integrated elevated station combines building and bridge structures, leading to notable unevenness in structural stiffness and mass distribution, both horizontally and vertically. This integration presents challenges like intricate load transfer paths, clarifying dynamic coupling from train vibrations, and a lack of unified design standards during the design phase. Many scholars have explored these issues, with extensive on-site experiments conducted on elevated stations. Cai et al. [2] investigated the vibration characteristics of elevated stations, examining vibrations in waiting halls and responses due to trains arriving, leaving, and passing through. Yu et al. [3] examined the effects of high-speed train loads on waiting room and business floors' vibrations, assessing vibration serviceability with data from Zhengzhou East Railway Station. Liu, Yang, et al. [4,5] conducted tests and analyses on environmental vibrations in various areas of elevated stations, including platforms and waiting halls, to explore the propagation of train-induced vibrations. Ba et al. [6] performed field tests to evaluate background and structural vibrations from varying speeds, measuring effective vibration acceleration at different points. Furthermore, high-speed trains impact nearby buildings with vibrations and noise [7–9]. Xia et al. [10,11] explored the mechanisms behind train-induced vibrations and noise at elevated stations, examining the effects of train type, speed, and proximity on nearby buildings. Sanayei M, Hesami S et al. [12,13] corroborated finite element models with field data, studying the impact of train speed, soil properties, and structural traits on building vibrations caused by trains. Li and his team performed empirical research on vibrations and noise at large high-speed railway stations due to passing trains, exploring structural vibration patterns and noise radiation during train operations [14].

Other researchers have proposed various numerical models for simulation calculations. Deng et al. [15] suggested taking the load time history from the train–bridge sub-model dynamic calculations as external excitation acting on the bridge–station sub-model. They conducted time history analysis to compute the dynamic response caused by trains passing through the elevated station and evaluated the station's vibration serviceability. Xu, Xie, Yang, Cui, et al. established a train–station vibration analysis system to explore station dynamics and train-induced vibration patterns, focusing on station structure comfort [16–19]. Salih Alan and colleagues researched vibration control for train-induced vibrations at an Ankara station. They simulated the station–foundation interaction with springs and dampers in a finite element model, aligning with Turkish environmental noise laws [20]. Zhang et al. [21] used rigid body dynamics for a train subsystem model and the mode superposition method for a structural model. They explored the analysis method for the train–station structure coupling system under braking force. Zhang et al. devised a numerical model for analyzing vibrations in large-scale integrated building–bridge structures (IBBS), focusing on high-speed railway stations and evaluating vibration mitigation effectiveness [22]. Yang et al. introduced a two-step time-frequency prediction method for superstructures to predict and analyze train-induced vibrations in buildings above subway tunnels [23].

Numerous researchers have explored the vibration characteristics of station structures during train operations, including aspects of safety and comfort. Additionally, they have documented the propagation and attenuation of train-induced vibrations [24–29]. Yet, the dynamic response of multi-line "building-bridge integrated" elevated stations remains less studied. To tackle these challenges, this study examines a typical multi-line "building-bridge integrated" elevated station. It treats the train, track, and station structures as an integrated unit, formulating spatially coupled vibration equations for this system. These equations are solved through time-domain analysis, leading to the development of a finite element model for the multi-line elevated station. The research delves into the dynamic characteristics of such stations and the propagation of train-induced vibrations, providing design insights for multi-line "building-bridge integrated" elevated station structures.

2. The Multi-Line Elevated Railway Station

Figure 1 displays the elevated station's structure, designed for 100 km/h with 15 tracks. The "U" in Figure 1 represents the "upward direction" and "D" represents the "downward direction". Its structural system combines steel-reinforced concrete columns, bidirectional prestressed concrete beams, and cast-in-place concrete slabs into a spatial frame. The station comprises three levels: a 9.09-m-high platform floor, a 6.90-m-high track-bearing floor, and a 0.70-m-high hall floor. It is uniquely segmented into three parts: perpendicular to the tracks, with a one-span, three-span, and two-span frame structure, respectively. Parallel to the tracks, each section features a three-span structure, and each span is 21 m long. In Figure 1, '#' represents the track. Tracks 3#, 4#, 11#, and 12# are supported by bridge structures, isolated from the main frame by seismic joints, making them independent. Thus, this study focuses on the vibrations from trains on the elevated station's tracks, not including bridges.

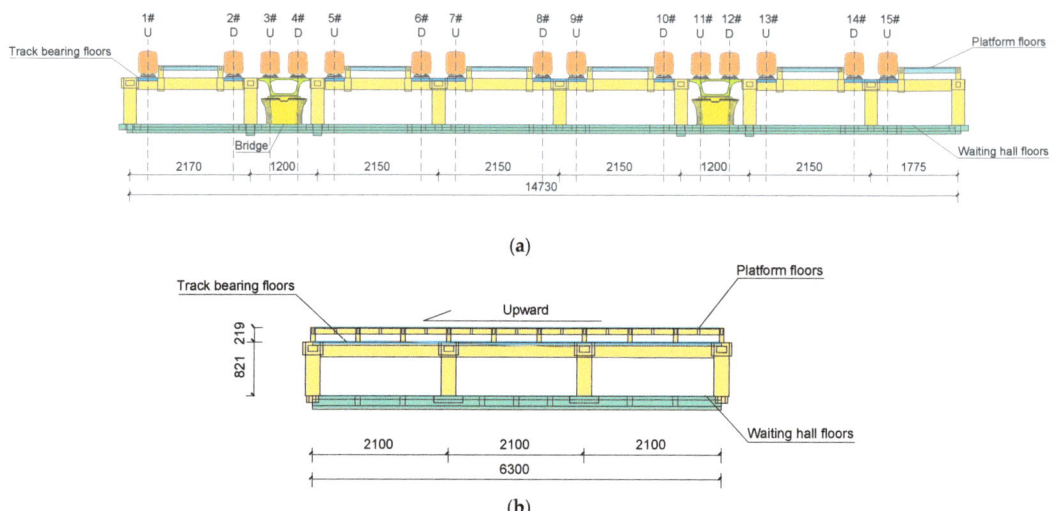

Figure 1. Schematic diagram of station structure (unit: mm): (a) lateral view of station; (b) elevation view of station.

3. Train–Track–Station Spatially Coupled Model

This paper utilizes GSAP, a dynamic simulation analysis finite element software developed by the author, to accurately model the track–station structure. The large scale of the station structure, coupled with its numerous degrees of freedom (DOFs), makes solving for such a multitude of DOFs quite challenging. To address this challenge and enhance computational efficiency, the paper adopts the principle of total potential energy with stationary value in elastic system dynamics and the "Set in right position" rule for formulating matrices as proposed by Zeng of Central South University [30,31]. It treats the train, track, and station structures as an integrated system, establishes a coupled vibration equation for the train–track–station, and employs time-domain analysis and computer numerical simulation for solution.

3.1. Train Spatial Vibration Model

3.1.1. Assumptions of the Train Model

To enhance computational efficiency while ensuring the accuracy of the train spatial model, this paper simplifies the train model based on its construction features, making the following assumptions [32]:

(1) The train body, bogies, and wheelsets are all assumed to be rigid bodies, considered only to undergo minor vibrations;

(2) The train travels at a constant speed on the track, with the longitudinal motion of the train not considered;

(3) The creep force of the wheel rail and the springs in the train model are linear, and the damping of the train model is viscous.

(4) The vertical displacement of the wheelset and track on the station remains constant.

3.1.2. Establishment of the Train Model

Based on the above assumptions, the train consists of one body, two bogies, and four wheelsets. The vehicle body and bogie exhibit five DOFs—lateral, vertical, roll, yaw, and pitch—while the wheelset's motion is described by two DOFs, namely in the lateral and roll. Figure 2 illustrates the specific vehicle vibration analysis model, and Table 1 details the DOFs of the vehicle.

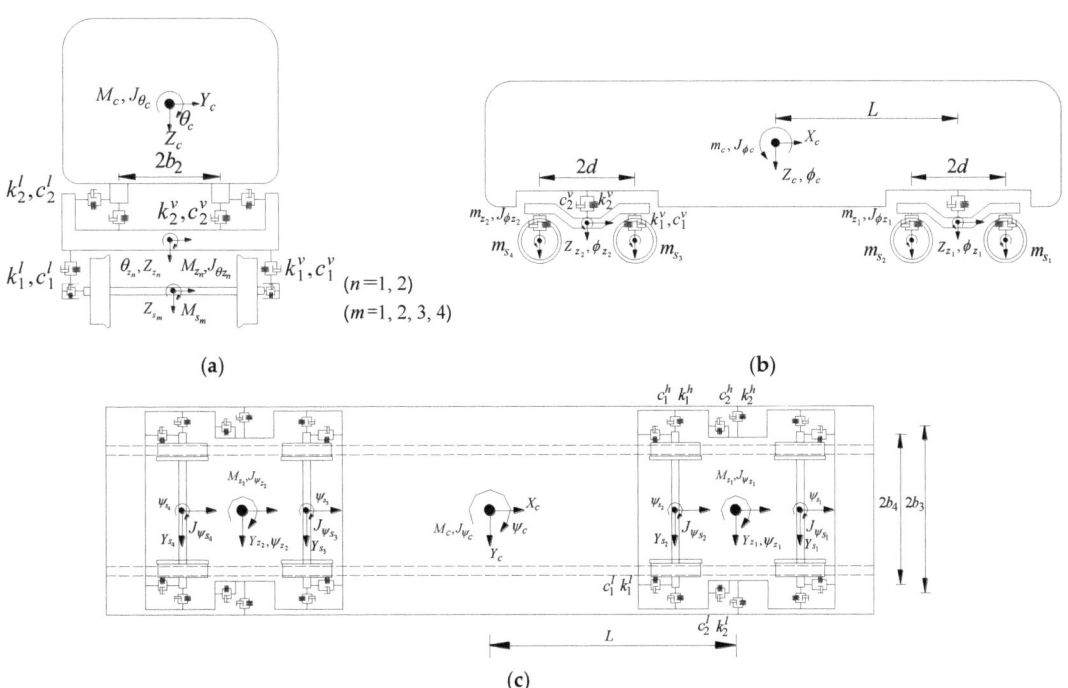

Figure 2. The model of the train: (**a**) side view; (**b**) elevation view; (**c**) plan view.

Table 1. Definition of DOF Variables for each vehicle.

Components	Lateral	Vertical	Yaw	Pitch	Roll
Body	Y_c	Z_c	ψ_c	ϕ_c	θ_c
Front bogie	Y_{Z_1}	Z_{Z_1}	ψ_{Z_1}	ϕ_{Z_1}	θ_{Z_1}
Rear bogie	Y_{Z_2}	Z_{Z_2}	ψ_{Z_2}	ϕ_{Z_2}	θ_{Z_2}
1st wheelset	Y_{S_1}	/	ψ_{S_1}	/	/
2nd wheelset	Y_{S_2}	/	ψ_{S_2}	/	/
3rd wheelset	Y_{S_3}	/	ψ_{S_3}	/	/
4th wheelset	Y_{S_4}	/	ψ_{S_4}	/	/

The other parameters in the Figure 2 are defined as follows: M_{Z_1}, M_{Z_2}, M_C represent the mass of the front bogie, rear bogie, and vehicle body. $J_{\phi Z_1}$, $J_{\phi Z_2}$, $J_{\phi C}$ denote the roll moments of inertia for the front bogie, rear bogie, and car body, while $J_{\psi Z_1}$, $J_{\psi Z_2}$, $J_{\psi C}$ represent the yaw moments of inertia for the same components, respectively. $J_{\psi S_1}$, $J_{\psi S_2}$, $J_{\psi S_3}$, $J_{\psi S_4}$ are the yaw moments of inertia for the 1st, 2nd, 3rd, and 4th wheelsets, respectively. k_1^v, c_1^v, k_1^h, c_1^h, k_1^l, c_1^l correspond to the vertical, horizontal, and longitudinal stiffness and damping of the primary suspension. k_2^v, c_2^v, k_2^h, c_2^h, k_2^l, c_2^l signify the same parameters for the secondary suspension.

The 23 displacement parameters of the train model are as shown in Equation (1).

$$\{\delta\}_v^e = \{Y_{s_1}\ Y_{s_2}\ Y_{s_3}\ Y_{s_4}\ \psi_{s_1}\ \psi_{s_2}\ \psi_{s_3}\ \psi_{s_4}\ Y_{z_1}\ Y_{z_2}\ Z_{z_1}\ Z_{z_2}\ \psi_{z_1}\ \psi_{z_2}\ \phi_{z_1}\ \phi_{z_2}\ \theta_{z_1}\ \theta_{z_2}\ Y_c\ Z_c\ \psi_c\ \phi_c\ \theta_c\}^T \quad (1)$$

The position of the train on the track constantly changes. Initially, the total vibration potential energy of a single vehicle section must be calculated. By applying the principle of total potential energy with stationary value in elastic system dynamics and the "Set in right position" rule for formulating matrices, the $[M_v]$ (mass), $[K_v]$ (Stiffness), and $[C_v]$ (Damping) matrices essential for the train simulation model in this paper can be generated:

$$[M_v]\{\ddot{\delta}_v\} + [C_v]\{\dot{\delta}_v\} + [K_v]\{\delta_v\} = \{P_v\} \quad (2)$$

where $\{\ddot{\delta}_v\}$, $\{\dot{\delta}_v\}$, $\{\delta_v\}$, respectively, represent the arrays of acceleration, velocity, and displacement parameters.

3.2. Track–Station Spatial Vibration Model

The finite element analysis model of this paper focuses on the track–station structure's main framework, omitting the canopy. It utilizes two-node spatial beam elements for beams and columns, and four-node shell elements for floor slabs. The core modeling strategy entails generating mass and stiffness matrices for each element from material parameters in a local coordinate system. Implementing the "Set in right position" rule for formulating matrices allows for the integration of all element matrices into comprehensive mass and stiffness matrices for the track–station model.

3.2.1. Establishment of Mass and Stiffness Matrices

1. Beam element;

A spatial beam element features two end nodes, i and j. Each node encompasses two translational degrees and one rotational degree of freedom (DOFs) within the local coordinate system, summing up to six DOFs per beam element. The local coordinate system originates at node i, with the axis from node i to node j constituting the x-axis. Introducing a point k, the plane spanned by i, j, and k constitutes the X-Y plane, with the Z-axis orientation determined by the right-hand rule. Thus, the mass and stiffness matrices of a spatial beam element in this local coordinate system are as follows:

$$[M_b]^e = \begin{bmatrix} \frac{\rho Al}{3} & & & & & & & & & & & \\ 0 & \frac{13\rho Al}{35} & & & & & & & & & & \\ 0 & 0 & \frac{13\rho Al}{35} & & & & & & & & & \\ 0 & 0 & 0 & \frac{\rho Al I_\rho}{3A} & & & & \text{symmetry} & & & & \\ 0 & 0 & -\frac{11\rho Al^2}{210} & 0 & \frac{\rho Al^3}{105} & & & & & & & \\ 0 & \frac{11\rho Al^2}{210} & 0 & 0 & 0 & \frac{\rho Al^3}{105} & & & & & & \\ \frac{\rho Al}{6} & 0 & 0 & 0 & 0 & 0 & \frac{\rho Al}{3} & & & & & \\ 0 & \frac{9\rho Al}{70} & 0 & 0 & 0 & \frac{13\rho Al^2}{420} & 0 & \frac{13\rho Al}{35} & & & & \\ 0 & 0 & \frac{9\rho Al}{70} & 0 & -\frac{13\rho Al^2}{420} & 0 & 0 & 0 & \frac{13\rho Al}{35} & & & \\ 0 & 0 & 0 & \frac{\rho Al I_\rho}{6A} & 0 & 0 & 0 & 0 & 0 & \frac{\rho Al I_\rho}{3A} & & \\ 0 & 0 & \frac{13\rho Al^2}{420} & 0 & -\frac{\rho Al^3}{140} & 0 & 0 & 0 & \frac{11\rho Al^2}{210} & 0 & \frac{\rho Al^3}{105} & \\ 0 & -\frac{13\rho Al^2}{420} & 0 & 0 & 0 & -\frac{\rho Al^3}{140} & 0 & -\frac{11\rho Al^2}{210} & 0 & 0 & 0 & \frac{\rho Al^3}{105} \end{bmatrix} \quad (3)$$

$$[K_b]^e = \begin{bmatrix} \frac{EA}{l} & & & & & & & & & & & \\ 0 & \frac{12EI_z}{(b_z+1)l^3} & & & & & & & & & & \\ 0 & 0 & \frac{12EI_y}{(b_y+1)l^3} & & & & & & & & & \\ 0 & 0 & 0 & \frac{GJ}{l} & & & & \text{symmetry} & & & & \\ 0 & 0 & \frac{-6EI_y}{(1+b_y)l^2} & 0 & \frac{(b_y+4)EI_y}{(1+b_y)l} & & & & & & & \\ 0 & \frac{6EI_z}{(1+b_z)l^2} & 0 & 0 & 0 & \frac{(b_z+4)EI_z}{(1+b_z)l} & & & & & & \\ -\frac{EA}{l} & 0 & 0 & 0 & 0 & 0 & \frac{EA}{l} & & & & & \\ 0 & \frac{-12EI_z}{(1+b_z)l^3} & 0 & 0 & 0 & \frac{-6EI_z}{(1+b_z)l^2} & 0 & \frac{12EI_z}{(1+b_z)l^3} & & & & \\ 0 & 0 & \frac{-12EI_y}{(b_y+1)l^3} & 0 & \frac{6EI_y}{(b_y+1)l^2} & 0 & 0 & 0 & \frac{12EI_y}{(b_y+1)l^3} & & & \\ 0 & 0 & 0 & -\frac{GJ}{l} & 0 & 0 & 0 & 0 & 0 & \frac{GJ}{l} & & \\ 0 & 0 & \frac{-6EI_y}{(1+b_y)l^2} & 0 & \frac{(2-b_y)EI_y}{(1+b_y)l} & 0 & 0 & 0 & \frac{6EI_y}{(b_y+1)l^2} & 0 & \frac{(4+b_y)EI_y}{(1+b_y)l} & \\ 0 & \frac{6EI_z}{(1+b_z)l^2} & 0 & 0 & 0 & \frac{(2-b_z)EI_z}{(1+b_z)l} & 0 & \frac{-6EI_z}{(1+b_z)l^2} & 0 & 0 & \frac{(4+b_z)EI_z}{(1+b_z)l} \end{bmatrix} \quad (4)$$

$$\left.\begin{array}{l} b_y = \frac{12kEI_y}{GA_z l^2} = \frac{24(\mu+1)A}{A_z}\left(\frac{r_y}{l}\right)^2 \\ b_z = \frac{12kEI_z}{GA_y l^2} = \frac{24(\mu+1)A}{A_y}\left(\frac{r_z}{l}\right)^2 \\ I_y = \int z^2 dA, \; I_z = \int y^2 dA \end{array}\right\} \quad (5)$$

where l is the length of the beam element; A is the cross-sectional area of the beam element; ρ is the density of the beam element material; I_ρ, I_y, I_z are the moments of inertia of the beam element's cross-section about the x, y, z axes, respectively; r_y, r_z are the radius of the gyration of the beam element's cross-section about the y, z axes, respectively; and A_y and A_z are the shear areas of the beam element along the local coordinate system's y, z axes, respectively.

2. Shell element;

A shell element consists of four nodes, corresponding to the four corners of a rectangle, with each node having two translational degrees of freedom, totaling eight degrees of freedom per shell element. Equations (6) and (7) display the mass and stiffness matrices of the shell element, where a and b represent half the length of the sides of the plate element, ρ is the material density, and t is the thickness of the plate element.

$$[M] = \frac{ab\rho t}{9} \begin{bmatrix} 4 & 0 & 2 & 0 & 1 & 0 & 2 & 0 \\ 0 & 4 & 0 & 2 & 0 & 1 & 0 & 2 \\ 2 & 0 & 4 & 0 & 2 & 0 & 1 & 0 \\ 0 & 2 & 0 & 4 & 0 & 2 & 0 & 1 \\ 1 & 0 & 2 & 0 & 4 & 0 & 2 & 0 \\ 0 & 1 & 0 & 2 & 0 & 4 & 0 & 2 \\ 2 & 0 & 1 & 0 & 2 & 0 & 4 & 0 \\ 0 & 2 & 0 & 1 & 0 & 2 & 0 & 4 \end{bmatrix} \quad (6)$$

$$[K] = \begin{bmatrix} k_{11} & k_{12} & k_{13} & k_{14} \\ k_{21} & k_{22} & k_{23} & k_{24} \\ k_{31} & k_{32} & k_{33} & k_{34} \\ k_{41} & k_{42} & k_{43} & k_{44} \end{bmatrix} \quad (7)$$

The expression for the sub-stiffness matrix within the stiffness matrix in Equation (7) is shown in Equation (8).

$$[k_{ij}] = \iint t[B_i]^T[D][B_j]dxdy \tag{8}$$

3.2.2. Establishment of the Damping Matrix

Calculating the damping matrix for a structure is notably more challenging than determining its stiffness and mass matrices. Due to its simplicity and convenience, Rayleigh damping is widely adopted in structural dynamic analysis. Consequently, it is commonly applied in engineering projects for deriving the damping matrix. The formula for Rayleigh damping used in this study is detailed in Equation (9).

$$[C] = \alpha[M] + \beta[K] \tag{9}$$

In Equation (9), $[M]$ and $[K]$ represent the structure's mass matrix and stiffness matrix, respectively. α and β are proportional coefficients, which are constants. Once the finite element model of the bridge is constructed, α and β can be obtained through modal analysis, as illustrated in Equations (10) and (11).

$$\alpha = \frac{2\omega_1\omega_2}{\omega_1^2 - \omega_2^2}(\omega_1\zeta_2 - \omega_2\zeta_1) \tag{10}$$

$$\beta = \frac{2(\omega_1\zeta_1 - \omega_2\zeta_2)}{\omega_1^2 - \omega_2^2} \tag{11}$$

ω_1 and ω_2 represent the first and second natural frequency of the station finite element model, respectively, while ζ_1 and ζ_2 correspond to the first and second modal damping of the finite element model. Therefore, once the overall stiffness and mass matrices of the bridge–station analysis model are established, the overall damping matrix can be obtained using the aforementioned method.

3.2.3. Finite Element Model of Station

Figure 3 illustrates the finite element model of the station, featuring 2574 nodes, 2533 beam elements, and 1615 shell elements, all constructed from linear elastic materials. The elasticity modulus and Poisson's ratio adhere to specified standards, with the concrete structure's damping ratio established at 2%. Table 2 outlines the components' material strength classes.

Figure 3. Finite element model of multi-line "building-bridge integrated" elevated station.

Table 2. Material strength classes for station components.

Components	Strength Class
Platform level beams, columns, and floor	C50
Track-bearing layer longitudinal beams, transverse beams	C60
Track-bearing layer columns, floor	C50
ground level beams, concrete columns, and floor	C50

3.3. Establishment of the Train–Track–Station System Coupled Vibration Equation

At any given time t, deriving the coupled vibration equation for the train–track–station system requires calculating the total potential energies of both the track–station and the train. The sum of these energies forms the system's total potential energy. The train–track–station system is viewed as an integrated unit, with self-excitation due to track irregularities and external excitation from vehicular loads. Utilizing Qingyuan Zeng's previously mentioned theory allows for the formulation of the system's vibration equation at time t:

$$[K]\{\delta\} + [C]\{\dot{\delta}\} + [M]\{\ddot{\delta}\} = \{P\} \tag{12}$$

where $[K]$, $[C]$, $[M]$ represent the stiffness, mass, and damping matrices of the train–track–station system at time t, respectively; $\{\delta\}$, $\{\dot{\delta}\}$, $\{\ddot{\delta}\}$ correspond to the displacement, velocity, and acceleration arrays for the system at time t; and $\{P\}$ is the load array acting on the system at time t, consisting of wheel-rail contact forces due to track irregularities and the self-weight of the vehicle.

In Equation (12), the right-side load array includes only the train's gravitational forces. To solve for the dynamic response of the train–track–station system during operation, excitations from track irregularities and deformations must replace the vibration parameters on the left side of Equation (12). Accordingly, the displacement parameter $\{\delta\}$ in Equation (12) is divided into k known and n unknown parameters, leading to $\{\delta\}$'s expression as follows:

$$\{\delta\} = \{\delta_k \; \delta_n\}^T \tag{13}$$

Equation (12) can be derived as:

$$\begin{bmatrix} K_{kk} & K_{nk} \\ K_{kn} & K_{nn} \end{bmatrix} \begin{Bmatrix} \delta_k \\ \delta_n \end{Bmatrix} + \begin{bmatrix} C_{kk} & C_{nk} \\ C_{kn} & C_{nn} \end{bmatrix} \begin{Bmatrix} \dot{\delta}_k \\ \dot{\delta}_n \end{Bmatrix} + \begin{bmatrix} M_{kk} & M_{nk} \\ M_{kn} & M_{nn} \end{bmatrix} \begin{Bmatrix} \ddot{\delta}_k \\ \ddot{\delta}_n \end{Bmatrix} = \begin{Bmatrix} 0 \\ P \end{Bmatrix} \tag{14}$$

Expanding Equation (14), we obtain:

$$[K_{nn}]\{\delta_n\} + [C_{nn}]\{\dot{\delta}_n\} + [M_{nn}]\{\ddot{\delta}_n\} = \{P\} - [K_{nk}]\{\delta_k\} - [C_{nk}]\{\dot{\delta}_k\} - [M_{nk}]\{\ddot{\delta}_k\} \tag{15}$$

$$[K_{kk}]\{\delta_k\} + [C_{kk}]\{\dot{\delta}_k\} + [M_{kk}]\{\ddot{\delta}_k\} + [M_{kn}]\{\ddot{\delta}_n\} + [C_{kn}]\{\dot{\delta}_n\} + [K_{kn}]\{\delta_n\} = 0 \tag{16}$$

Equation (16) is a linearly dependent matrix equation and must be eliminated. Since all terms on the right side of Equation (15) are known, it constitutes a solvable vibration matrix equation for the train–track–station system with a free term. Therefore, the dynamic response of the train–track–station coupling system can be determined based on this.

3.4. Solution of the Train–Track–Station System Coupled Vibration Equation

For linear structures, mode superposition methods, such as Duhamel integration and Fourier transform, are commonly used for dynamic calculation and analysis. These methods offer simplicity in calculation, and provide an intuitive understanding of the contribution of each vibration mode to structural vibration.

This study addresses the train–track–station system, a complex time-varying entity. Changes in spring and damping properties mean the system's vibration equation coefficients constantly shift, making mode superposition methods impractical. Instead, a step-by-step integration method within the time domain emerges as a viable analytical approach for the motion equations. This method divides the system's vibration duration under external loads into numerous intervals, keeping parameters static within each. The system's dynamic response at the beginning of a time interval is used as the initial condition for vibration, allowing the system's response at the end of that interval to be determined and used as the initial condition for the next interval. By repeating this calculation process, the vibration response for all time intervals can be determined. This paper employs the step-by-step integration method for solution.

3.5. Verification of the Spatial Model

This study's finite element simulation model and computational approaches align with those described in Reference [33]. That research examines the influence parameters of the dynamic response of the elevated station with the "integral station-bridge system". Designed for 120 km/h speeds, the station utilizes a complete cast-in-place spatial frame structure. Structurally, it includes a hall level, track-bearing level, and platform level, organized from the bottom upwards. The overall dimensions are 21.62 m in height, 120 m in length, and 31.80 m in width, with three transverse rows of columns spaced 10.5 m apart. Longitudinally, it comprises 10 spans of 12 m each, devoid of expansion joints. Figure 4 displays the station's finite element model.

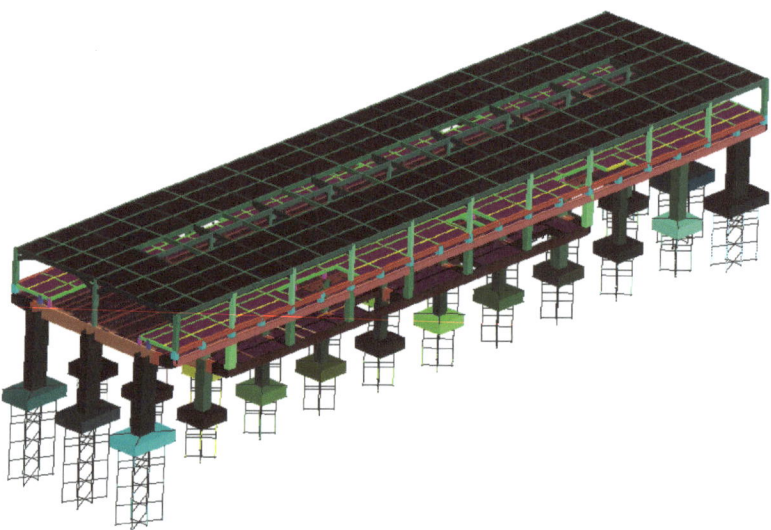

Figure 4. Finite element model of station.

Numerous studies indicate that the vibration response of station structures is primarily influenced by the vertical loads from trains, with the station's vibration predominantly characterized by vertical responses [2,6,33]. To validate the model's credibility and the computational results' reliability, this study focused on the hall and platform levels for vertical response measurements, comparing the observed data with numerical simulations. An unloaded six-car train passing at 110 km/h in a downward direction served as the on-site condition for measurement. The schematic diagram of the lateral arrangement of measuring points is shown in Figure 5. Points A1 and A2 were designated at the hall level, with points B1 and B2 at the platform level. Figure 6 illustrates the vertical acceleration time–history curves at these points, while Table 3 compares the measured and simulated acceleration peak values. The close match in peak values, being of the same magnitude order, confirms the model's spatially coupled vibration calculation accuracy.

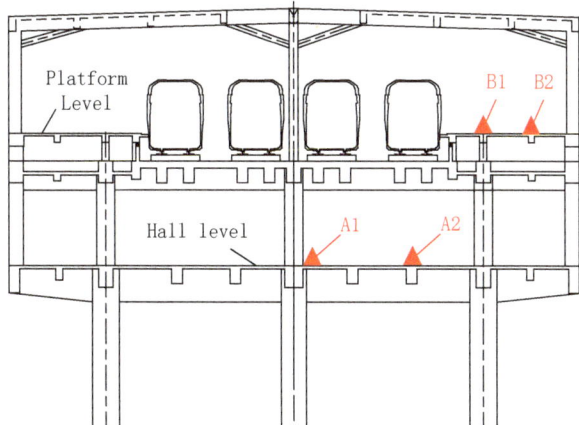

Figure 5. Schematic diagram of lateral arrangement of measuring points.

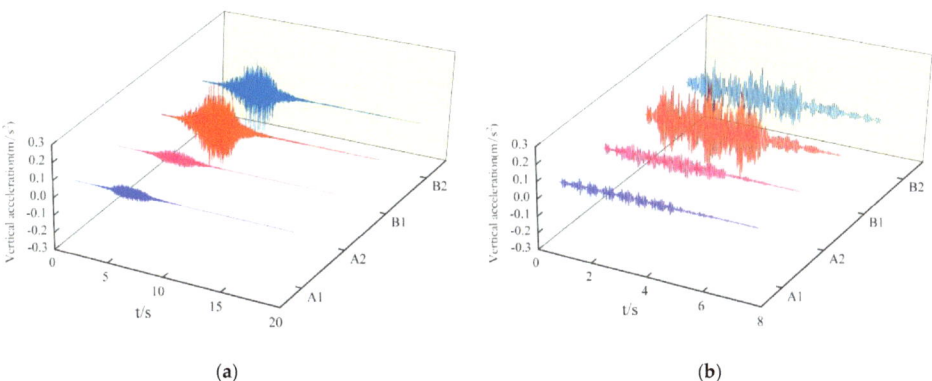

Figure 6. Vertical acceleration time–history curve of each measuring point: (a) measured time–history curve; (b) calculated time–history curve.

Table 3. Comparison of peak acceleration.

Position	Point	Peak Acceleration (mm/s^2)	
		Measured Value	Calculated Value
hall level	A1	42.46	32.98
	A2	50.96	38.37
platform level	B1	123.86	152.57
	B2	107.75	101.50

4. Calculation and Analysis of Vibration in the Train–Track–Station Spatially Coupled System

4.1. The Natural Vibration Characteristics of Station

The station's natural frequency and mode shapes provide an intuitive representation of its structural stiffness. Utilizing the finite element analysis model developed earlier, the study examined the station structure's natural vibration characteristics, detailed in Table 4.

Table 4. The natural vibration characteristics of elevated station.

Finite Element Model	Mode Frequency (Hz)	Vibration Characteristics
	1st Mode 3.310 Hz	Vertical bending mode (Front view)
	12th Mode 4.810 Hz	Lateral translation mode (Side view)
	15th Mode 4.994 Hz	Longitudinal translation mode (Front view)

4.2. Dynamic Response Results and Analysis of the Station under Different Single-Track Conditions

Based on the finite element model established earlier, a coupled vibration analysis of the train–track–station system is conducted. The train model uses the China Railway High-speed 3 (CRH3). It consists of 16 cars, including 8 motor cars and 8 trailer cars. The German low-disturbance spectrum is used to simulate additional track irregularities during train operation. The analysis in this section focuses on single-track train conditions. Out of the station's 11 tracks, 7 with unique operating conditions were selected due to the similarities in the conditions among some tracks. The seven single-track train conditions correspond to tracks 1#, 2#, 5#, 6#, 7#, 13#, and 14# in Figure 1a. The specific train formation and operational conditions are shown in Table 5 below. The structural dynamic response is calculated for a single-track fully loaded train passing through the station in the down direction at speeds of 60~100 km/h.

Table 5. The train formations and operational conditions.

Train Model	Train Formation	Speed (km/h)	Working Condition
CRH3	$(M^1 + T^1 + T + M) \times 4$	60, 70, 80, 90, 100	Single-track train (1#, 2#, 5#, 6#, 7#, 13#, 14#)

[1] The M is motor car, and the T is trailer car.

To illustrate the variation in structural responses under different single-track train conditions and speeds, Figure 7 presents curves of lateral and vertical movements and accelerations for seven distinct single-track scenarios, for both the track-bearing and platform floors. All structural vibration displacement values are measured relative to the initial equilibrium position.

Figure 7 reveals that for a single-track CRH3 train traveling at speeds of 60~100 km/h, both the track-bearing and platform floors experience greater vertical than lateral displacements and accelerations at all speeds. Lateral movements and accelerations exhibit minimal changes as the speed increases. Overall, vertical displacements and accelerations at the track-bearing floor increase with speed. At the platform floor, vertical displacement increases with speed, while vertical acceleration first decreases then increases.

A single-track bearing floor contains one track, while a double-track bearing floor contains two tracks. As depicted in Figure 1, tracks 1#, 2#, 5#, and 13# reside on a single-track floor, with tracks 6#, 7#, and 14# situated on a double-track floor. At identical speeds, the platform floor's lateral and vertical movements and accelerations—collectively termed as lateral and vertical responses—are more pronounced when a train traverses track 5# than on tracks 6# and 7#. Similarly, at equal speeds, track 13# elicits greater lateral and vertical responses on the platform floor than track 14#. Moreover, at the same velocity, a single-track train induces notably lower lateral and vertical responses on a double-track bearing floor than on a single-track bearing floor.

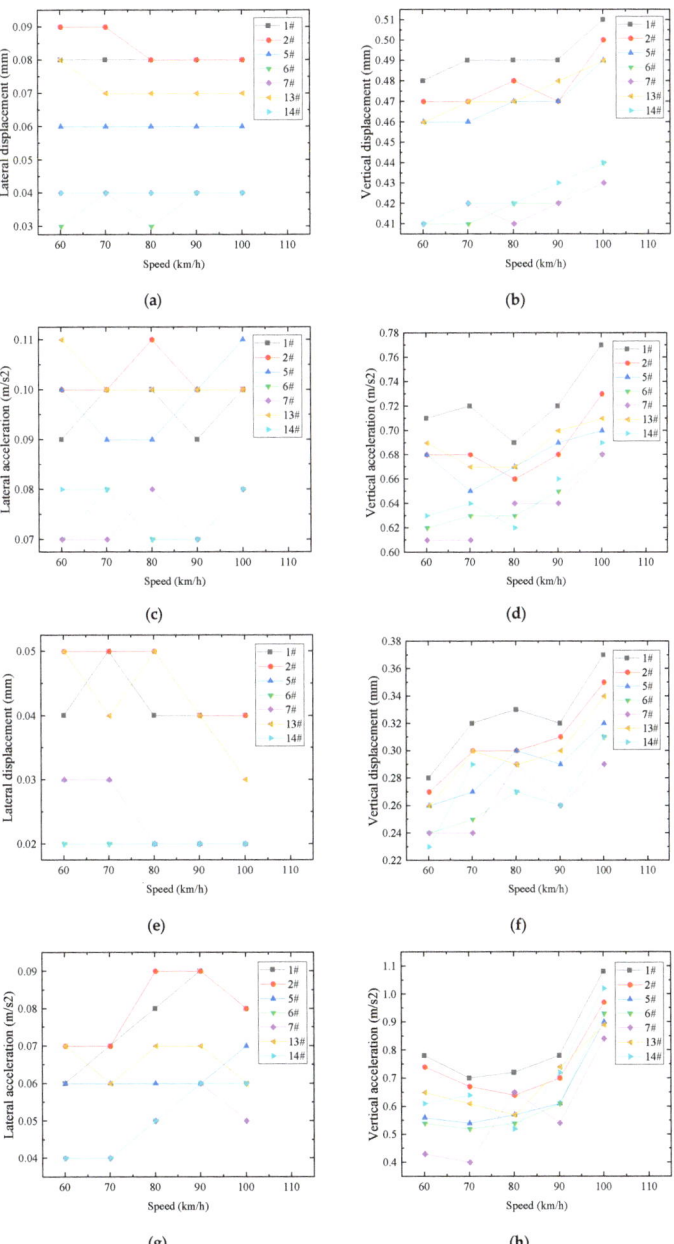

Figure 7. Maximum dynamic response of track-bearing floors and platform floors at the speed of 60~100 km/h under seven types of single-track train conditions: (**a**) lateral displacements of the track-bearing floors; (**b**) vertical displacements of the track-bearing floors; (**c**) lateral acceleration of the track-bearing floors; (**d**) vertical acceleration of the track-bearing floors; (**e**) lateral displacements of the platform floors; (**f**) vertical displacements of the platform floors; (**g**) lateral acceleration of the platform floors; (**h**) vertical acceleration of the platform floors.

Considering the span of frame structures, tracks 1# and 2# are on single-span frames, 5#, 6#, and 7# are on triple-span frames, and 13# and 14# are on double-span frames. In a general trend, the platform floor's lateral and vertical responses from trains on the double-span 13# track are less than on the single-span 1# and 2# tracks but greater than on the triple-span 5# track.

4.3. The Impact of Multi-Track Train Operations on Station Dynamic Response

To assess the station's structural dynamic response to varying numbers of simultaneously operating trains, this study concurrently passed double, quadruple, and sextuple CRH3 trains through the station at 100 km/h. The train count was symmetrically increased from the central triple-span frame structure, as shown in Table 6, which details specific train formations and operational conditions. The vibration response collection points started at the intersection of cross-track and along-track central axes at the edge of the track-bearing floor, near track 1#. These points extended along the cross-track direction. For the track-bearing floor, collection points were positioned at the center. For the platform floor, collection points were located at both side edges and the center, as illustrated in Figure 8.

Table 6. The train formations and operational conditions.

Train Model	Train Formation	Speed (km/h)	Working Condition
CRH3	$(M^1 + T^1 + T + M) \times 4$	100	Double trains (7#, 8#)
CRH3	$(M^1 + T^1 + T + M) \times 4$	100	Quadruple trains (6#, 7#, 8#, 9#)
CRH3	$(M^1 + T^1 + T + M) \times 4$	100	Sextuple trains (5#, 6#, 7#, 8#, 9#, 10#)

^{1} The M is motor car, and the T is trailer car.

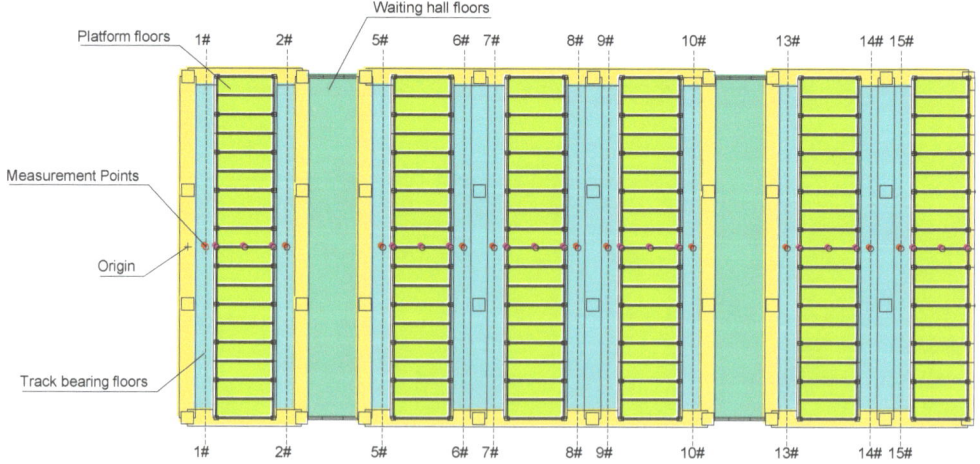

Figure 8. Schematic diagram of lateral arrangement of vibration measuring points.

To demonstrate the impact of the number of operating trains on the vibration response of the track-bearing and platform floors, as well as the response distribution across the cross-track direction, the calculated lateral and vertical dynamic responses are plotted as curves, as shown in Figures 9 and 10.

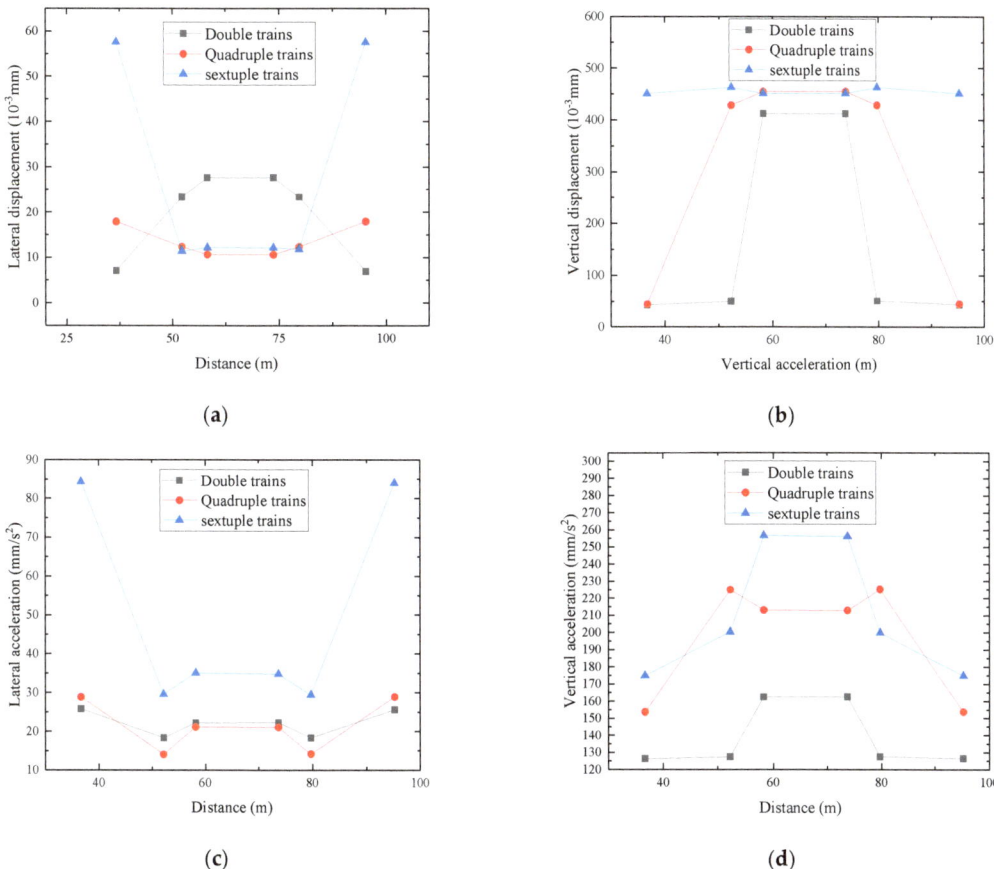

Figure 9. Maximum dynamic response at the mid-span of station track-bearing floors under different number of operating train lines: (**a**) lateral displacements of the track-bearing floors; (**b**) vertical displacements of the track-bearing floors; (**c**) lateral acceleration of the track-bearing floors; (**d**) vertical acceleration of the track-bearing floors.

Figure 9 shows that with two trains, the lateral displacement in the track-bearing floor decreases from the mid-span towards the edges. With four and six trains, the lateral displacement at the frame edges exceeds that at mid-span. This occurs because, with more trains, the symmetric dynamic loads are partially offset at the structure's center. At the same time, the track-bearing floor edges experience vibration amplification due to superposition. Moreover, when multiple trains are in operation, the lateral acceleration in the track-bearing layer is lower at mid-span compared to the boundaries. Overall, the trend shows that as the number of trains increases, both the vertical displacement and acceleration in the track-bearing floor rise, with vertical responses markedly surpassing lateral ones.

Figure 10 reveals that with an increase in the number of operating trains, the lateral displacement and acceleration at the platform floor of the frame structure follow a similar pattern, displaying troughs at the center and amplified vibration responses near the edges. Overall, as the number of operating trains increases, there is a trend of increasing vertical displacement and acceleration at the platform floor, with vertical responses being significantly greater than lateral responses.

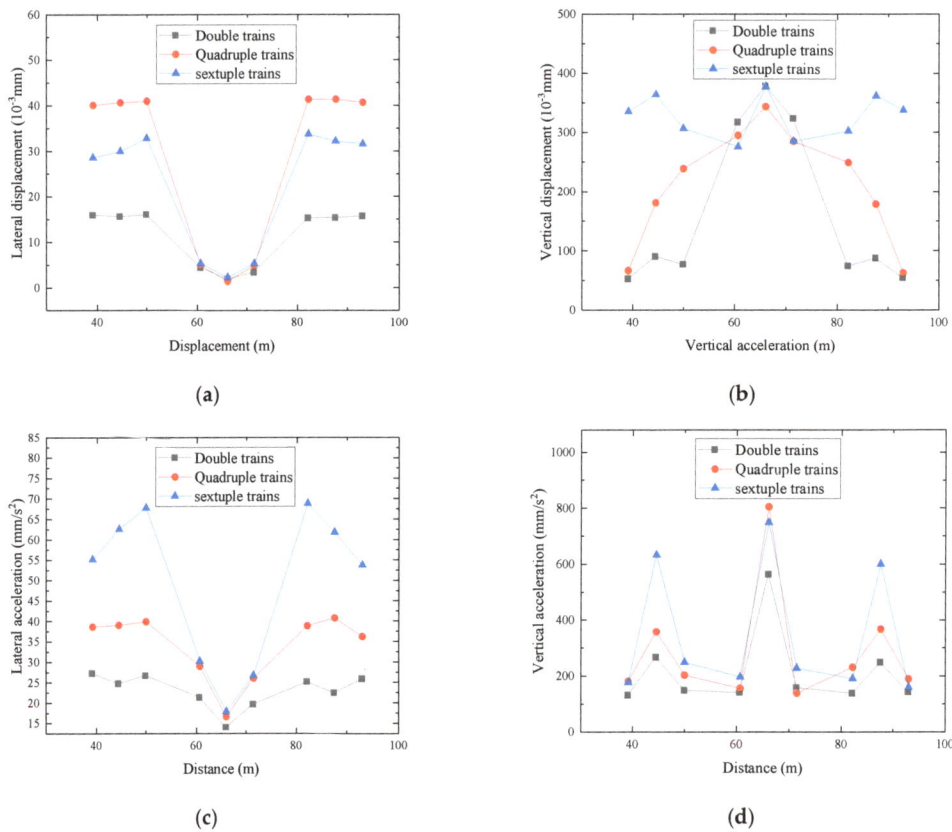

Figure 10. Maximum dynamic response at the mid-span of station platform floors under different number of operating train lines: (**a**) lateral displacements of the platform floors; (**b**) vertical displacements of the platform floors; (**c**) lateral acceleration of the platform floors; (**d**) vertical acceleration of the platform floors.

5. Conclusions

This paper examines a typical multi-line "building-bridge integration" elevated station. It establishes a spatially coupled vibration equation and a finite element analysis model for the train–track–station system. These are based on the principle of total potential energy with stationary value in elastic system dynamics and the "Set in right position" rule for formulating matrices. The study analyzes the dynamic response characteristics of multi-line elevated station structures and how these responses vary with different numbers of operational train tracks. It yields the following conclusions:

(1) When a single train passes through the station, the track-bearing and platform floors show only slight increases in lateral displacement and acceleration with speed, which are smaller than the vertical movements and accelerations at the same speed. Generally, the vertical displacement and acceleration of the track-bearing floor, as well as the vertical displacement of the platform floor, increase with speed. Notably, the platform floor's vertical acceleration first decreases and then increases as the speed rises.

(2) The track-bearing layer is classified into a single-track bearing floor and double-track bearing floor, depending on the number of tracks. When moving at the same speed, a single train on a single-track bearing floor induces a stronger dynamic response in

(3) Station structures can be classified into single-span, double-span, and triple-span frame types. With train speed as the sole variable, the dynamic response of the platform floor generated by a single train on the single-track bearing floor of multi-span frame structures decreases sequentially as the number of spans increases.

(4) This study uses the methodology described in the references for developing spatially coupled models and solving vibration equations. Minor differences between the peak acceleration values measured on-site at the platform and hall levels, compared to their theoretical counterparts, confirm the spatial model's reliability and the computational data's accuracy. This demonstrates the method's effectiveness in closely approximating the vibration response characteristics of real structures.

(5) As the number of operating train lines increases, the track-bearing and platform floor layers exhibit a compensation of lateral dynamic responses at the center of the three-span structure, while an amplification of vibration due to superposition occurs at the frame edges. Both the track-bearing and platform floors show that vertical responses exceed lateral responses, with an increasing trend as the number of operational train tracks rises.

This study explored the structural dynamic response of multi-line "building-bridge integration" elevated stations. An increase in the number of train lines necessitates larger station structures, making the propagation of dynamic responses more complex. This research provides design references for future large-scale "building-bridge integration" elevated stations and data for future station design standards. In this paper, larger responses were observed in the double-track bearing floors and at the edges of multi-span structures within the station structure, identifying them as critical areas needing attention in future station designs. Moreover, this study focused only on the vibration responses of the platform and track-bearing floors of station structures, without examining waiting hall responses. Future research should extend to include the investigation of train-induced responses in waiting halls.

Author Contributions: Conceptualization, X.G.; Methodology, X.G.; Software, X.G.; Validation, X.G. and S.W.; Formal analysis, X.G. and S.W.; Investigation, S.W.; Resources, X.G.; Data curation, X.G.; Writing—original draft, X.G. and S.W.; Writing—review & editing, S.W.; Visualization, S.W.; Supervision, X.G.; Project administration, X.G.; Funding acquisition, X.G. All authors have read and agreed to the published version of the manuscript.

Funding: This work is supported by the National Natural Science Foundation of China (U1934207).

Data Availability Statement: Data are contained within the article.

Conflicts of Interest: The authors declare no conflict of interest.

References

1. Luo, J. Study on Design Ideas of Elevated Station. *Build. Technol. Dev.* **2020**, *47*, 15–16.
2. Cai, C.; Deng, S.; He, X.; Zhou, Y.; Yu, K.; Wu, Y. Vibration measurement and analysis of an elevated station of Dongguan-Shenzhen intercity railway. *J. Railw. Sci. Eng.* **2022**, *19*, 1395–1403.
3. Yu, Y.L.; He, W. Comfortable Evaluation of Bridge-Station Combined Large Span High Speed Railway Station under High Speed Train Load. *Appl. Mech. Mater.* **2013**, *477–478*, 727–731. [CrossRef]
4. Liu, Q.M.; Zhang, X.; Zhang, Z.J.; Li, X.Z. In Situ Experimental Study of Environmental Vibration Induced by CRH on Elevated Railway Station. *Adv. Mater. Res.* **2013**, *639–640*, 930–934. [CrossRef]
5. Yang, N.; Guo, T.; Sun, G. Train-induced vibration on elevated railway station. *J. Cent. South. Univ.* **2013**, *20*, 3745–3753. [CrossRef]
6. Ba, Z.; Jiao, P.; Liang, J.; Liu, G.; Gao, Y. Actual Measurement and Analysis of Station Vibration Induced by High-Speed Train. *J. Tianjin Univ.* **2020**, *53*, 1211–1217.
7. Nielsen, J.C.O.; Anderson, D.; Gautier, P.; Iida, M.; Nelson, J.T.; Thompson, D.; Tielkes, T.; Towers, D.A.; de Vos, P. Ground-Borne Vibration due to Railway Traffic: A Review of Excitation Mechanisms, Prediction Methods and Mitigation Measures. In *Noise and Vibration Mitigation for Rail Transportation Systems*; Springer: Berlin/Heidelberg, Germany, 2015; Volume 126, pp. 253–287.
8. Connolly, D.P.; Marecki, G.P.; Kouroussis, G.; Thalassinakis, I.; Woodward, P.K. The growth of railway ground vibration problems—A review. *Sci. Total Environ.* **2016**, *568*, 1276–1282. [CrossRef] [PubMed]
9. Kaynia, A.M.; Madshus, C.; Zackrisson, P. Ground Vibration from High-Speed Trains: Prediction and Countermeasure. *J. Geotech. Geoenviron* **2000**, *126*, 531–537. [CrossRef]

10. Xia, H.; Zhang, N.; Cao, Y.M. Experimental study of train-induced vibrations of environments and buildings. *J. Sound. Vib.* **2005**, *280*, 1017–1029. [CrossRef]
11. Xia, H.; Gao, F.; Wu, X.; Zhang, N.; de Roeck, G.; Degrande, G. Running train induced vibrations and noises of elevated railway structures and their influences on environment. *Front. Struct. Civ. Eng.* **2009**, *3*, 9–17. [CrossRef]
12. Sanayei, M.; Moore, J.A.; Brett, C.R. Measurement and prediction of train-induced vibrations in a full-scale building. *Eng. Struct.* **2014**, *77*, 119–128. [CrossRef]
13. Hesami, S.; Ahmadi, S.; Ghalesari, A.T. Numerical modeling of train-induced vibration of nearby multi-story building: A case study. *Ksce J. Civ. Eng.* **2016**, *20*, 1701–1713. [CrossRef]
14. Li, X.-Z.; Gao, W.; Lei, K.-N.; Hu, X.-H.; Liang, L. Vibration and Noise Measurement of Railway Station Hall Induced by High-Speed Trains. *J. Southwest Jiaotong Univ.* **2020**, *55*, 1067–1075. [CrossRef]
15. Deng, Z. Structural Vibration Analysis and Vibration Control Study on The New Changsha Railway Station. Ph.D. Thesis, Central South University, Changsha, China, 2010.
16. Xu, M.-D.; Wei, M.-Y.; Ye, J.-N.; Yang, Y.-Z.; Xie, W.-P. Vibration serviceability of the subway elevated station of "combination of bridge and construction" type. *J. Railw. Sci. Eng.* **2022**, *19*, 230–238. [CrossRef]
17. Xie, W.-P.; Ma, C.-P.; Zhou, D.-J.; Cao, L.-C.; Hua, Y.-M. Research on Vibration Response and Serviceability of Four Lane Elevated Station of "Combination of Bridge and Construction". *J. Wuhan Univ. Technol.* **2022**, *44*, 31–37.
18. Na, Y.; Min, W.; Shuai, Z. Vehicle-induced Vibration Response Analysis and Comfort Evaluation of "Building-Bridge Integration"Structure. *J. Hunan Univ.* **2016**, *43*, 96–104. [CrossRef]
19. Cui, C.; Lei, X.; Zhang, L. Study on Vehicle-induced Vibration Response Characteristics of an Integrated Traffic Hub. *Noise Vib. Control* **2017**, *37*, 144–150+172. [CrossRef]
20. Alan, S.; Caliskan, M. Vibration isolation design of railroad tracks within Ankara high speed train station. *J. Acoust. Soc. Am.* **2015**, *137*, 2343. [CrossRef]
21. Zhang, N.; Xia, H.; Cheng, Q.; De Roeck, G. Analysis method for a vehicle structure coupled system under braking force. *J. Vib. Shock* **2011**, *30*, 138–143. [CrossRef]
22. Zhang, X.; Ruan, L.; Zhao, Y.; Zhou, X.; Li, X. A frequency domain model for analysing vibrations in large-scale integrated building–bridge structures induced by running trains. *Proc. Inst. Mech. Eng. Part F J. Rail Rapid Transit.* **2020**, *234*, 226–241. [CrossRef]
23. Yang, J.; Zhu, S.; Zhai, W.; Kouroussis, G.; Wang, Y.; Wang, K.; Lan, K.; Xu, F. Prediction and mitigation of train-induced vibrations of large-scale building constructed on subway tunnel. *Sci. Total Environ.* **2019**, *668*, 485–499. [CrossRef] [PubMed]
24. Li, X. Dynamic Effect Analysis of Vehicle-induced with Frame-Type "Integration Construction-Bridge" Structure. Master's Thesis, Chang'an University, Xi'an, China, 2019.
25. Feng, D.; Yu, Z.; Xu, H.; Yang, J. Test and Analysis of Train-Induced Vibration Comfort of Railway Station Structure. In *First International Conference on Rail Transportation 2017*; American Society of Civil Engineers: Reston, VA, USA, 2017; pp. 288–296.
26. Gao, R.; Li, C. Traffic-induced Structural Vibrations and Noises in Elevated Railway Stations. In *IABSE Symposium Report*; International Association for Bridge and Structural Engineering: Zürich, Switzerland, 2003; pp. 1–7.
27. Zhao, Y.; Zhang, Y.; Xu, J.; Yu, P.; Sun, B. Shaking table test on seismic performance of integrated station-bridge high-speed railway station. *Structures* **2022**, *46*, 1981–1993. [CrossRef]
28. Xie, W.; Zhang, J.; Liang, H.; Zhang, Z.; Lu, Y.; Wei, P. Analysis of human-induced vibration response in elevated waiting hall of Beijing Fengtai High-Speed Railway Station. *Structures* **2023**, *2023*, 105348. [CrossRef]
29. Cui, K.; Su, L. Dynamic finite element analysis of an elevated station-track structure coupled system under resonance. *Teh. Vjesn.* **2019**, *26*, 449–456.
30. Zeng, Q. The principle of total potential energy with stationary value in elastic system dynamics. *J. Huazhong Univ. Sci. Technol.* **2000**, *28*, 1–3+14.
31. Zeng, Q.Y.; Yan, G.P. The "set-in-right-position" rule forming structural method for space analysis of truss bridges. *J. China Railw. Soc.* **1986**, *8*, 48–59.
32. Zeng, Q.Y.; Guo, X.R. *Theory and Application of Vibration Analysis of Time-Varying Systems on Train Bridges*; China Railway Publishing House: Beijing, China, 1999.
33. Guo, X.; Liu, J.; Wu, Y.; Jiang, Y. Research on the influencing parameters of the dynamic responses of the elevated station with "integral station-bridge system". *J. Railw. Sci. Eng.* **2023**, *20*, 671–681. [CrossRef]

Disclaimer/Publisher's Note: The statements, opinions and data contained in all publications are solely those of the individual author(s) and contributor(s) and not of MDPI and/or the editor(s). MDPI and/or the editor(s) disclaim responsibility for any injury to people or property resulting from any ideas, methods, instructions or products referred to in the content.

Article

Effect of Soil Anisotropy on Ground Motion Characteristics

Yuhong Xie [1,2], Zhou Cao [3] and Jian Yu [1,2,*]

[1] Department of Geotechnical Engineering, Tongji University, Shanghai 200092, China; 2310070@tongji.edu.cn
[2] Key Laboratory of Geotechnical and Underground Engineering of Ministry of Education, Tongji University, Shanghai 200092, China
[3] Nuclear Power Operations Research Institute, Shanghai 200120, China; caozhou@cnnp.com.cn
* Correspondence: 002yujian@tongji.edu.cn

Citation: Xie, Y.; Cao, Z.; Yu, J. Effect of Soil Anisotropy on Ground Motion Characteristics. Buildings 2023, 13, 3017. https://doi.org/10.3390/buildings13123017

Academic Editor: Mengmeng Lu

Received: 11 October 2023
Revised: 29 November 2023
Accepted: 1 December 2023
Published: 3 December 2023

Copyright: © 2023 by the authors. Licensee MDPI, Basel, Switzerland. This article is an open access article distributed under the terms and conditions of the Creative Commons Attribution (CC BY) license (https:// creativecommons.org/licenses/by/ 4.0/).

Abstract: Soil transverse isotropy results in different stiffness characteristics in horizontal and vertical directions. However, the effect is usually neglected in seismic motion analysis. In this study, an equivalent linear anisotropic soil model was established based on the finite element method, and we investigated the impact of anisotropic parameters on ground motion at the site under various seismic wave inputs. It was found that the anisotropic parameters have a more significant effect on seismic waves, with the dominant frequency being closer to the fundamental frequency of the site. As an example, the soil dynamic parameters in Shanghai Yangshan Port were calibrated by a series of bending elements, resonance columns, and cyclic triaxial tests. The influences of anisotropy on the peak ground acceleration (PGA) and response spectrum were studied for Yangshan Port. Additionally, the standard design response spectra considering the soil anisotropy were provided. A comparison reveals that the existing isotropic design response spectrum may lead to dangerous seismic design for the structures at Yangshan port.

Keywords: transverse isotropy; ground motion characteristics; standard design response spectrum

1. Introduction

To improve the seismic design of structures, it is necessary to conduct a site-specific seismic motion analysis to understand the characteristics of ground motion at the site. Earthquakes have resulted in significant economic losses and casualties. Earthquakes have the potential to inflict significant destruction upon various types of structures, encompassing inland buildings [1], hydraulic structures [2], and subterranean constructions [3]. One of the important causes of the structural damage caused by earthquakes is soil amplification. This is illustrated in damage assessments of buildings in the countries of Turkey [4–6], Greece [7], Iran [8], Mexico [9], Korea [10], Pakistan [11], and Nepal [12]. Numerous scholars have extensively researched the seismic resilience of engineering structures [13–16]. The analysis of structures for seismic resistance utilizes a range of methodologies, encompassing response spectrum analysis, the base shear method (quasi-static method), and time history analysis. The response spectrum method is extensively utilized as the predominant seismic analysis approach in engineering practice. The response spectrum is obtained by calibrating ground motion characteristics at the specific site. Consequently, numerous scholars have researched the seismic design of structures based on the seismic motion characteristics at the site [16–18].

Soil dynamic constitutive models can be categorized into total stress models and effective stress models, considering the aspects of stress transmission and inter-particle contact [19]. The effective stress model offers superior capability in addressing the issue of seismic liquefaction [20–22]. Regarding the issue of seismic ground motion response, researchers commonly employ the total stress model for analysis. The total stress models include the elastic–plastic model [23–30], nonlinear model [22,31,32], and equivalent linear model [33]. Traditional elastoplastic models such as the Mohr–Coulomb model [34] and the

Drucker–Prager model [35] can be integrated with the boundary interface theory [25–27], kinematic hardening theory [23,24], and nested yield surface theory [29,30] to account for the effects of cyclic loading. Although the elastoplastic model theory effectively describes the hysteresis characteristics and nonlinearity of soils, its computational complexity and significant workload make it inconvenient for engineering applications. Researchers employ nonlinear models [31,32] based on the Masing criteria to depict the nonlinearity and hysteresis characteristics of soils. While these nonlinear models simplify the workload associated with elastoplastic models, they still entail a certain level of complexity. The linear model offers the advantages of a low workload and low complexity, allowing for the consideration of soil nonlinearity and hysteresis characteristics through the application of an equivalent concept. Hence, the most widely utilized approach at present is the equivalent linear model, which has evolved from linear models. Schnabel [33] initially pioneered the development of the frequency domain equivalent linear model within the SHAKE program. Idriss et al. [36] modified the expression of damping in the SHAKE program and developed the SHAKE91 program. With the development of commercial software, the SHAKE and SHAKE91 programs have been integrated into software such as EERA 1998 and DEEPSOIL v7.0.33. The SAHKE program demonstrates limited proficiency in effectively addressing anisotropic challenges.

Soil anisotropy has aroused great interest in recent years. Zhang et al. [37,38] and Teng [39] investigated the impact of soil anisotropy on excavation-induced effects in excavations. Wei et al. [40], Soe et al. [41], and Zhang et al. [42] investigated the impact of soil anisotropy on tunnel design and construction. Peric et al. [43] and Ai et al. [44,45] investigated the impact of anisotropy on the design of pile foundations. It has been found that there is anisotropy in the soil under small strain conditions. Bentil [46] has delved into the anisotropy of the small-strain shear modulus by conducting bending element experiments. The consideration of soil anisotropy in the seismic response analysis of soil layers has been scarce among scholars, primarily due to the complexities involved in studying anisotropy, seismic loads, and their associated intricacies. Considering the lack of relevant research and to maintain the coherence and rigor of scientific inquiry, this study investigates the impact of anisotropy on the seismic response of the site using a finite element method with an anisotropic time-domain equivalent linear model.

2. Soil Dynamic Characteristics

2.1. Linear Viscoelastic Model

The viscoelastic Kelvin model (a spring connected in parallel with a sticky pot) is used to reflect the hysteresis of the soil under cyclic loading. The stress–strain relationship is described by Equation (1):

$$\tau = G\gamma + \eta_G \dot{\gamma} \tag{1}$$

where G is the shear modulus; τ is the shear stress; γ is the shear strain; and η_G is the shear viscosity coefficient, as given in Equation (2):

$$\eta_G = \frac{2GD}{\omega} \tag{2}$$

where D is the damping ratio and ω is the circular frequency.

2.2. Modulus and Damping Models

The key to the effective linearization method is to determine the relationship between the shear modulus ratio and the damping ratio with the shear strain. Hyperbolic models are widely used to describe nonlinear soil behavior under cyclic loading, such as the Pyke model and the Stokoe model [47]. In this study, the improved Stokoe model is selected to fit the relationship between the shear modulus ratio and shear strain, as given in Equation (3):

$$\frac{G}{G_0} = \frac{1}{1 + \left(\frac{\gamma}{\gamma_r}\right)^a} \tag{3}$$

where G_{norm} is the normalized shear modulus, r_r is the reference strain, and α is the fitting parameter; the definition of r_r is different from the Harden–Drnevich model ($r_r = \tau_{max}/G_{max}$).

Zhang et al.'s [48] formula is adopted to describe the relationship between D and G_{norm} as follows:

$$D = K_1 G_{norm}^2 + K_2 G_{norm} + K_3 \qquad (4)$$

where K_1, K_2, K_3 are the model fitting parameters.

2.3. Effects of Anisotropy

The notation for an anisotropic material used herein is the y-axis (the vertical direction) represents the direction of the anisotropy, and the x, z-plane is the plane of isotropy. The stress–strain increment equation for an anisotropic material can be written as follows (5) [49].

$$\begin{Bmatrix} \delta\varepsilon_x \\ \delta\varepsilon_y \\ \delta\varepsilon_z \\ \delta\gamma_{xy} \\ \delta\gamma_{yz} \\ \delta\gamma_{zx} \end{Bmatrix} = \begin{bmatrix} \frac{1}{E_h} & -\frac{v_{vh}}{E_v} & -\frac{v_{hh}}{E_h} & 0 & 0 & 0 \\ -\frac{v_{hv}}{E_h} & \frac{1}{E_v} & -\frac{v_{hv}}{E_h} & 0 & 0 & 0 \\ -\frac{v_{hh}}{E_h} & -\frac{v_{vh}}{E_v} & \frac{1}{E_h} & 0 & 0 & 0 \\ 0 & 0 & 0 & \frac{1}{G_{hv}} & 0 & 0 \\ 0 & 0 & 0 & 0 & \frac{1}{G_{vh}} & 0 \\ 0 & 0 & 0 & 0 & 0 & \frac{1}{G_{hh}} \end{bmatrix} \begin{Bmatrix} \delta\sigma_x \\ \delta\sigma_y \\ \delta\sigma_z \\ \delta\sigma_{xy} \\ \delta\sigma_{yz} \\ \delta\sigma_{zx} \end{Bmatrix} \qquad (5)$$

where E_v and E_h are Young's moduli in the vertical and horizontal directions, respectively; V_{hh} and V_{vh} are Poisson's ratios for horizontal strains from a horizontal and vertical strain, respectively; V_{hv} is Poisson's ratio for vertical strains from a horizontal strain; G_{vh} and G_{hv} are the shear moduli in the vertical plane; and G_{hh} is the shear modulus in the horizontal plane.

The anisotropy ratios AR_G and AR_E for the shear modulus and Young's modulus are, respectively, defined as:

$$AR_G = \frac{G_{hh}}{G_{vh}} \qquad (6)$$

$$AR_E = \frac{E_h}{E_v} \qquad (7)$$

Under the undrained condition, these Poisson's ratios [50] need to satisfy the additional relationships ($v_{vh} = 0.5, v_{hh} + v_{hv} = 1$). Therefore, one can further obtain the following equations [50]:

$$v_{hh} = 1 - v_{hv} = 1 - AR_E \cdot v_{vh} \qquad (8)$$

$$v_{hv} = \frac{E_h}{E_v} \cdot v_{vh} = AR_E \cdot v_{vh} \qquad (9)$$

$$G_{hh} = \frac{AR_E \cdot E_v}{2(2 - AR_E \cdot v_{vh})} \qquad (10)$$

$$G_{vh} = G_{hv} = \frac{1}{AR_G} \frac{AR_E \cdot E_v}{2(2 - AR_E \cdot v_{vh})} \qquad (11)$$

3. Simulation of Time-Domain Equivalent Linear Model for Anisotropic Soil Layers

This study takes the actual recorded seismic waves as input conditions, degenerates the model to isotropy, and compares it with EERA for verification. This study focuses on exploring the ground motion response of anisotropic sites with different sedimentary characteristics (different anisotropic parameters).

3.1. Input Ground Motion and Finite Element Model

In Eurocode 8 [51], sites are classified according to the average shear wave velocity in the upper 30 m thick soil profile. Therefore, the soil layer thickness is selected as 30 m. Also, three seismic waves are chosen: the Ei-Centro Wave, Shanghai Wave, and Kobe Wave. The amplitudes of these seismic waves were adjusted to have a peak acceleration of 0.1 g. Figure 1 illustrates the three seismic waves. As shown in Figure 2, a two-dimensional plane strain soil finite element model is established based on ABAQUS 6.14. The type of the finite element is CPE4, and the estimated size of the mesh is 1 m × 1 m. The mesh size meets the requirement of less than 1/10 wavelength. The boundary adopts the infinite element boundary, and seismic waves are input to the base. Therefore, the wave vibrates in the x direction and propagates in the y direction.

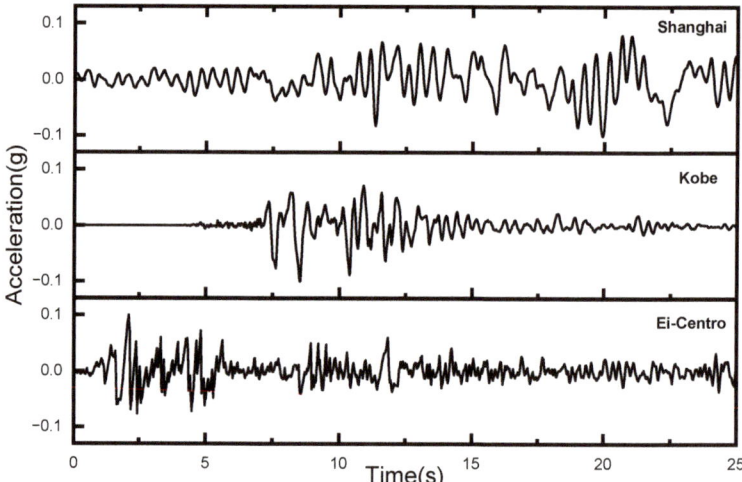

Figure 1. Seismic wave time history curve.

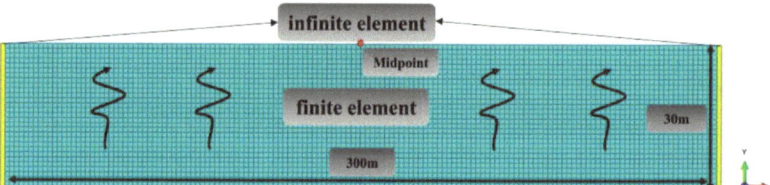

Figure 2. Finite element model.

3.2. Isotropic Soil Layer

This subsection is used to verify the validity of the established finite element method with an anisotropic time-domain equivalent linear model in the isotropic case studies. The shear modulus model chosen here is the Seed–Idriss mode [52], and the damping curve model used is the Idriss mode [36]. Table 1 presents the model parameters. The properties of the soil layers are shown in Table 2.

Table 1. Model parameters.

γ_γ (10^{-4})	α	K_1	K_2	K_3
5.87	0.93	0.26	−0.51	0.26

Table 2. Isotropic soil example.

Seismic Wave	Thickness (m)	V_s (m/s)	ρ (g/cm^3)
Ei-Centro; Shanghai; Kobe	30	100 150 200 300	2

Figure 3 illustrates the comparison of the peak ground acceleration (PGA) with depth, considering three seismic records, namely the Ei-Centro Wave, Shanghai Wave, and Kobe Wave, along with four soil conditions. They exhibit good agreements with the results obtained from the classical seismic analysis code EERA.

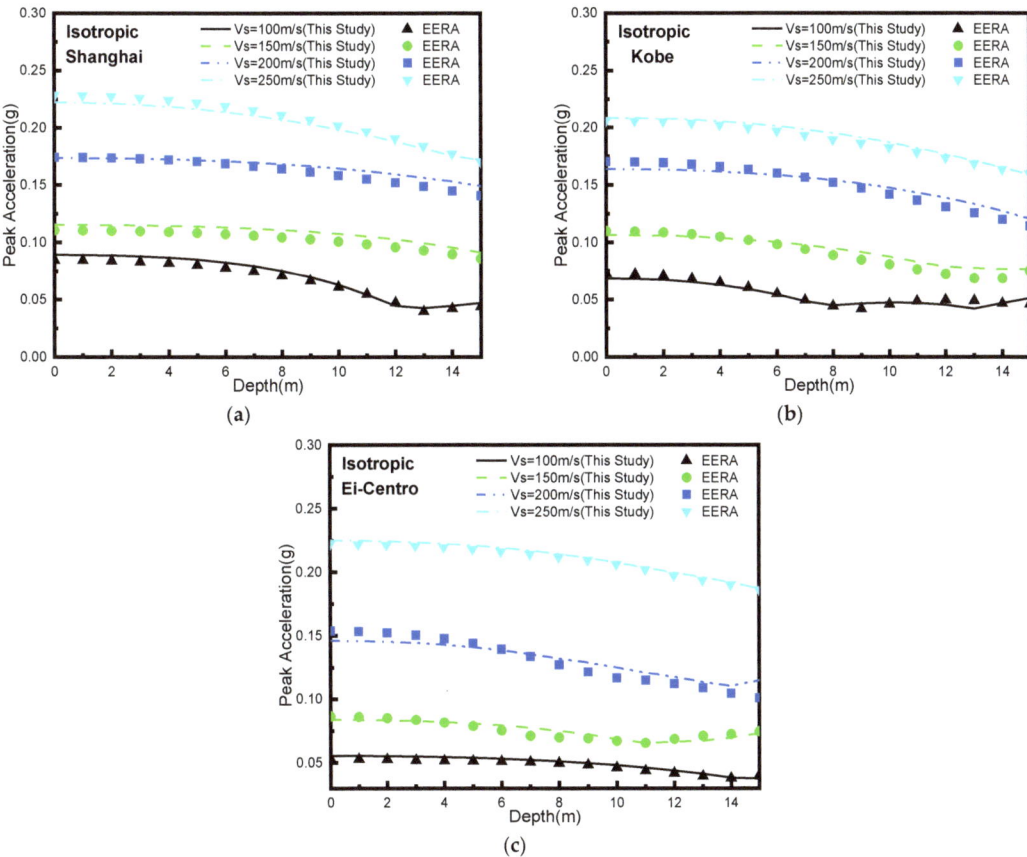

Figure 3. Comparison of results for homogeneous soil layer with isotropic: (**a**) Shanghai; (**b**) Kobe; (**c**) Ei-Centro.

3.3. The Influence of Anisotropic Parameters

It should be noted that this study assumes a consistent relationship between $G_{vh} - \gamma$ and $G_{hh} - \gamma$. Diffserent anisotropic parameter (AR_E = 1.00, 1.30, 1.60; AR_G = 1.05, 1.30, 1.55, 1.80, 2.05) were selected for a 30-meter-thick layer of soft soil with a shear wave velocity of 150 m/s. Detailed parameters are listed in Table 3.

Table 3. Example parameters.

Seismic Wave	AR_E	AR_G	Thickness (m)	V_s (m/s)	ρ (g/cm^3)
EI-Centro; Shanghai; Kobe	1.00; 1.30; 1.60;	1.05; 1.30; 1.55; 1.80; 2.05;	30	150	2

PGA_{aniso}/PGA_{iso} is the ratio between the peak ground acceleration obtained at an anisotropic site and that obtained at an isotropic site. Figure 4 shows that, with AR_E increasing, PGA_{aniso}/PGA_{iso} gradually decreases. Conversely, as AR_G increases, PGA has an obvious increase. Especially when inputting the Kobe wave, the PGA of anisotropic conditions exceeds that of isotropic conditions up to 14% with AR_G increasing. Consequently, it is considered that soil anisotropy is essential in analyzing site conditions for seismic response.

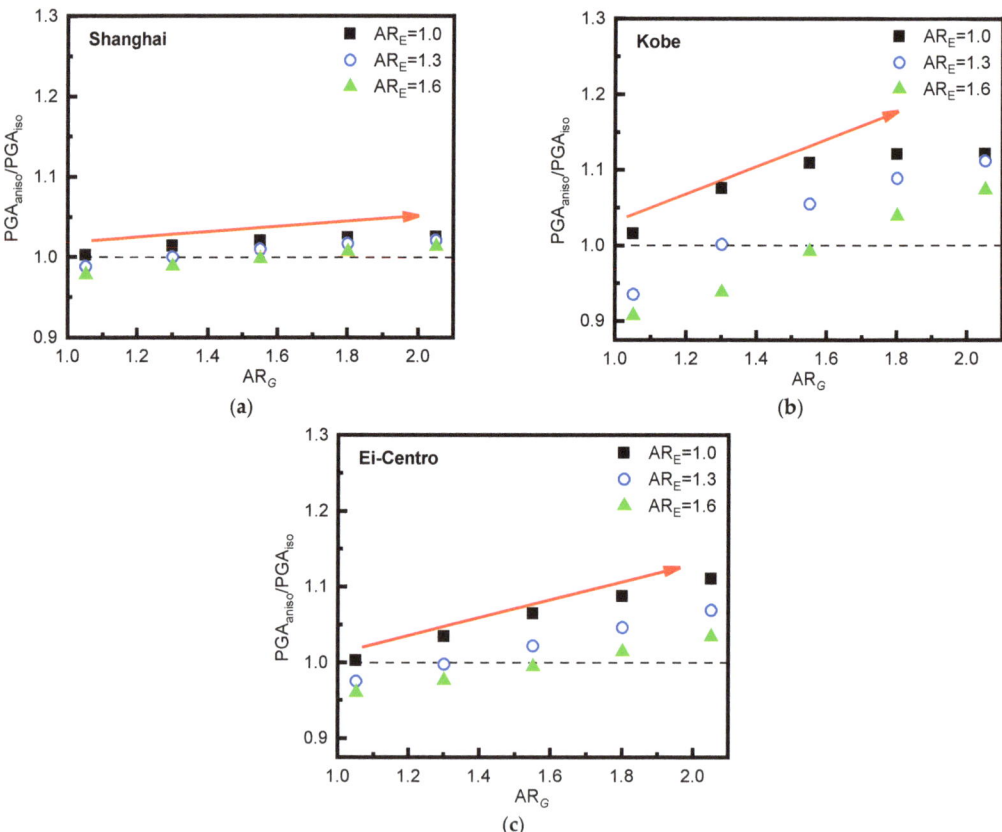

Figure 4. Normalized peak ground acceleration scatter plot: (a) Shanghai; (b) Kobe; (c) Ei-Centro.

To assess the impact of anisotropy on the seismic response, the dominant frequency of the seismic waves and the fundamental frequency of the site are further discussed as listed in Table 4, where the fundamental frequency of the site is calculated using Equation (12) [53]:

$$f_g = \frac{1}{T_g} = V_{Si}/4\sum_{i=1}^{n} H_i \qquad (12)$$

where the symbol "f_g" represents the fundamental frequency of the site; "T_g" refers to the site's characteristic period; and the variable "H_i" denotes the thickness of individual soil layers, whereas "V_{Si}" represents the shear wave velocity specific to each respective soil layer.

Table 4. Site fundamental frequency and seismic wave dominant frequency.

Seismic Wave	Dominant Frequency of Seismic Wave (Hz)	f_g (Hz)
Shanghai	0.92	
Kobe	1.2	1.25
Ei-Centro	1.16	

As shown in Figure 4, the Ei-Centro and Kobe waves have significant influences on the PGA_{aniso}/PGA_{iso}, while the Shanghai wave has a relatively slight effect. From Table 4, compared with the dominant frequency of the Shanghai wave, those of the Kobe wave and Ei-Centro waves are closer to the fundamental frequency of the site. The waves with dominant frequencies closer to the fundamental frequencies of the sites may result in having a more significant effect in terms of anisotropy. Therefore, detailed investigations on the effects of the soil anisotropy on the seismic response of the Shanghai Yangshan Port site are conducted as an example.

4. Seismic Response of Shanghai Yangshan Port

4.1. Input Seismic Wave

The seismic characteristics of Yangshan Port are examined in this study through the analysis of artificial seismic wave data from Shanghai. According to the site classification method specified in the "Code for Seismic Design of Buildings" (GB 50011-2010) [54], the Yangshan Port area belongs to the fourth category of site. Based on the linear elastic soil layer, the ground acceleration time history is inverted to bedrock to obtain the bedrock acceleration time history [55].

4.2. Calculation Model and Parameters of Soil Layer

Most of the topsoil in Yangshan port is soft clay [56], and there is sand soil in the lower layer. Through the geological investigation, the site of Yangshan Port consists of four typical soil layers: clay layer, silty clay layer, muddy silty clay, and sand soil layer. According to Hou [57], the AR_E values for this area range from 1.6 to 2.40. Li [58] reported AR_G values ranging from 1.08 to 1.39 for Shanghai soil, while Ng [59] stated that the values were between 1.07 and 1.38. Detailed borehole data for the soil layers are provided in Table 5.

Table 5. Soil layer information of Yangshan port.

Number	Soil	Bottom Depth (m)	V_s (m/s)	ρ (g/cm^3)
1	Muddy silty clay	17.6	140	
2	Clay	21.6	180	1.8
3	Silty clay	31.2	230	
4	Sand	43.2	290	2

Resonance column and cyclic triaxial tests were conducted to obtain $G_{norm} - \gamma$ and $D - \gamma$ curves for clay, silty clay layer, muddy silty clay, and sand. The design confining pressures are 150 kPa, 200 kPa, and 250 kPa for clay, silty clay layer, and muddy silty clay.

The design confining pressures are 200 kPa, 250 kPa, and 300 kPa for sand. Figure 5 shows the details of the cyclic triaxial test and the resonant column test. The experimental results pertaining to silty clay are specifically analyzed in this study. Figure 6 illustrates the S-wave output signals of the bending element in the silty clay specimen subjected to a confining pressure of 150 kPa across different input frequencies. Furthermore, the G_0 value derived from the bending element tests is approximately 1.1 to 1.2 times greater than the results obtained from the resonant column tests, which aligns with the outcomes reported by Yang et al. [60] and Gu et al. [61,62]. The reason behind this is that the results of the bending element tests specifically relate to the localized stiffness of the shear wave propagation path, while the resonant column offer insights into the overall stiffness characteristics of the specimens [60–62]. This confirms the accuracy of the resonant column test. Simultaneously, Yang et al. [60], Gu et al. [61,62], and Youn et al. [63] propose to determine G_0 based on the resonant column test.

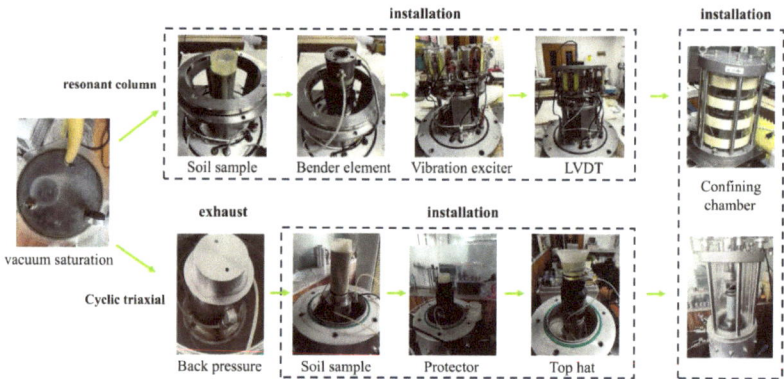

Figure 5. Cyclic triaxial test and resonant column test details.

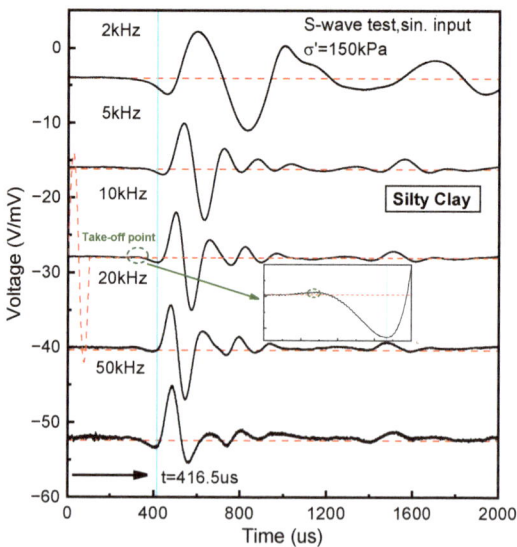

Figure 6. BE S-wave signals in silty clay specimen.

The cyclic triaxial experiments were performed employing strain control. To mitigate

stage was limited to six. Figure 7 depicts the outcomes of the cyclic triaxial tests conducted on the silty clay. The modulus degrades as the strain increases, while it amplifies with the elevated confining pressure. Through fitting analysis, we obtained the $G_{norm} - \gamma$ and $D - \gamma$ curves for clay in the Yangshan Port site. Figure 8 depicts the curve of best fit for the silty clay specimen. For the other soil layers, the model parameters calibrated from the resonance column and the cyclic triaxial tests are also provided in Table 6.

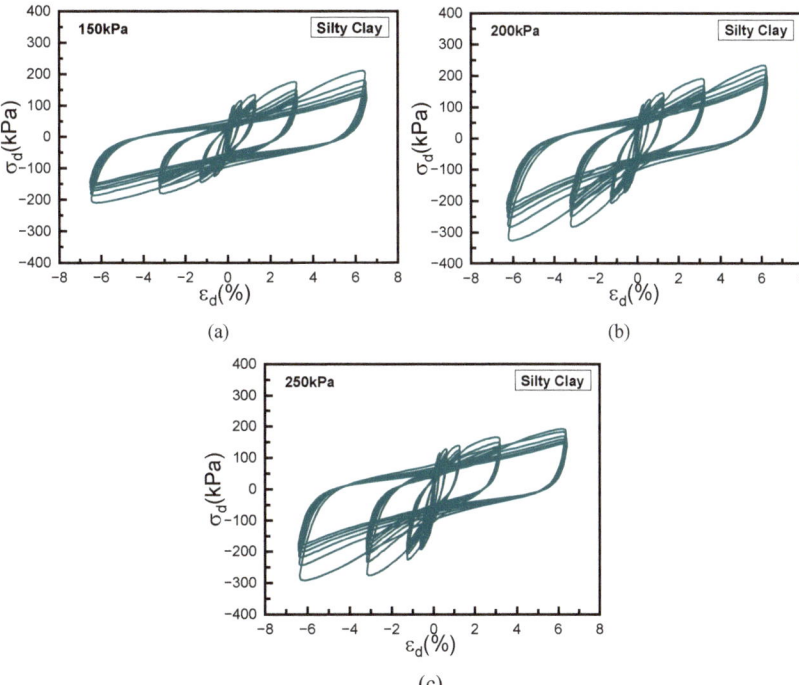

Figure 7. Partial results of cyclic triaxial testing (silty clay): (**a**) 150 kpa; (**b**) 200 kpa; (**c**) 250 kpa.

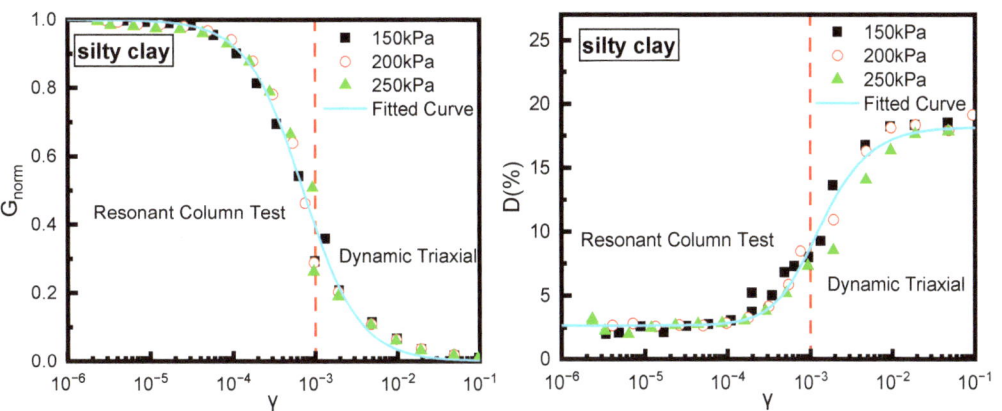

Figure 8. (**a**) $G_{norm} - \gamma$; (**b**) $D - \gamma$.

Table 6. The soil parameters of the Yangshan Port site.

Soil	γ_γ (10^{-4})	α	K_1	K_2	K_3
Clay	8.1	1.28	0.16	−0.32	0.19
Silty Clay	7.17	1.27	0.14	−0.29	0.18
Muddy silty Clay	7.32	1.31	0.10	−0.26	0.18
sand	6.09	1.07	0.12	−0.26	0.16

4.3. Results

Taking into account the previous discussion regarding the suppressive effect of AR_E and the enhancing effect of AR_G on PGA performance, we initially selected a value of 1.6 for AR_E and a value of 1.4 for AR_G to simulate the unfavorable conditions. In this case, study, the Shanghai wave is adjusted to have peak ground acceleration of 0.1 g. Figures 9 and 10 show the changes in peak acceleration along the depth and the ground peak acceleration spectrum, respectively. It is evident that when the proposed calculation model reduces to the isotropic scenario, it aligns well with the results obtained from EERA. However, when considering soil anisotropy, the PGA increases by 19.70%; the peak value of the ground response spectrum undergoes a significant increase of 28.82% at 1 Hz.

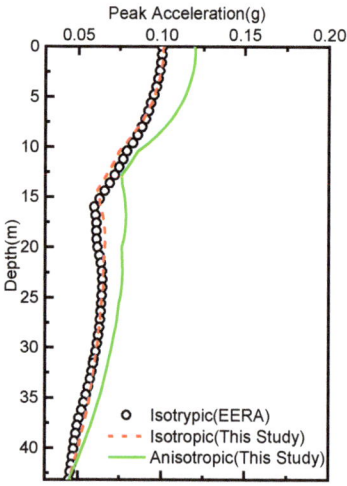

Figure 9. Peak acceleration along depth.

In order to discuss the influence of anisotropy on the ground acceleration frequency, the ground motion response of the Ei-Centro waves and Kobe waves input at Yangshan port is calculated. The result can be seen in Figure 10. It can be found that anisotropy has little influence on the frequency position of the ground Fourier acceleration spectrum peak.

To further explore the design response spectrum of Yangshan Port, the ground motion responses of the site under different seismic intensities (0.1 g, 0.15 g, 0.2 g, 0.25 g, 0.3 g) are studied. Based on the 5% damped ground acceleration response spectrum of different intensities, the least squares method by Andreotti [64] et al. was used to calibrate the design response spectrum. The shape function of the design response spectrum by Chinese code "GB 51247-2018" is as follows [65]:

$$\beta(T) = \begin{cases} 1 + (\beta_{max} - 1)\frac{T}{T_0} & 0 \le T \le T_0 \\ \beta_{max} & T_0 \le T \le T_g \\ \beta_{max}\left(\frac{T_g}{T}\right) & \end{cases} \quad (13)$$

where T_0 is the period associated with the initial inflection point, established at 0.1 s. T_g signifies the characteristic period. Since this study focuses on a certain soil layer in Yangshan Port, the characteristic period (T_g) can be calculated according to Equation (12), which is 0.92 s. T_m is the cutoff period, χ represents the attenuation index, and β_{max} denotes the plateau value.

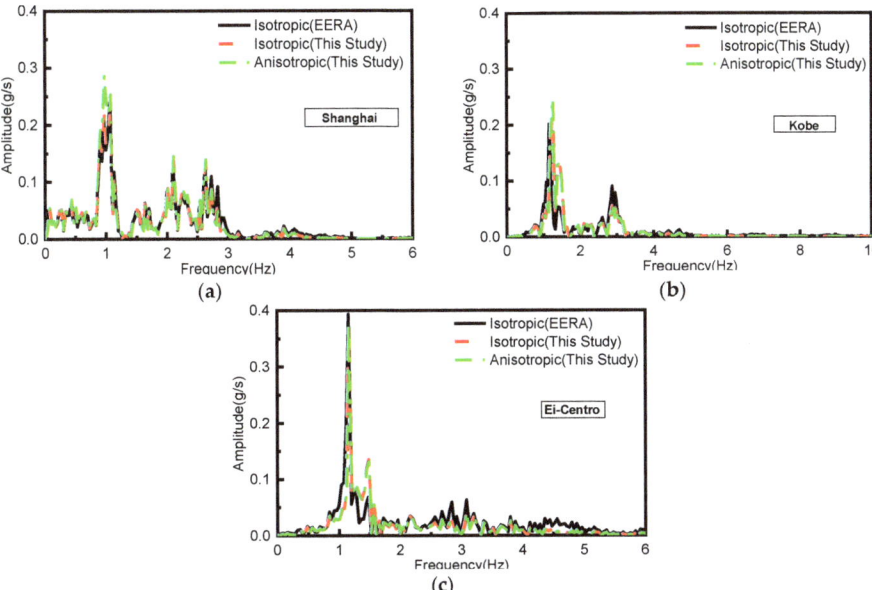

Figure 10. Ground Fourier acceleration spectrum: (**a**) Shanghai; (**b**) Kobe; (**c**) Ei-Centro.

As well as that given in Equation (13) (termed as "Shape Function One"), another two typical commonly used shape functions shown in Figure 11 are also collected for calibrating the design response spectrum for Yangshan Port. Shape Function Two was provided by Deoda and Adhikary (2020) [66], where T_D is defined as the value defining the beginning of the constant displacement response range of the spectrum. Shape Function Three is employed by both NZS 1170.5 [67] and Eurocode 8 [51].

The gray lines in Figure 11 are the ground response spectrums at different intensities with consideration of soil anisotropy. The upper and lower bounds of the gray area are the 16th to 84th percentiles, respectively, showing the region with a probability range of 16% to 84% for the occurrence of seismic events. The red line represents the average of the response spectrum curve. Utilizing Shape Function One, the 5% damped design response spectrum for Yangshan Port is calibrated and represented by the blue line. The design response spectrums based on Shape Function Two and Three are depicted as the magenta and orange lines, respectively.

For Shape Function One suggested by GB 51247-2018, the plateau value (β) is 2.97, and the attenuation index (χ) is 0.5. The Pearson correlation coefficient (r value) is 0.63. Additionally, utilizing Shape Function Two results in a plateau value (β) of 3.00 and an attenuation index (χ) of 0.5. This yields a Pearson correlation coefficient (r value) of 0.57. Similarly, when employing Shape Function Three, the plateau value (β) is 3.00, while the attenuation index (χ) is 0.5, with a significantly higher Pearson correlation coefficient (r value) of 0.71. Three shape functions provide a similar plateau value (β), but Shape Function Three exhibits a better overall fitting performance.

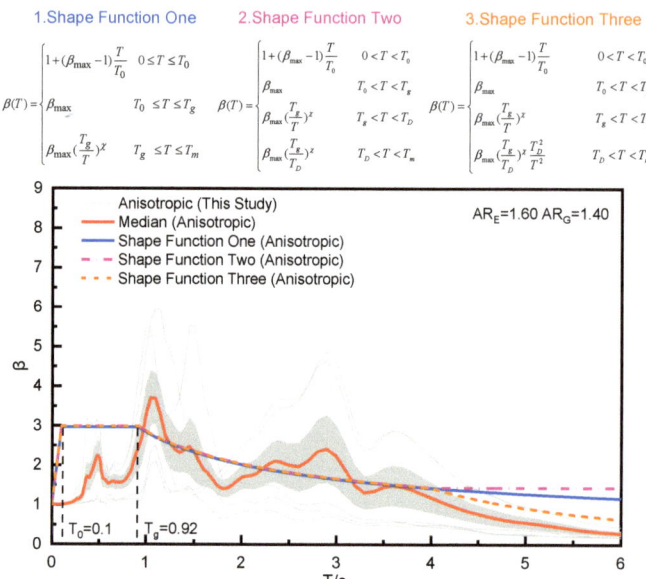

Figure 11. The 5% damped design response spectrum for Yangshan Port ($AR_E = 1.60$, $AR_G = 1.40$).

To further investigate the influence of anisotropy on ground response spectra, a series of cross-parameter studies were conducted focusing on anisotropy-related parameters (elastic modulus ratio $AR_E = 1.60, 2.00, 2.40$; shear modulus ratio $AR_G = 1.10, 1.25, 1.40$), resulting in a total of nine combinations. Following the method illustrated in Figure 11, Figure 12 presents nine average response spectrum curves. Utilizing the mean values of these spectrums, the root mean square error (RMSE) across the entire period range for the nine response spectrum curves was calculated and is depicted by the gray line in Figure 12. It is observed that the maximum RMSE among the nine response spectrum curves reaches 34.5% at lower frequencies.

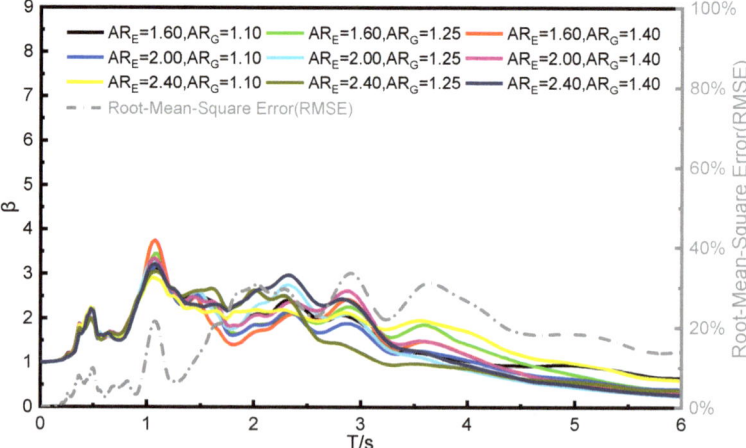

Figure 12. The 5% damped design response spectrum for Yangshan Port (different anisotropic parameters).

Finally, the proposed 5% damped standard design response spectrum for Yangshan Port has been developed from the previously described nine combinations of anisotropic parameters and five seismic intensity levels. The design response spectrum for the anisotropic site is refined using Shape Function Three, which is depicted as an orange line in Figure 13. This calibration involves a plateau value (β) of 3.00 and an attenuation index (χ) of 0.50. When compared to the isotropic spectrum, the anisotropic spectrum demonstrates a notably elevated design response spectrum. Factoring in anisotropy has led to an increase of 18% in the plateau value (β) for the standard design response spectrum. Consequently, utilizing the isotropic design response spectrum for the seismic design of structures at Shanghai Yangshan Port may result in dangerous results.

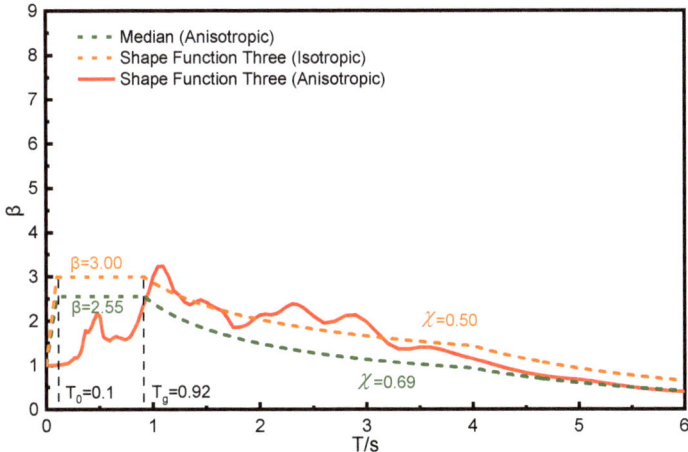

Figure 13. Proposed 5% damped standard design response spectrum for Yangshan Port.

Figure 14 illustrates the application of the response spectrum for a single-degree-of-freedom frame model. When utilizing the response spectrum, two key aspects must be contemplated. Firstly, the fundamental frequency ($f = 2\pi\sqrt{m/k}$) is derived from the concentration of mass and lateral stiffness. Secondly, the design base shear is computed through $F = m\beta \times PGA$.

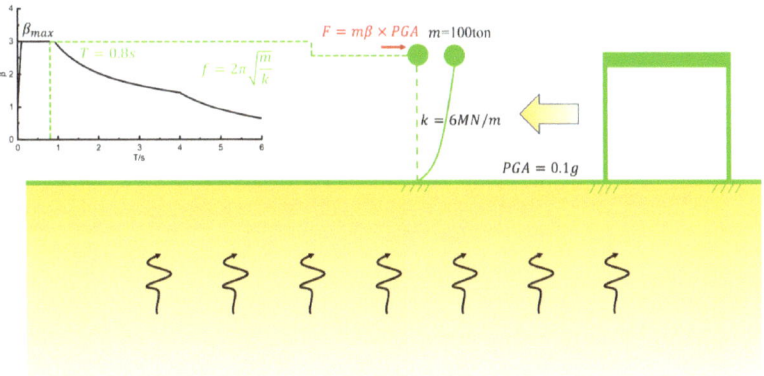

Figure 14. Application of response spectrum for single-degree-of-freedom systems.

For instance, let us consider a single-story frame with a total seismic weight of 980 kN and a total column lateral stiffness k of 6 MN/m. The structure has a damping ratio of 5%.

The basic acceleration (PGA) by design is 0.1 g. For the conventional response spectrum, the design base shear amounts to 250 kN. However, when the response spectrum is applied with consideration of anisotropy, the design base shear amounts to 294 kN, marking an increase of 18%.

5. Conclusions

This study establishes an equivalent linear ground motion model for anisotropic sites to analyze the ground motion response characteristics for layered soils with various anisotropy parameters. Further investigations were conducted to study the ground motion response features of anisotropic sites when subjected to different seismic wave inputs. It was found that the anisotropy ratios for the shear modulus (AR_G) have a promoting effect on the peak ground acceleration (PGA), while the anisotropic ratio of undrained Young's modulus (AR_E) has an inhibitory effect on peak ground acceleration (PGA) in the site earthquake response problem. The impact of anisotropy on the ground motion of the site becomes more significant when seismic waves have dominant frequencies closer to the fundamental frequencies of the sites. In the ground motion response problem, anisotropy cannot be ignored when the dominant frequencies of the seismic waves are closer to the fundamental frequencies of the sites.

This study further takes the Yangshan Port site as an example to calibrate the seismic motion parameters ($G - \gamma$ and $D - \gamma$) of the Yangshan Port soil layer based on cyclic triaxial tests and resonant column tests. The results show that the combination of the resonant column test and cyclic triaxial test can better describe the dynamic characteristics of soil from small strains ($10^{-6} \sim 10^{-3}$) to large strains ($10^{-3} \sim 10^{-1}$). Based on the calibrated dynamic parameters, the ground motion characteristics of the anisotropic site of Yangshan Port were studied. It was found that the consideration of site anisotropy leads to significant increases in both the peak ground acceleration (PGA) and the peak ground Fourier acceleration spectrum for the Yangshan Port site. Anisotropy has little influence on the frequency position of the ground Fourier acceleration spectrum peak. The frequency region where the amplitude of the ground Fourier acceleration spectrum increases significantly is close to the site fundamental frequency (1 Hz). This demonstrates that site anisotropy can potentially lead to a severe underestimation of the acceleration response in regions proximate to the site's fundamental frequency.

A series of cross-parameter studies regarding anisotropic parameters were also conducted to investigate the influence of anisotropy on ground response spectra. When taking site anisotropy into account, the calibrated design response spectrum plateau value (β) is 3.00. Conversely, when anisotropy is not considered, the calibrated design response spectrum plateau value (β) stands at 2.55. The design response spectrum calibrated for an anisotropic site surpasses that for an isotropic site. Thus, when performing the seismic design of buildings, utilizing the design response spectrum derived from isotropic site ground motion response calibration might present higher risks.

Author Contributions: Conceptualization, Z.C. and J.Y.; Methodology, Y.X. and J.Y.; Software, Y.X.; Resources, Z.C.; Writing – original draft, Y.X.; Writing – review & editing, J.Y.; Visualization, Y.X.; Supervision, Z.C. and J.Y.; Funding acquisition, Z.C. and J.Y. All authors have read and agreed to the published version of the manuscript.

Funding: This work was supported by the National Key Research and Development Program of China (No. 2021YFB2600700) and the Natural Science Foundation of Shanghai (No. 23ZR1468500).

Data Availability Statement: Data are availableble upon request.

Conflicts of Interest: The authors declare no conflict of interest.

References

1. Tong, X.; Hong, Z.; Liu, S.; Zhang, X.; Xie, H.; Li, Z.; Bao, F. Building-damage detection using pre-and post-seismic high-resolution satellite stereo imagery: A case study of the May 2008 Wenchuan earthquake. *ISPRS J. Photogramm. Remote Sens.* **2012**, *68*, 13–27. [CrossRef]
2. Zhang, M.; Jin, Y. Building damage in Dujiangyan during Wenchuan earthquake. *Earthq. Eng. Eng. Vib.* **2008**, *7*, 263–269. [CrossRef]
3. An, X.; Shawky, A.A.; Maekawa, K. The collapse mechanism of a subway station during the Great Hanshin earthquake. *Cem. Concr. Compos.* **1997**, *19*, 241–257. [CrossRef]
4. Kelam, A.A.; Karimzadeh, S.; Yousefibavil, K.; Akgün, H.; Askan, A.; Erberik, M.A.; Ciftci, H. An evaluation of seismic hazard and potential damage in Gaziantep, Turkey using site specific models for sources, velocity structure and building stock. *Soil Dyn. Earthq. Eng.* **2022**, *154*, 107129. [CrossRef]
5. Polat, O.; Çeken, U. Investigation of the local soil effects at the new strong-motion array (MATNet) in Hatay-K. Maras Region, Turkey. In Proceedings of the AGU Fall Meeting, San Francisco, CA, USA, 15–19 December 2014. abstract S13F–07.
6. Wardman, J.; Officer, C.C. The 2023 Kahramanmaraş Earthquake Sequence: A Counterfactual Contemplation—On 6 February 2023 a Magnitude Mw7. 8 and Mw7. 6 Earthquake Doublet Occurred in the East Anatolian Fault Zone Resulting in Catastrophic Destruction and Loss of Life in Southeast-Central Turkey and Northern Syria. Knowing What We Know Now, How Might the Events Have Played Out Differently? Available online: https://www.maxinfo.io/blog/the-2023-kahramanmaras-earthquake-sequence-a-counterfactual-perspective (accessed on 1 December 2023).
7. Askan, A.; Gülerce, Z.; Roumelioti, Z.; Sotiriadis, D.; Melis, N.S.; Altindal, A.; Margaris, B. The Samos Island (Aegean Sea) M7. 0 earthquake: Analysis and engineering implications of strong motion data. *Bull. Earthq. Eng.* **2022**, *20*, 7737–7762. [CrossRef]
8. Daraei, A.; Hama Ali, H.F.; Qader, D.N.; Zare, S. Seismic retrofitting of rubble masonry tunnel: Evaluation of steel fiber shotcrete or inner concrete lining alternatives. *Arab. J. Geosci.* **2022**, *15*, 1074. [CrossRef]
9. Preciado, A.; Santos, J.C.; Silva, C.; Ramírez-Gaytán, A.; Falcon, J.M. Seismic damage and retrofitting identification in unreinforced masonry Churches and bell towers by the september 19, 2017 (Mw = 7.1) Puebla-Morelos earthquake. *Eng. Fail. Anal.* **2020**, *118*, 104924. [CrossRef]
10. Aaqib, M.; Park, D.; Adeel, M.B.; Hashash, Y.M.; Ilhan, O. Simulation-based site amplification model for shallow bedrock sites in Korea. *Earthq. Spectra* **2021**, *37*, 1900–1930. [CrossRef]
11. Adeel, M.B.; Nizamani, Z.A.; Aaqib, M.; Khan, S.; Rehman, J.U.; Bhusal, B.; Park, D. Estimation of V S30 using shallow depth time-averaged shear wave velocity of Rawalpindi–Islamabad, Pakistan. *Geomat. Nat. Hazards Risk* **2023**, *14*, 1–21. [CrossRef]
12. Bhusal, B.; Aaqib, M.; Paudel, S.; Parajuli, H.R. Site specific seismic hazard analysis of monumental site Dharahara, Kathmandu, Nepal. *Geomat. Nat. Hazards Risk* **2022**, *13*, 2674–2696. [CrossRef]
13. Joyner, M.D.; Gardner, C.; Puentes, B.; Sasani, M. Resilience-Based seismic design of buildings through multiobjective optimization. *Eng. Struct.* **2021**, *246*, 113024. [CrossRef]
14. Morandi, P.; Butenweg, C.; Breis, K.; Beyer, K.; Magenes, G. Latest findings on the behaviour factor q for the seismic design of URM buildings. *Bull. Earthq. Eng.* **2022**, *20*, 5797–5848. [CrossRef]
15. Leyva, H.; Bojórquez, J.; Bojórquez, E.; Reyes-Salazar, A.; Carrillo, J.; López-Almansa, F. Multi-objective seismic design of BRBs–reinforced concrete buildings using genetic algorithms. *Struct. Multidiscip. Optim.* **2021**, *64*, 2097–2112. [CrossRef]
16. Jiang, J.; El Nggar, H.M.; Xu, C.; Zhong, Z.; Du, X. Effect of ground motion characteristics on seismic fragility of subway station. *Soil Dyn. Earthq. Eng.* **2021**, *143*, 106618. [CrossRef]
17. Mo, T.; Wu, Q.; Li, D.Q.; Du, W. Influence of ground motion characteristics (velocity pulse and duration) on the pile responses in liquefiable soil deposits. *Soil Dyn. Earthq. Eng.* **2022**, *159*, 107330. [CrossRef]
18. Zhang, X.; Tang, L.; Ling, X.; Chan, A.H.C.; Lu, J. Using peak ground velocity to characterize the response of soil-pile system in liquefying ground. *Eng. Geol.* **2018**, *240*, 62–73. [CrossRef]
19. Wang, Z.Z.; Zhang, J.; Huang, H. Interpreting random fields through the U-Net architecture for failure mechanism and deformation predictions of geosystems. *Geosci. Front.* **2024**, *15*, 101720. [CrossRef]
20. Dammala, P.K.; Kumar, S.S.; Krishna, A.M.; Bhattacharya, S. Dynamic soil properties and liquefaction potential of northeast Indian soil for non-linear effective stress analysis. *Bull. Earthq. Eng.* **2019**, *17*, 2899–2933. [CrossRef]
21. Markham, C.S.; Bray, J.D.; Macedo, J.; Luque, R. Evaluating nonlinear effective stress site response analyses using records from the Canterbury earthquake sequence. *Soil Dyn. Earthq. Eng.* **2016**, *82*, 84–98. [CrossRef]
22. Chen, G.; Wang, Y.; Zhao, D.; Zhao, K.; Yang, J. A new effective stress method for nonlinear site response analyses. *Earthq. Eng. Struct. Dyn.* **2021**, *50*, 1595–1611. [CrossRef]
23. Li, Z.; Liu, H. An isotropic-kinematic hardening model for cyclic shakedown and ratcheting of sand. *Soil Dyn. Earthq. Eng.* **2020**, *138*, 106329. [CrossRef]
24. Elia, G.; Rouainia, M. Investigating the cyclic behaviour of clays using a kinematic hardening soil model. *Soil Dyn. Earthq. Eng.* **2016**, *88*, 399–411. [CrossRef]
25. Moghadam, S.I.; Taheri, E.; Ahmadi, M.; Ghoreishian Amiri, S.A. Unified bounding surface model for monotonic and cyclic behaviour of clay and sand. *Acta Geotech.* **2022**, *17*, 4359–4375. [CrossRef]
26. Cheng, X.; Du, X.; Lu, D.; Ma, C.; Wang, P. A simple single bounding surface model for undrained cyclic behaviours of saturated clays and its numerical implementation. *Soil Dyn. Earthq. Eng.* **2020**, *139*, 106389. [CrossRef]

27. Shi, Z.; Huang, M. Intergranular-strain elastic model for recent stress history effects on clay. *Comput. Geotech.* **2020**, *118*, 103316. [CrossRef]
28. Huang, X.W.; Guo, J.; Li, K.Q.; Wang, Z.Z.; Wang, W. Predicting the thermal conductivity of unsaturated soils considering wetting behavior: A meso-scale study. *Int. J. Heat Mass Transf.* **2023**, *204*, 123853. [CrossRef]
29. Elbadawy, M.A.; Zhou, Y.G.; Liu, K. A modified pressure dependent multi-yield surface model for simulation of LEAP-Asia-2019 centrifuge experiments. *Soil Dyn. Earthq. Eng.* **2022**, *154*, 107135. [CrossRef]
30. Mroz, Z. On the description of anisotropic workhardening. *J. Mech. Phys. Solids* **1967**, *15*, 163–175. [CrossRef]
31. Hardin, B.O.; Drnevich, V.P. Shear modulus and damping in soils: Measurement and parameter effects (terzaghi leture). *J. Soil Mech. Found. Div.* **1972**, *98*, 603–624. [CrossRef]
32. Hardin, B.O.; Drnevich, V.P. Shear modulus and damping in soils: Design equations and curves. *J. Soil Mech. Found. Div.* **1972**, *98*, 667–692. [CrossRef]
33. Schnabel, P.B. *SHAKE, a Computer Program for Earthquake Response Analysis of Horizontally Layered Sites*; Report No. EERC 72-12; University of California: Berkeley, CA, USA, 1972.
34. Wang, J.P.; Xu, Y. A non-stationary earthquake probability assessment with the Mohr–Coulomb failure criterion. *Nat. Hazards Earth Syst. Sci.* **2015**, *15*, 2401–2412. [CrossRef]
35. Luo, C.; Yang, X.; Zhan, C.; Jin, X.; Ding, Z. Nonlinear 3D finite element analysis of soil–pile–structure interaction system subjected to horizontal earthquake excitation. *Soil Dyn. Earthq. Eng.* **2016**, *84*, 145–156. [CrossRef]
36. Idriss, I.M.; Sun, J.I. A Computer program for conducting equivalent linear seismic response analysis of horizontally layered soil deposits. In *Users Manual for SHAKE91*; University of California, Berkeley: Berkeley, CA, USA, 1992.
37. Zhang, R.; Wu, C.; Goh, A.T.; Böhlke, T.; Zhang, W. Estimation of diaphragm wall deflections for deep braced excavation in anisotropic clays using ensemble learning. *Geosci. Front.* **2021**, *12*, 365–373. [CrossRef]
38. Zhang, R.; Li, Y.; Goh, A.T.; Zhang, W.; Chen, Z. Analysis of ground surface settlement in anisotropic clays using extreme gradient boosting and random forest regression models. *J. Rock Mech. Geotech. Eng.* **2021**, *13*, 1478–1484. [CrossRef]
39. Teng, F. A Simplified Expression for Ground Movements Induced by Excavations in Soft Clay. In Proceedings of the 2nd International Symposium on Asia Urban GeoEngineering, Changsha, China, 24–27 November 2017; Springer: Singapore, 2018; pp. 93–115.
40. Wei, F.; Wang, H.; Zeng, G.; Jiang, M. Seepage flow around twin circular tunnels in anisotropic ground revealed by an analytical solution. *Undergr. Space* **2023**, *10*, 1–14. [CrossRef]
41. Soe, T.E.E.; Ukritchon, B. Three-dimensional undrained face stability of circular tunnels in non-homogeneous and anisotropic clays. *Comput. Geotech.* **2023**, *159*, 105422. [CrossRef]
42. Zhang, W.; Li, Y.; Wu, C.; Li, H.; Goh, A.T.C.; Liu, H. Prediction of lining response for twin tunnels constructed in anisotropic clay using machine learning techniques. *Undergr. Space* **2022**, *7*, 122–133. [CrossRef]
43. Perić, D.; Tran, T.V.; Miletić, M. Effects of soil anisotropy on a soil structure interaction in a heat exchanger pile. *Comput. Geotech.* **2017**, *86*, 193–202. [CrossRef]
44. Ai, Z.Y.; Ye, J.M.; Zhao, Y.Z. The performance analysis of energy piles in cross-anisotropic soils. *Energy* **2022**, *255*, 124549. [CrossRef]
45. Ai, Z.Y.; Ye, J.M. Thermo-mechanical analysis of pipe energy piles in layered cross-isotropic soils. *Energy* **2023**, *277*, 127757. [CrossRef]
46. Bentil, O.T.; Zhou, C.; Zhang, J.; Liu, K.K. Cross-anisotropic stiffness characteristics of a compacted lateritic clay from very small to large strains. *Can. Geotech. J.* **2023**. [CrossRef]
47. Stokoe, K.H.; Darendeli, M.B.; Andrus, R.D.; Brown, L.T. Dynamic soil properties: Laboratory, field and correlation studies. In *Earthquake Geotechnical Engineering*; CRC Press: Boca Raton, FL, USA, 1999; pp. 811–845.
48. Zhang, M.; Liao, W.M.; Wang, Z.J. Statistical analysis of the relationship between dynamic shear modulus ratio and damping ratio and shear strain of cohesive soil. *Earthq. Eng. Eng. Vib.* **2013**, *33*, 256–262.
49. Nishimura, S. Cross-anisotropic deformation characteristics of natural sedimentary clays. *Géotechnique* **2014**, *64*, 981–996. [CrossRef]
50. Teng, F.C.; Ou, C.Y.; Hsieh, P.G. Measurements and numerical simulations of inherent stiffness anisotropy in soft Taipei clay. *J. Geotech. Geoenviron. Eng.* **2014**, *140*, 237–250. [CrossRef]
51. EN 1998-5; Eurocode 8: Design of Structures for Earthquake. CEN (European Committee for Standardization): Brussels, Belgium, 2004.
52. Seed, H.B. *Soil Moduli and Damping Factors for Dynamic Response Analyses*; Report EERC 70-10; University of California: Berkeley, CA, USA, 1970.
53. Forte, G.; Chioccarelli, E.; De Falco, M.; Cito, P.; Santo, A.; Iervolino, I. Seismic soil classification of Italy based on surface geology and shear-wave velocity measurements. *Soil Dyn. Earthq. Eng.* **2019**, *122*, 79–93. [CrossRef]
54. GB 50011-2010; Code for Seismic Design of Buildings. China Architecture Building Press: Beijing, China, 2010. (In Chinese)
56. Zhang, J.Z.; Huang, H.W.; Zhang, D.M.; Phoon, K.K. Experimental study of the coupling effect on segmental shield tunnel lining under surcharge loading and excavation unloading. *Tunn. Undergr. Space Technol.* **2023**, *140*, 105199. [CrossRef]

57. Hou, Y.M.; Wang, J.H.; Zhang, L.L. Finite-element modeling of a complex deep excavation in Shanghai. *Acta Geotech.* **2009**, *4*, 7–16. [CrossRef]
58. Li, Q.; Ng, C.W.W.; Liu, G.B. Determination of small-strain stiffness of Shanghai clay on prismatic soil specimen. *Can. Geotech. J.* **2012**, *49*, 986–993. [CrossRef]
59. Ng, C.W.W.; Li, Q.; Liu, G.B. Experimental study on the measurement of anisotropic shear modulus of undisturbed Shanghai soft clay using bending elements. *Chin. J. Geotech. Eng.* **2013**, *35*, 150–156.
60. Yang, J.; Liu, X. Shear wave velocity and stiffness of sand: The role of non-plastic fines. *Géotechnique* **2016**, *66*, 500–514. [CrossRef]
61. Gu, X.; Yang, J.; Huang, M. Laboratory measurements of small strain properties of dry sands by bender element. *Soils Found.* **2013**, *53*, 735–745. [CrossRef]
62. Gu, X.Q.; Yang, J.; Huang, M.S.; Gao, G.Y. Bending element, resonant column and cyclic torsional shear tests for the determination of shear modulus of sand. *Chin. J. Geotech. Eng.* **2016**, *38*, 740–746.
63. Youn, J.U.; Choo, Y.W.; Kim, D.S. Measurement of small-strain shear modulus G max of dry and saturated sands by bender element, resonant column, and torsional shear tests. *Can. Geotech. J.* **2008**, *45*, 1426–1438. [CrossRef]
64. Andreotti, G.; Famà, A.; Lai, C.G. Hazard-dependent soil factors for site-specific elastic acceleration response spectra of Italian and European seismic building codes. *Bull. Earthq. Eng.* **2018**, *16*, 5769–5800. [CrossRef]
65. *GB 51247-2018*; Standard for Seismic Design of Hydraulic Structures. China Architecture Building Press: Beijing, China, 2018. (In Chinese)
66. Deoda, V.R.; Adhikary, S. A preliminary proposal towards the revision of Indian seismic code considering site classification scheme, amplification factors and response spectra. *Bull. Earthq. Eng.* **2020**, *18*, 2843–2889. [CrossRef]
67. *NZS 1170.5*; Structural Design Actions, Part 5: Earthquake Actions. Standards New Zealand: Wellington, New Zealand, 2004.

Disclaimer/Publisher's Note: The statements, opinions and data contained in all publications are solely those of the individual author(s) and contributor(s) and not of MDPI and/or the editor(s). MDPI and/or the editor(s) disclaim responsibility for any injury to people or property resulting from any ideas, methods, instructions or products referred to in the content.

Article

Dynamic Response of Transmission Tower-Line Systems Due to Ground Vibration Caused by High-Speed Trains

Guifeng Zhao, Meng Wang, Ying Liu and Meng Zhang *

School of Civil Engineering, Zhengzhou University, Zhengzhou 450001, China; gfzhao@zzu.edu.cn (G.Z.)
* Correspondence: zhangmeng@zzu.edu.cn

Abstract: With the continuous expansion of the scale of power grid and transportation infrastructure construction, the number of crossovers between transmission lines and high-speed railways continues to increase. At present, there is a lack of systematic research on the dynamic characteristics of transmission tower-line structures crossing high-speed railways under vehicle-induced ground vibration. This article focuses on the phenomenon of accidents such as line drops when crossing areas in recent years and establishes a high-speed train track foundation soil finite element model in ABAQUS that considers track irregularity. The three-dimensional vibration characteristics and attenuation law of train ground vibration are analyzed. Acceleration data for key points are also extracted. A separate finite element model of the transmission tower-line system is established in ANSYS, where acceleration is applied as an excitation to the transmission tower-line system, and the coupling effect between the tower and the line is considered to analyze its dynamic response. Subsequently, modal analysis is conducted on the tower-line system, providing the vibration modes and natural frequencies of the transmission tower-line structure. The effects of factors such as train speed, soil quality, and distance from the tower to the track on the dynamic response of the transmission tower-line system under vehicle-induced ground vibration are studied. The results show that the speed range (300 km/h–400 km/h) and track distance range (4.5 m–30 m) with the greatest impacts are obtained. The research results can provide a reference for the reasonable design of transmission tower-line systems in high-speed railway sections.

Keywords: high-speed train; ground vibration; transmission tower-line structure; numerical simulation; dynamic response

Citation: Zhao, G.; Wang, M.; Liu, Y.; Zhang, M. Dynamic Response of Transmission Tower-Line Systems Due to Ground Vibration Caused by High-Speed Trains. *Buildings* **2023**, *13*, 2884. https://doi.org/10.3390/buildings13112884

Academic Editors: Carmelo Gentile and Fabio Rizzo

Received: 4 August 2023
Revised: 13 November 2023
Accepted: 16 November 2023
Published: 18 November 2023

Copyright: © 2023 by the authors. Licensee MDPI, Basel, Switzerland. This article is an open access article distributed under the terms and conditions of the Creative Commons Attribution (CC BY) license (https://creativecommons.org/licenses/by/4.0/).

1. Introduction

Power grid systems and railway systems are important material foundations of modern society and an important part of lifeline engineering systems. In the past 20 years, high-speed railways have developed rapidly in many countries due to their advantages of safety, efficiency, low energy consumption, and large transportation capacity. At the same time, high-voltage overhead transmission, as the main mode of power supply in countries around the world, has also developed significantly in the past few decades. With the continuous increase in social electricity demand and transportation demand, the power grid and transportation infrastructure are constantly being upgraded and constructed, which brings about the inevitable problem of crossings of transmission tower-line systems and high-speed railways (Figure 1). In the crossover area of the two, once an accident, such as disconnection, string drop, or even tower collapse, occurs, it may cause a large-scale power supply interruption. Therefore, ensuring the long-term safe operation of the power grid system and the railway system across both sections is a matter of great concern to both the power sector and the railway sector.

Figure 1. Cases of overhead transmission lines crossing high-speed railways.

With the increase in high-speed railway construction mileage and the improvement of high-speed train operating speed [1], the environmental vibration problem caused by high-speed trains has become increasingly prominent. At present, many scholars have conducted relatively comprehensive research on the generation mechanism of ground vibration caused by high-speed trains and the law of vibration propagation. For the generation mechanism of vehicle-induced ground vibration, the research mainly focuses on three aspects: theoretical analysis models, field tests, and finite element numerical simulations. Typical theoretical analysis models include the Winkler foundation beam theory [2], the Timoshenko elastic foundation beam model [3], and the basic model of wheel–rail interaction considering unsprung mass and track stiffness [4]. Because the ground vibration caused by trains usually propagates near the surface [5], research mainly focuses on the propagation law of ground vibration, including the simulation of roadbeds and ground [6], the simulation of track irregularity [7], and the theoretical analysis of large coupling vibration problems [8]. In recent years, researchers have studied the problems of vehicle-induced ground vibration using finite element numerical simulation and field test methods. For example, Xia et al. [9] established a comprehensive model considering train-track-foundation dynamic interactions. Factors such as the quasistatic axle load and dynamic excitation between the wheel and rail are analyzed. The results show that the ground vibration characteristics are closely related to the train speed and soil characteristics. With increasing track distance, the ground acceleration tends to decrease. Erkal et al. [10] measured triaxial vibrations of road and rail traffic on and around a typical residential masonry building in Istanbul and its response to adjacent ground-born vibrations through numerical modeling. The results show that train-induced vibrations caused the walls of the building to experience tensile stresses up to 23% of the masonry tensile strength. Motazedian et al. [11] found that the durations and amplitudes of the train-induced seismic waves at soil sites increased dramatically compared to those at the reference bedrock site. On the other hand, very large soil amplifications have been observed based on local earthquake recordings, with a very different source mechanism than train-induced seismic waves. Niu et al. [12] studied the ground vibration caused by the operation of the Datong–Xi'an high-speed railway through field tests. The results show that the ground vibration caused by a high-speed train is a periodic excitation, and the vertical vibration acceleration of the ground decreases with increasing distance from the vibration source.

Because the vibration caused by trains will be transmitted to the surrounding soil layer through the track, roadbed, etc., and then cause secondary vibration of the adjacent structures, some scholars have also conducted studies on the dynamic response of such structures under the action of high-speed trains. Chen et al. [13] experimentally studied the site dynamic response of a bridge and its surrounding environment on the Wuhan–Guangzhou high-speed railway. The results show that the vertical vibration acceleration of the bridge is generally between 0.07 and 0.25 m/s^2. With increasing train speed, the ground vibration gradually increases. Hesami et al. [14] used a two-dimensional finite element method to analyze the influence of train vibration on residential buildings near the

Qaemshahr railway. The train–ground dynamic model is preliminarily verified by the measured data. The results show that the vibration level decreases significantly with increasing distance from the track centerline to the building. Zhou et al. [15] sampled vibration data from the proposed site near the railway, and the measured ground acceleration was used as the excitation for the proposed building. The law of some trains' impact on the vibration of nearby buildings was obtained. Erkan et al. [16] studied the ground vibration caused by high-speed trains and its impact on surrounding residential areas through a large amount of field work and many field measurements in Türkiye. The above studies mainly focus on the impact of ground vibration caused by high-speed trains on adjacent high-rise buildings, bridge structures, and residential areas.

In the study of train vibration, one of the very important factors is the excitation source target interaction system. When an excitation source (such as mechanical vibration or vehicle dynamic load) acts on the ground, the soil will generate and transmit excitation energy and interact with the target structure. Conversely, the response of the structure, such as vibration and dynamic forces, will also be transmitted along the soil, affecting the source of motivation. One important method is to use transfer functions to analyze the impact of excitation sources on ground motion [17] and, finally, the impact on the analyzed object. Due to the high stability requirements of the large Hadron collider (HL-LHC) for the orbit, Schaumann et al. [18] conducted some research to characterize the actual ground motion in the large Hadron collider tunnel and summarized the observations made on the LHC beams. Farahani et al. [19] developed a numerical model based on the modal analysis results of buildings to address the impact of vibrations caused by trains on residents in the vicinity of railway lines. The double confirmation analysis method is used to identify the modal parameters of buildings: obtain the transfer function through the dynamic response of modal analysis; reproduce the vibration acceleration of different floors of the building from on-site measurement records; and apply it to the building foundation.

Considering the large-span and high-rise structural characteristics of high-voltage overhead transmission tower-line systems, wind load is usually the dominant load. A large number of studies have been conducted in such areas, including the design wind loads [20], wind-induced vibration [21–23], and galloping [24,25] of transmission tower-line systems, which provide effective technical support for the rational design and maintenance of high-voltage transmission tower-line systems. In fact, the transmission tower-line system across the high-speed railway will also be affected by the environmental vibration caused by high-speed trains. Taking China as an example, Feng et al. [26] reported that with the continuous development of power transmission capacity and railway transportation capacity, the proportion of the crossing of transmission tower-line systems and high-speed railways will continue to increase. Yin et al. [27] numerically studied the dynamic response of transmission lines under the action of high-speed trains and verified it by field tests. Zhang et al. [28] and Liu [29] analyzed the transient force and typical dynamic response of an overhead transmission tower-line structure under the action of high-speed train-induced wind. The results show that when the train passes through the overhead transmission tower-line structure, the ultimate force of the transmission line has a significant quadratic function relationship with the train speed.

In summary, the phenomenon of crossings between overhead transmission tower-line systems and high-speed railways will continue to increase in the future, but there is currently a lack of systematic research on the dynamic response of transmission tower-line structures across high-speed railways under vehicle-induced ground vibration. To this end, a high-speed train track foundation soil finite element model is established in ABAQUS that considers track irregularity. The three-dimensional vibration characteristics and attenuation law of train ground vibration are analyzed. A separate finite element model of the transmission tower-line system was established in ANSYS. Factors such as different soil qualities, different train speeds, and different distances to the track are discussed. This ... maintenance of transmission tower-line systems across high-speed railways.

2. Establishment and Verification of the Finite Element Model

To study the effect of ground vibration caused by high-speed trains on the transmission tower-line system, the acceleration time history of the ground surface during the process of the high-speed train passing through the transmission tower-line system was obtained by a numerical simulation method, and then it was used to perform a dynamic analysis of the transmission tower-line system.

2.1. Establishment and Verification of the Train-Track-Foundation-Soil Coupling Model

This section mainly focuses on the theoretical methods and parameters related to the establishment of the finite element model of the train-track-foundation soil. Because ABAQUS 2021 can better simulate the nonlinear contact problem between wheels and rails and the explicit integration algorithm in ABAQUS can solve highly nonlinear quasistatic problems, complex contact problems, and high-speed dynamic loads, the explicit dynamic integration method (ABAQUS/Explicit) is used in this study. A schematic diagram of the model is shown in Figure 2.

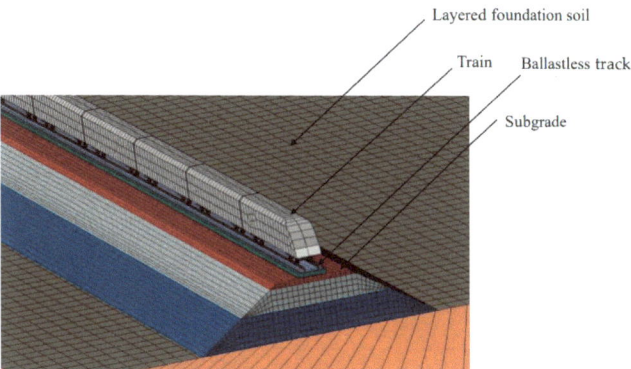

Figure 2. Train-track-foundation-soil coupling model.

2.1.1. Model of the Train

The motion of the vehicle system in the vertical longitudinal plane can be considered a multirigid body system, and the vehicle body model with a secondary spring mass is used. The following assumptions are used for the model [2]. Only the wheel–rail vertical dynamic effect is considered in the model. The car body, bogie, and wheelset are regarded as rigid bodies, and the influence of the deformation of these components on the overall model is not considered. The wheelset and the bogie are connected by a series of springs and damping elements; the connection between the bogie and the car body is composed of a second series of springs and damping elements. The masses of the car body, bogie, and wheelset are simplified as centralized mass considerations. The single vehicle part of the overall model has 16 degrees of freedom, such as the ups and downs of the car body and the three-way nodding, the ups and downs of the front and rear bogies and the three-way nodding, and the ups and downs of the four-wheel sets.

Taking an ICE3-type train [30] as an example, a finite element model of train-track-foundation-soil is established to simulate the propagation and attenuation of vibration waves generated by wheel–rail action in the soil layer. The driving distance is 444 m. The speed of the train selected in this section is 250 km/h. The schematic diagram is shown in Figure 3. In this figure, M_c and J_c are the mass and moment of inertia of the car body, respectively; M_t and J_t are the mass and moment of inertia of the bogie, respectively; M_ω is the quality of the wheelset; K_{s1} and C_{s1} are the primary suspension mass and damping, respectively; and K_{s2} and C_{s2} are the suspension mass and damping of the secondary series, respectively. The above parameters and other geometric parameters are listed in Table 1.

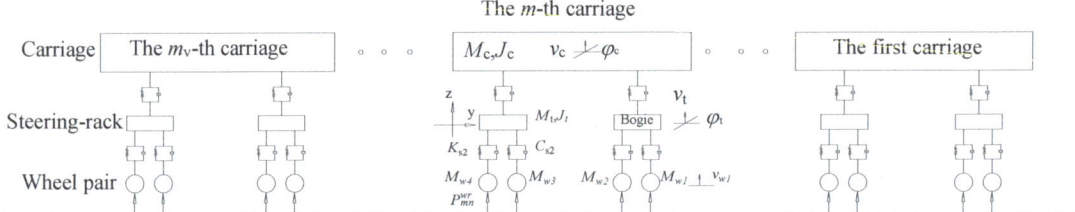

Figure 3. Secondary suspension vehicle model and parameters.

Table 1. Mechanical and geometric parameters of train vehicles.

Vehicle Parameters	Size
M_c (kg)	47,900
J_c (kg/m^2)	Lateral is 8.224×10^6 Vertical is 8.232×10^6 Longitudinal is 2.751×10^5
M_t (kg)	1381
J_t (kg/m^2)	Lateral is 1695 Vertical is 2844 Longitudinal is 1378
M_ω (kg)	1400
K_{s1} (N/m)	1.87×10^6
C_{s1} (N·s/m)	5×10^5
K_{s2} (N/m)	1.72×10^5
C_{s2} (N·s/m)	1.92×10^5
Tire size (m)	0.46
Distance from coupler to coupler (m)	2.50
Wheel base $2d$ (m)	2.50

2.1.2. Train-Track Model

The CRTS I-type ballastless track is adopted to build the train-track model. The track components include steel rails, sleepers, elastic fasteners, track slabs, CAE mortar filling layers, and concrete bases. In this paper, a simplified track model is used, and its cross-section is shown in Figure 4. The gauge of the two rails is 1.435 m, and the track fastener spacing is 0.6 m. The parameters of the sleeper, track slab, CAE mortar, and concrete base are shown in Table 2.

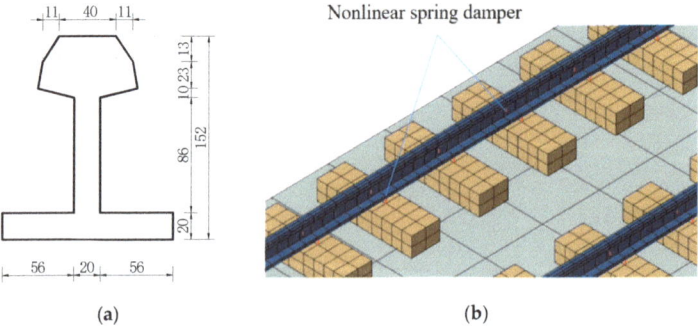

(a) (b)

Figure 4. Details of the track section. (**a**) Simplified track cross-section size diagram (mm). (**b**) Finite element model of ballastless track.

Table 2. Parameters of the track structure.

Structure Layer Name	Width (m)	Depth (m)
Track board	2.4	0.20
CA mortar bed	2.4	0.05
Concrete base	3	0.30
Rail bearing	0.25	0.16

To prevent the occurrence of an hourglass phenomenon where the unit lacks stiffness and cannot resist deformation, each unit adopts an hourglass control. The actual connection between the rail and the sleeper is a fastener, which is used to limit the vertical displacement of the rail through tension. Therefore, in ABAQUS, the fastener is simulated with a nonlinear spring-damper element that can only be tensioned, as shown in Figure 4b. The actual track board is connected by many standard-length track boards, but considering the strong longitudinal connection between the track boards, this model will not model the standard track board and then consider the longitudinal connection but, instead, will model the overall structure of the track board and other structures in the longitudinal direction. The convex retaining platform and other lateral limiting devices that play a longitudinal limiting role will not be physically modeled, and their limiting effect will be replaced by specifying boundary conditions for the track structure. The material parameters of the track structure are listed in Table 3.

Table 3. Track structure material parameters.

Structure Layer Name	Density (kg/m^3)	Elastic Modulus (Pa)	Poisson's Ratio
Steel rail	7800	2.06×10^{11}	0.25
Rail bearing	2500	3.60×10^{10}	0.20
Track board	2600	3.50×10^{10}	0.17
CA mortar bed	1800	9.20×10^{6}	0.40
Concrete base	2500	2.40×10^{10}	0.20

2.1.3. Track Irregularities

Track irregularity refers to the deviation between the track contact surface used to support and guide the wheels along the length direction of the track and the theoretical smooth track, which is the main excitation that causes the change in the wheel–rail action and then the coupled vibration of the entire train-track-foundation-soil system. The track irregularity spectrum of each country is divided into two levels: low interference and high interference. The low-interference level is suitable for high-speed railways above 250 km/h. For China's trunk railways, the more typical statistical spectrum functions that can characterize the irregularity characteristics include various speed levels, such as 120 km/h, 160 km/h, and 200 km/h [31].

For the high-speed train considered in this study, the above various track irregularity spectra cannot be better adapted to the working conditions of this study. To address this issue, Xu et al. [32] compared and analyzed the track irregularity spectrum at all levels and noted that the distribution of the low-interference track irregularity spectrum in Germany is similar to the standard spectrum of the 200 km/h speed-up line in China and can be used in the speed-up line spectrum in China. Additionally, the simulated power spectral density function of the track irregularity of the 350 km/h high-speed rail line at the design speed is also obtained based on the above track spectral density function and the sample data of the Shanghai–Nanjing passenger dedicated line. In view of the need to predict the impact of higher train speeds on the ground vibration caused by high-speed trains, the track irregularity power spectral density function used in this study is the German low-interference track irregularity spectrum suitable for speeds greater than 250 km/h and the simulated irregularity spectrum corresponding to a speed of 350 km/h [32].

Due to the unevenness of the track level and other directions, it contributes less to the excitation between the wheel and rail [33], so, in this study, only the level track irregularity is considered. Using $S_v(\Omega)d\Omega = S_v(f)df$, the spectral functions of the German low-interference high–low irregularity spectrum, and the 350 km/h high–low irregularity spectrum varying with time and frequency can be obtained as follows:

(1) German low-interference track irregularity spectrum:

$$S_v(f) = S_v\left(\frac{2\pi f}{v}\right) \cdot \frac{2\pi}{v} = \frac{A_v \cdot f_c^2 v}{2\pi(f^2 + f_r^2)(f^2 + f_c^2)} \quad (1)$$

where f_r is the spatial cutoff frequency, Ω_r is the corresponding time truncation frequency ($f_r = v\Omega_r/2\pi$), f_c is the spatial cutoff frequency, and Ω_c is the corresponding time truncation frequency ($f_c = v\Omega_c/2\pi$).

(2) Simulated track irregularity spectrum for 350 km/h [32]:

$$S_v(f) = \frac{a(f^2 v^{-3}) + b(fv^{-2})}{(1 + bfv^{-1} + cf^3 v^{-3})} \quad (2)$$

With the help of the MATLAB program and inverse Fourier transform method, the discrete data of the amplitude of track irregularity changing with time can be obtained based on the above-mentioned power spectral density function that changes with time and frequency. The results are shown in Figure 5a,b. In comparison with the current literature, it is found that the simulated amplitude of the track irregularities according to the German low-interference spectrum is very close to the data calculated by Chen et al. [34], and the amplitude of the track irregularities obtained by the Shanghai–Nanjing 350 km/h spectrum is almost the same as the amplitude in the literature [32]. To verify the precision of the simulated method in this study, the simulated results and analytical values of the two kinds of irregularity spectra are also compared, as shown in Figure 6.

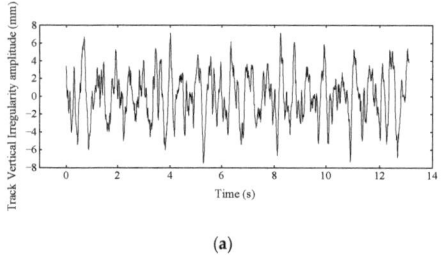 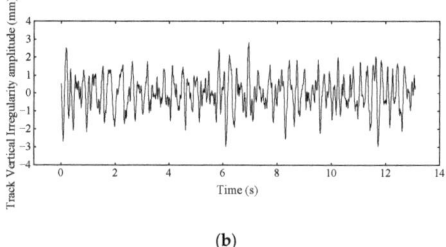

(a) (b)

Figure 5. Vertical irregularity amplitude simulation time series. (a) Simulated time series of low-interference high–low irregularity amplitude in Germany. (b) The simulated 350 km/h track irregularity amplitude time series.

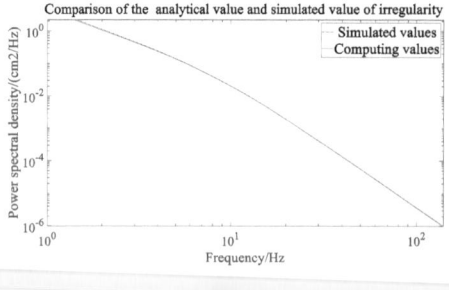 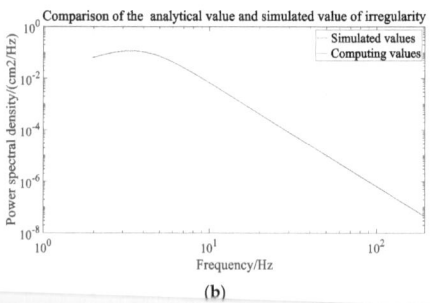

(b)

Figure 6. Comparison of the simulated and analytical values of the irregularity power spectrum. (a) The spectrum of German low interference. (b) The spectrum of 350 km/h.

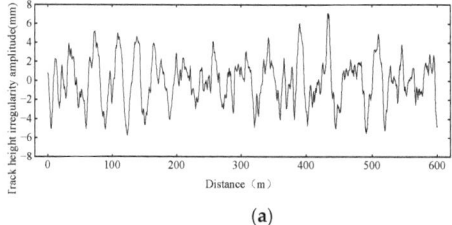
(a)

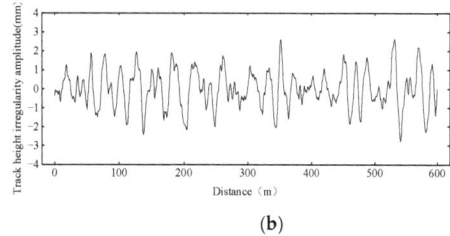
(b)

Figure 7. The relationship between the amplitude vertical irregularity and the forward distance of the train. (**a**) The relationship between the amplitude of German low-interference vertical irregularity and the forward distance of the train. (**b**) The relationship between the amplitude of 350 km/h irregularity amplitude and the forward distance of the train.

2.1.4. Subgrade and Foundation Soil Model

(1) Subgrade model parameters

According to the "Code for Design of High-speed Railway" [35], the subgrade section of the model in this study is based on the standard cross-sectional dimensions of single-line embankments for medium-ballasted tracks, as shown in Figure 8. The specification stipulates that the surface of the subgrade should be filled with graded gravel, and considering the large deformation of the subgrade, for the accuracy of the model, the Drucker–Prager plastic material constitutive model is used for each subgrade. The parameters of each layer of the roadbed structure are listed in Table 4.

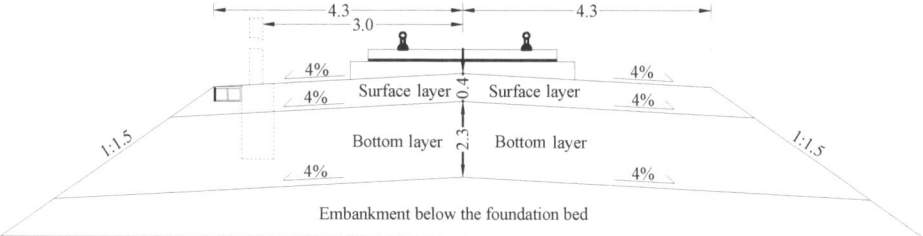

Figure 8. Single-line embankment ballastless track standard cross-section diagram.

Table 4. Geometric and mechanical parameters of each layer of the subgrade structure.

Names of Each Foundation Bed	Thickness (m)	Dynamic Elastic Modulus (MPa)	Poisson's Ratio	Density (kg/m³)	Cohesion (Pa)	Internal Friction Angle (°)	Damping Ratio
Surface layer of foundation bed	0.4	120	0.3	2184	7×10^4	27	0.045
Bottom layer of subgrade bed	2.3	70	0.3	1939	5×10^4	23	0.039
Embankment	3.6	50	0.35	1837	4×10^4	20	0.035

The triaxial test parameters that need to be input into the definition of the Drucker–Prager plastic material constitutive model are obtained by the following formula:

$$\tan \beta = \frac{6 \sin \varphi}{3 - \sin \varphi} \quad (3)$$

$$K = \frac{3 - \sin \varphi}{3 + \sin \varphi} \quad (4)$$

$$\sigma_c = \frac{2c \cos \varphi}{1 - \sin \varphi} \tag{5}$$

where φ is the friction angle in the Coulomb constitutive model (see Table 4), c is cohesion, K is the flow stress ratio ($0.078 \leq K \leq 1$), and the plastic parameters needed for each layer of the model subgrade structure are shown in Table 5.

Table 5. Drucker–Prager parameters for each layer of the subgrade structure.

Name of the Subgrade Structure of Each Layer	Angle of Friction (°)	Flow Stress Ratio K)	Expansion Angle (°)	Compression Yield Stress (Pa)	Absolute Plastic Strain
Surface layer of foundation bed	27	0.855	0	177847.90	0
Bottom foundation bed	23	0.876	0	122247.00	0
Embankment	20	0.892	0	95136.97	0

(2) Parameters of the foundation soil model

Usually, the loading speed and different strain levels on the soil will directly lead to the state of elasticity, elastoplasticness, or failure of the soil [36]. The dynamic strain of soil caused by rail transit is generally very small, and generally less than 1×10^{-5}. At this time, the soil is almost completely in the elastic stage. Therefore, the following assumptions [2] are adopted for the soil model in this study. The foundation soil is assumed to be a layered elastic body, and the material of each layer of soil is consistent and simplified as isotropic. The atomic and molecular motions and internal pores of soil particles in the soil are not considered, and continuous functions can be used to describe the changing laws of physical quantities such as soil stress, deformation, and displacement. The initial stress of the soil is neglected.

According to the literature [34], the soil below the subgrade is divided into two types: soft soil and hard soil. Among them, the soft soil is analyzed by using a representative three-layer soil in a soft soil area in Shanghai. The distinction between soft and hard soil is based on the shear wave velocity of the soil. Both soft and hard soil materials adopt linear elastic constitutive models. In this study, the shear wave velocity of the soil is calculated by using the following formula:

$$V_s = \sqrt{\frac{E}{2\rho(1+\mu)}} \tag{6}$$

where E is the elastic modulus of the soil, μ is Poisson's ratio, and ρ is the soil density. The material parameters of soft and hard soil and the shear wave velocity obtained from the above formula are shown in Table 6.

The size of the foundation soil along the length of the train is 600 m, the length of the foundation soil in the vertical direction of the train is 150 m, and the thickness of the entire foundation soil is 60 m. In addition, since the vibration wave will produce a reflection effect when it propagates to the finite element boundary, the calculation accuracy of the vibration wave will be greatly reduced. To reduce this effect, the infinite element boundary is set in the INP file of ABAQUS for the surrounding and bottom of the foundation soil. The model is shown in Figure 9. The infinite part of the Earth's foundation is equivalent to a boundary. For the boundary at the bottom of the foundation, considering a certain depth, the influence of boundary impedance and scattering characteristics is very small, and the bottom belongs to the rock layer, so the bottom of the model adopts consolidation

soil surface to extract the response of surface points within the range of 4.5~49.5 m from the track center.

Table 6. Geometric and material parameters of soft and hard soil layers.

Soil Type	Name of Each Layer of Soil	Thickness (m)	Dynamic Elastic Modulus (MPa)	Poisson's Ratio	Shear Wave Velocity (m/s)	Density (kg/m³)	Damping Ratio
Soft soil	Silty clay	6	30	0.290	78.27	1898	0.050
	Silt clay	9	14	0.300	56.21	1704	0.050
	Sandy silt	24	74	0.310	123.66	1847	0.050
	Uniform elastic half-space soil layer	21	141	0.330	167.03	1900	0.023
Hard soil	Silty clay	6	124	0.302	158.40	1898	0.020
	Silt clay	9	111	0.310	157.82	1704	0.020
	Sandy silt	24	159	0.318	180.83	1847	0.020
	Uniform elastic half-space soil layer	21	141	0.330	167.03	1900	0.020

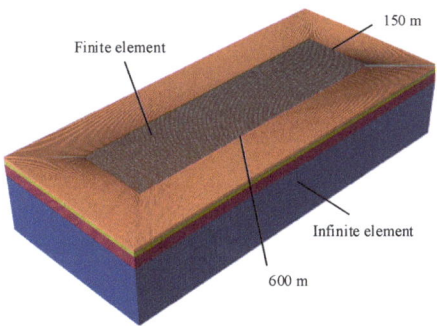

Figure 9. Schematic diagram of the foundation soil model with an infinite element boundary.

2.1.5. Calculation of Damping

Damping in ABAQUS/Explicit is mainly defined by the Rayleigh damping option, which can be determined by:

$$[C] = \alpha[M] + \beta[K] \tag{7}$$

where α, β is the proportionality constant related to the natural circular frequency of the structure and the damping ratio of the material [37], which can be determined by the following formula:

$$\begin{cases} \alpha = \dfrac{2(\xi_i\omega_j - \xi_j\omega_i)\omega_i\omega_j}{(\omega_j+\omega_i)(\omega_j-\omega_i)} \\ \beta = \dfrac{2(\xi_j\omega_j - \xi_i\omega_i)}{(\omega_j+\omega_i)(\omega_j-\omega_i)} \end{cases} \tag{8}$$

where ω_i and ω_j are the ith- and jth-order natural frequencies, respectively, and ξ_i and ξ_j are the damping ratios corresponding to the ith- and jth-order natural frequencies, respectively. In practical applications, due to the difficulty in determining the variation of ξ_i and ξ_j with natural frequencies, they are usually simplified as $\xi_i = \xi_j = \xi$. The Rayleigh damping can be obtained from Equations (7) and (9):

$$\begin{cases} \alpha = \dfrac{2\xi\omega_i\omega_j}{(\omega_j+\omega_i)} \\ \beta = \dfrac{2\xi}{(\omega_j+\omega_i)} \end{cases} \tag{9}$$

The entire foundation soil, including the subgrade section, is subjected to modal analysis, the calculation efficiency and model accuracy are considered comprehensively, and only the first 30 orders of natural circle frequencies of the foundation soil of different soil qualities are extracted. Since the vertical vibration of the foundation soil is the main concern in this study, only the participation coefficient of each order frequency in the vertical formation is extracted, as shown in Figure 10.

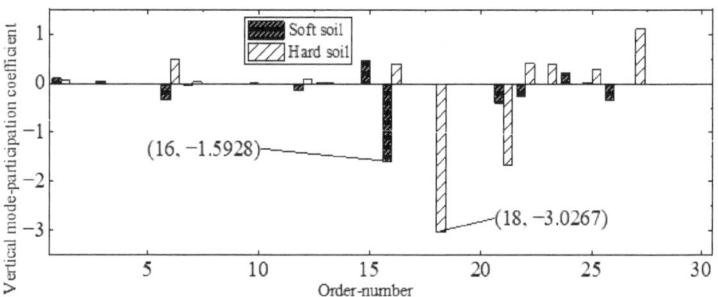

Figure 10. Each order natural frequency vertical mode-participation coefficient diagram.

Figure 10 indicates that the participation coefficient of the 16th-order natural frequency of soft soil is the largest in the vertical formation, which is 1.5928, and the participation coefficient of the 18th-order natural frequency of hard soil is the largest in the vertical formation, which is 3.0267. Therefore, the 16th-order natural frequency of soil and the 18th-order natural frequency of hard soil are selected as ω_j, and the first-order natural frequencies of soft and hard soils are selected as ω_i in Formulas (9) to calculate the Rayleigh damping coefficient. The calculated coefficients α and β are shown in Table 7.

Table 7. Rayleigh damping coefficient table of soft and hard soil.

Soil Quality	Name	ζ	ω_i	ω_j	α	β
Soft soil	Surface layer of foundation bed	0.045	3.0536	3.9663	0.1553	0.0128
	Bottom layer of subgrade bed	0.039	3.0536	3.9663	0.1346	0.0111
	Embankment	0.035	3.0536	3.9663	0.1208	0.0100
	Silty clay	0.050	3.0536	3.9663	0.1725	0.0142
	Silt clay	0.050	3.0536	3.9663	0.1725	0.0142
	Sandy silt	0.050	3.0536	3.9663	0.1725	0.0142
	Uniform elastic half-space soil layer	0.023	3.0536	3.9663	0.0794	0.0066
Hard soil	Surface layer of foundation bed	0.045	3.9628	6.3627	0.2198	0.0087
	Bottom layer of subgrade bed	0.039	3.9628	6.3627	0.1905	0.0076
	Embankment	0.035	3.9628	6.3627	0.1709	0.0068
	Silty clay	0.020	3.9628	6.3627	0.0977	0.0039
	Silt clay	0.020	3.9628	6.3627	0.0977	0.0039
	Sandy silt	0.020	3.9628	6.3627	0.0977	0.0039
	Uniform elastic half-space soil layer	0.020	3.9628	6.3627	0.0977	0.0039

2.1.6. Wheel–Rail Contact and Track-Subgrade Connection

Regarding the contact relationship between the wheel and rail, the typical Hertz nonlinear elastic contact theory is adopted in this study. The contact elastic action between the wheel and rail is simplified as a linear spring and is defined by the Hertz contact stiffness.

$$k_H = \frac{\overline{d}}{d\Delta z} = \frac{\overline{a}}{2G}$$

where the wheel–rail contact constant of the tapered tread is , and R is the wheel radius, with a value of 0.46 m in this study, so $G = 5.131 \times 10^{-8}$.

In ABAQUS/Explicit, nonlinear elastic contact is mainly achieved by setting the proportional relationship between contact pressure and interference according to Formula (3), which is set in ABAQUS/Explicit as the relationship between "softening" pressure and interference that conforms to the exponential law, as shown in Figure 11.

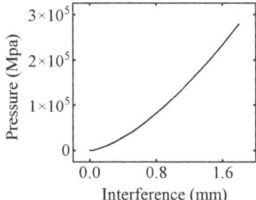

Figure 11. Pressure–interference diagram defined in ABAQUS/Explicit.

Considering that the sliding between the base of the track and the foundation is relatively small, the TIE connection is used. The so-called TIE connection binds the two surfaces that are in contact with each other. This processing method can better meet the deformation co-ordination relationship between the various parts of the track structure with a lower computational cost than the specified contact connection method.

2.1.7. Verification of the Finite Element Model

To verify the correctness of the model in this study, the same foundation soil size as in refs. [30,38] is used, soft soil type foundation soil is selected, and the train speed is 250 km/h. Considering the influence of the unevenness of the track on the ground vibration caused by the vehicle, the calculation and extraction are located in the middle of the model along the running direction of the track, and the distance from the vertical direction of the track is calculated and extracted. The monitoring points are demonstrated in Figure 12. Figure 13 shows a comparison of the vertical acceleration time history and the amplitude frequency between the present results and the results in ref. [30] at monitoring points of 4.5 m, 19.5 m, and 49.5 m.

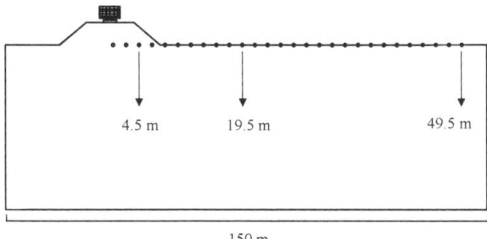

Figure 12. Train-track-foundation-soil coupling model monitoring point diagram.

As demonstrated in Figure 13, the present results are in good agreement with the results of ref. [30], and all of them are dominated by low-frequency responses, which is mainly due to the strong suppression of high-frequency vibrations by soft soils. To further verify the model in this study, working conditions (i.e., soft soil foundation and train speed 260 km/h) similar to those in ref. [38] are used to perform the analysis. The comparison of acceleration and displacement amplitude between the present study and the results of ref. [38] are listed in Table 8.

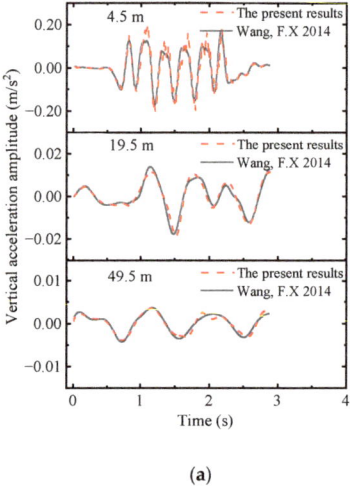 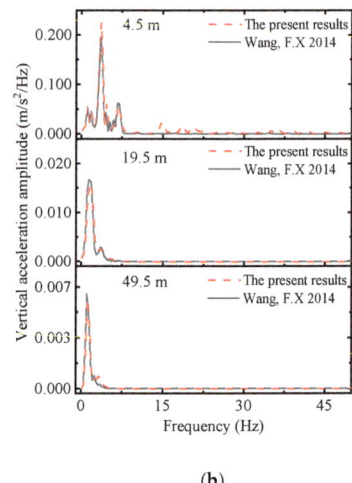

Figure 13. Comparison of vertical acceleration between the present results and the results in ref. [30]. (**a**) Vertical acceleration time history diagram. (**b**) Vertical acceleration amplitude frequency diagram.

Table 8. Comparison of acceleration and displacement amplitude between the present results and the results in ref. [38].

Data Sources	Monitoring Point	Vertical Acceleration Amplitude (m/s²)	Vertical Displacement Amplitude (mm)
Results of ref. [38]	Ground surface at a distance of 5 m from the track	0.15	1.3
Present results	Ground surface at a distance of 4.5 m from the track	0.18	1.4

Table 8 shows that both the acceleration amplitude and displacement amplitude results calculated by using the present model are close to the results in ref. [39]. In summary, the finite element model established in this study is reliable and can be used to simulate more engineering cases.

2.2. Analysis of Three-Dimensional Vibration Characteristics and Attenuation Law

To obtain the general vibration characteristics and attenuation law of the coupling model of train-track-foundation soil during train operation, this section selects a speed of 250 km/h, considers the high and low irregularity of the track (using German low-interference high- and low-irregularity spectrum), and analyzes the conditions when the foundation soil is soft soil. To reflect the variation in foundation soil vibration with the entire process of train travel, a series of monitoring points were selected in the middle of the entire soil model train travel direction at different distances perpendicular to the track direction, as shown in Figure 12. For the convenience of description, the direction perpendicular to the train travel is defined as the X direction, the vertical vibration direction is defined as the Z direction, and the train travel direction is defined as the Y direction. By extracting triaxial acceleration time history data at various monitoring points located at different distances (1.8 m–90 m) perpendicular to the track and plotting a triaxial acceleration–distance–amplitude waterfall chart, the vibration response characteristics of each measurement point

Figure 11 at a distance of 1.8 m–90 m from the track center on the ground surface. The time history

eration can better reflect the entire process of the train passing through the monitoring points. The change in acceleration amplitude in the middle part of the entire time domain can well reflect the process of the train passing through, and, due to the presence of wheels, the peaks in the triaxial acceleration time domain graph all exhibit periodic changes. The amplitude of the acceleration dynamic response in all three directions shows a decreasing trend as the distance to the center of the track increases, and the attenuation speed is first fast and then slow. At the same time, the periodic phenomenon of wave peaks caused by the wheel set effect gradually weakens as the distance to the track increases.

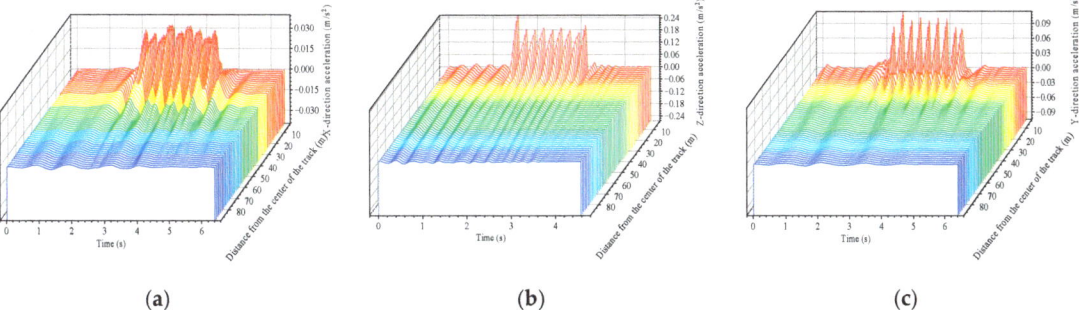

Figure 14. Time history of ground acceleration at different distances from the center of the track. (a) Time history of ground X-direction acceleration at different distances from the center of the track. (b) Time history of ground Z-direction acceleration at different distances from the center of the track. (c) Time history of ground Y-direction acceleration at different distances from the center of the track.

In terms of the amplitude of triaxial acceleration, the Z direction is the largest, followed by the Y direction, and the X direction is the smallest. In terms of the overall attenuation speed, the Z direction is the fastest, followed by the Y direction, and the X direction is the slowest. In addition, the attenuation speed of the triaxial acceleration also exhibits different patterns at different distances from the track: the attenuation speed of the Y and Z acceleration amplitudes within 20 m of the track center is significantly greater than that of the X direction, and the attenuation speed of the Z acceleration outside 20 m of the track sharply decreases and tends to flatten out. Within 20–40 m of the track, the attenuation speed of the Y direction acceleration is the highest among the three directions, while the attenuation speed of the X-direction acceleration is the slowest compared to the other two directions, and a rapid decrease in attenuation speed only occurs at approximately 40 m to the center of the track.

To study the vibration characteristics of the foundation soil in more detail, monitoring points were selected at the center of the roadbed surface and at distances of 4.5 m, 19.5 m, and 49.5 m from the track center, and their dynamic response data were analyzed in the time and frequency domains.

Figure 15 shows the triaxial acceleration time history and amplitude frequency at the center of the roadbed surface. Due to the proximity of the roadbed surface to the wheel–rail contact position, the amplitude of the triaxial acceleration dynamic response is significantly greater than that of the foundation soil surface. In addition, from the amplitude frequency of the triaxial acceleration at the roadbed, it can be seen that the frequency bandwidth of the roadbed surface is significantly greater than that of the foundation soil surface, and the triaxial acceleration frequency is significantly concentrated in the higher frequency range, indicating that the roadbed structure has a strong inhibitory effect on the vibration at higher frequencies.

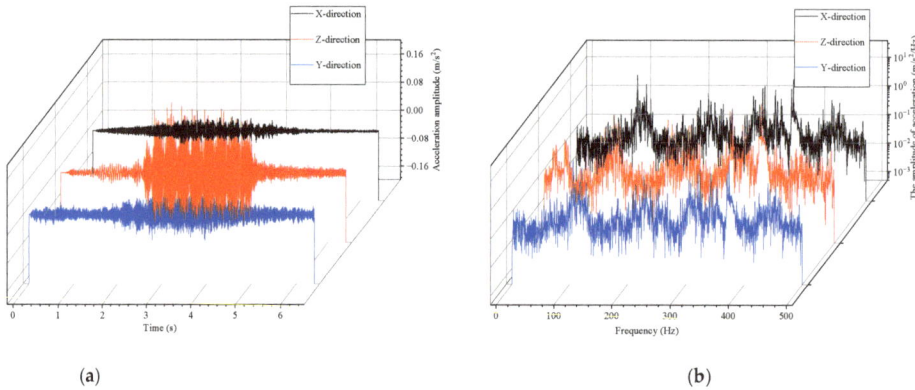

Figure 15. Three-dimensional dynamic response diagram of the roadbed surface. (**a**) The acceleration time history. (**b**) Amplitude frequency of the acceleration.

Figure 16 shows the time history and amplitude frequency of the triaxial acceleration on the foundation soil surface at distances of 4.5 m, 19.5 m, and 49.5 m from the center of the track. To provide a detailed explanation of the ground surface vibration characteristics and attenuation law, the following are described separately in terms of the time and frequency domains.

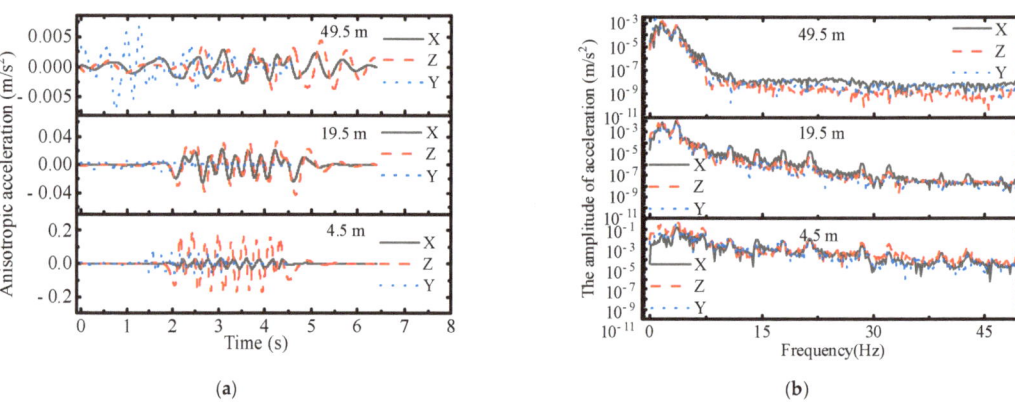

Figure 16. Time history and amplitude frequency of ground three-dimensional acceleration at distances of 4.5 m, 19.5 m, and 49.5 m from the track center. (**a**) The acceleration time history. (**b**) Amplitude frequency of the acceleration.

From the perspective of acceleration curves at different distances, the short-range acceleration time history curve can better reflect the impact effect of train passing. From the perspective of the dynamic response amplitude, all dynamic response values are relatively large at close range and show a gradual attenuation as the distance to the track center increases. From the perspective of triaxial acceleration, the dynamic amplitude of the Z-direction acceleration at 4.5 m is greater than that of the other two directions. At 19.5 m, the dynamic amplitude of the X and Y directions is slightly smaller than that of the Y direction, showing a tendency to catch up. At 49.5 m, the dynamic amplitude of the X and Y directions has already exceeded that of the Z direction. In terms of the propagation speed

its peak first, followed by the X direction, and the slowest in the Z direction.

the difference in the propagation velocity of the vibration wave in each direction increases with increasing distance to the center of the orbit.

2.3. Finite Element Model of the Transmission Tower-Line System

The transmission tower-line structure system is a large system composed of a series of single transmission towers and conducting (ground) lines. Previous research [36] has shown that the 'three-tower two-line system' is sufficient to meet the calculation requirements, and the calculation results are closer to the real situation. Therefore, the three-tower two-line model is also selected in this study.

2.3.1. Parameters of the Transmission Tower-Line System

The type of transmission tower is a 2A-ZM1 linear tower. The beam element is used to establish the finite element model of the transmission tower. The main parameters of the tower body are shown in Table 9. Table 10 lists the performance parameters of the established transmission conductance (ground) wire. The insulator model used in this paper is XP2-70. The tower and the ground are in a completely fixed form of restraint, and the cross arm of the transmission line, the insulator, and the conductor (ground) line are connected in a hinged manner. Finally, to make the model boundary more realistic, the insulator and conductor (ground) on both ends of the tower are restricted. The degrees of freedom of the nodes between the lines in the X direction. The finite element model of the three-tower two-line system in this section is shown in Figure 17.

Table 9. Main member parameters of the transmission tower.

Numbering	Tower Parts	Rod Specifications	Numbering	Tower Parts	Rod Specification
1	Tower leg main material	L80 × 7	8	Inner main material of upper crank arm	L45 × 4
2	Tower leg inclined material	L56 × 5	9	Outer main material of lower crank arm	L63 × 5
3	Tower leg diagonal brace	L40 × 4	10	Inner main material of lower crank arm	L56 × 5
4	Main material of tower body	L80 × 7	11	Tower leg top surface cross-section main material	L56 × 4
5	Tower body inclined material	L45 × 4, L40 × 4	12	Tower body top surface cross-section main material	L100 × 8
14	Cross-arm inclined material	L40 × 4	13	Outer main material of upper crank arm	L63 × 5
13	Cross-arm main material	L50 × 4	14	Tower body	L40 × 4

Table 10. Parameters of the transmission line.

Item	Cross-Sectional Area (mm^2)	Diameter (mm)	Line Density (kg/m)	Elastic Modulus (MPa)	Average Operating Tension (N)	Rupture Force ×0.95 (N)
LGJ—400/35	425.24	26.82	1.349	65,000	21,870	98,705
JLB40-150	148.07	15.75	0.6967	103,600	23,847	90,620

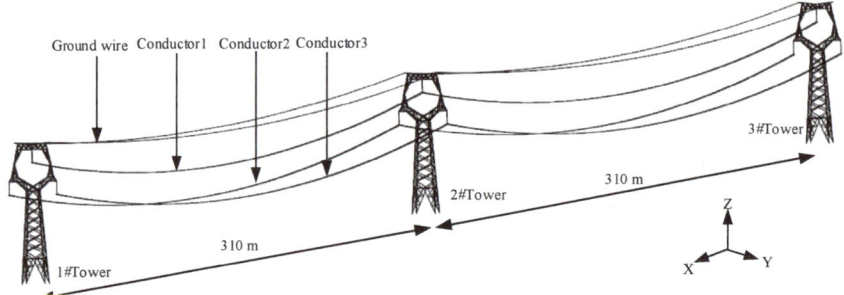

Figure 17. Finite element model of the transmission tower-line system.

2.3.2. Modal Analysis of the Transmission Tower-Line System

The interaction between transmission towers, transmission lines, and some armor clamps can affect the dynamic characteristics of individual components. Analyzing the dynamic characteristics of the tower-line coupling system is of great significance for studying the vibration characteristics of the tower-line system under vehicle-induced vibration. The partial vibration modes of the transmission tower-line structure and single transmission tower are shown in Figure 18.

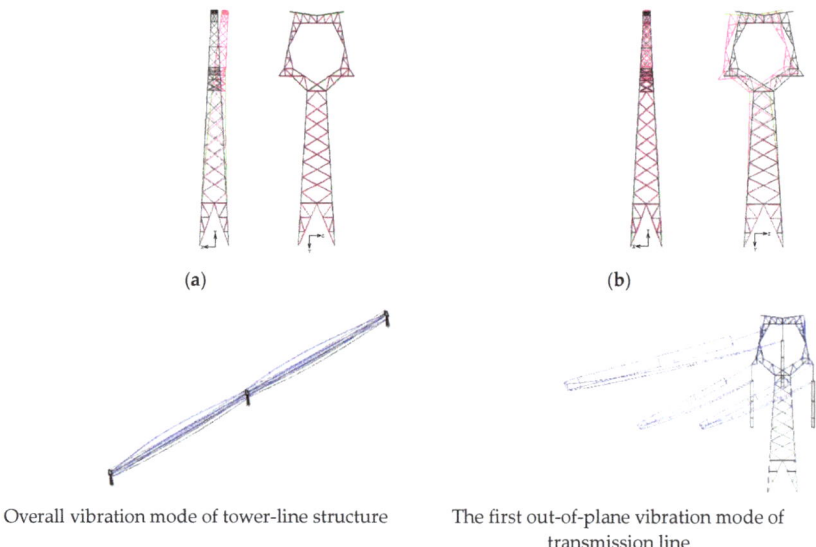

Figure 18. Partial vibration mode diagram of the transmission tower-line structure and single transmission tower. (**a**) First-order vibration mode of single tower (2.99 Hz). (**b**) Second-order vibration mode of single tower (3.73 Hz). (**c**) The first mode of vibration (0.200 Hz).

As indicated in Figure 18, the first and second natural frequencies of the transmission tower in the tower-line structure are 3.486 Hz and 3.487 Hz. Comparing the first and second natural frequencies of a single tower, it can be seen that the natural frequencies of transmission towers with tower-line structures exhibit significant amplification compared

Ref. [40] analyzes the dynamic response of transmission towers and single towers in a tower-line system. The study shows that under the same design wind speed, the stress of the main members of the tower in the tower-line system increases more than that of the single tower. The maximum stress of the multiple members approaches or reaches the design yield strength of the steel. However, in the corresponding single tower, the stress of the members is much less than the design yield strength of the steel, and the tower remains safe. Under the same design wind speed, the member stress increase in the tower-line system is mainly caused by the vibration of the transmission lines due to the coupling effect, whereas the stress increase in the single tower is mainly caused by its self-vibration. Under a 90° wind of varying speeds, the displacement of the tower top and the stress of the main members are greater than the results of the quasistatic analysis for the corresponding single tower, demonstrating that the amplifying effect of dynamic coupling on the response of the transmission tower cannot be neglected in the tower-line system. Therefore, this article analyzes a tower-line system.

3. Results and Discussion

3.1. Working Conditions

The case where the traveling direction of the train is perpendicular to the direction of the transmission tower line (X direction) is taken as an example to study the impact of different train speeds, soil conditions, and different distances to the track on the structural vibration of the transmission tower-line system. The schematic diagram is shown in Figure 19.

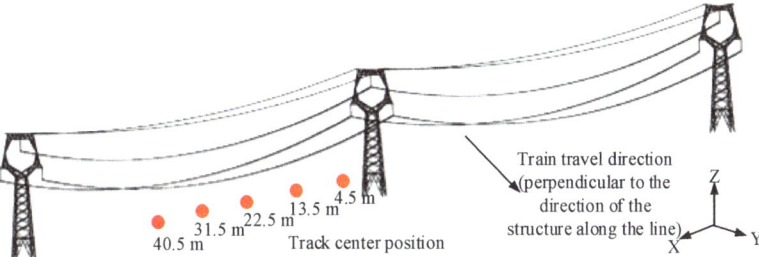

Figure 19. Schematic diagram of the train vibration source.

In this section, the soft and hard soil types introduced in Section 2 are selected, the train speed is considered to be 250 km/h, 300 km/h, 350 km/h, 400 km/h, and 450 km/h, and the distance to the track is 4.5 m, 13.5 m, 22.5 m, 31.5 m, and 40.5 m. The effective acceleration of ground vibration in the X, Y, and Z directions under different working conditions a_{rms} can be calculated by using Formula (11), and the results are shown in Figure 20.

$$a_{\text{rms}} = \sqrt{\overline{a^2(t)}} = \sqrt{\frac{\int_0^T a^2(t)dt}{T}} \qquad (11)$$

where a_{rms} is the effective acceleration, $a(t)$ is the acceleration at different times, and T is the duration of vibration action.

Figure 20 shows that the Z direction (vertical direction) is the largest in the effective value of the three-way ground vibration acceleration, and the energy is high, especially when the distance from the center of the track is short (4.5 m~13.5 m). For the law that the effective value of ground vibration acceleration changes with the speed of the train, there is a large difference between soft and hard soil foundations. When it is a soft soil foundation, the effective ground vibration acceleration at different distances from the track has the same value with increasing train speed. For the hard soil foundation, the effective acceleration increased slowly when the train speed was lower than 350 km/h and then increased rapidly.

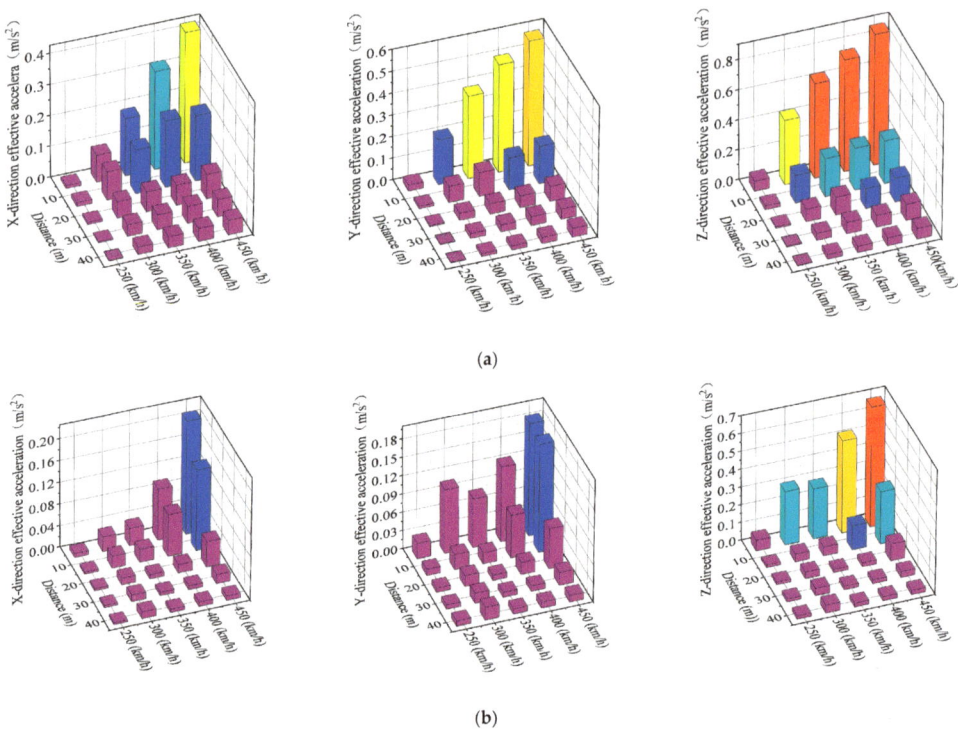

Figure 20. Effective acceleration under different working conditions. (**a**) Soft soil. (**b**) Hard soil.

3.2. Monitoring Points

To compare the dynamic response of the transmission tower-line structure under train-induced ground vibration for different vehicle speeds, soil qualities, and distances to the track, the axial stress of the main material of the tower legs at different heights and the displacement dynamics of the tower top in different directions under the above 50 working conditions were extracted. The monitoring points of the main material of the tower legs are shown in Figure 21a. Fifteen monitoring points evenly distributed on the main material of the tower legs are selected, and the monitoring points of the tower top displacement are shown in Figure 21b.

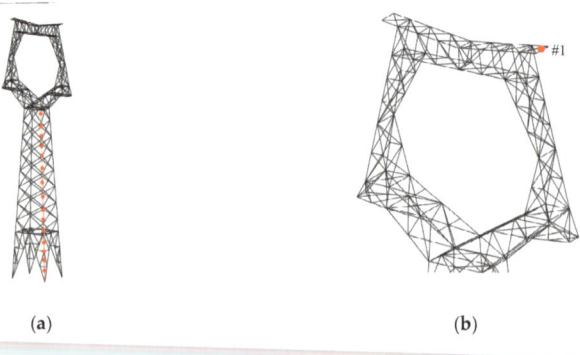

monitoring points. (**b**) Transmission tower top monitoring points.

According to the distribution law of ground vibration acceleration under the above 50 working conditions, the acceleration under various working conditions is not a simple linear distribution, and, when considering the influence of the three mixed factors of soil quality, vehicle speed, and distance, the factors that need to be considered are very complicated. Therefore, in this section, the method of controlling variables is used to analyze the dynamic response data of the transmission tower in detail. Finally, based on the dynamic response data of 50 working conditions, a three-dimensional surface diagram of the dynamic response of the transmission tower with changes in vehicle speed, soil quality, and distance is fitted.

3.3. Influence of Different Soil Qualities and Different Track Distances on the Vibration Response of the Transmission Tower-Line System

A speed of 250 km/h is taken as the control variable, and the dynamic response of the tower top displacement and the main material stress under the condition of a speed of 250 km/h are analyzed, which is affected by the soil and the distance to the track. From the modal analysis of the transmission tower-line structure in Section 2.3.2, it can be seen that the main mode shape of the transmission tower that appears for the first time is that the transmission tower bends in the X direction (in-plane), which is consistent with the bending direction of the first-order mode shape of the single transmission tower. This shows that the in-plane stiffness of the transmission tower is smaller, and it is more susceptible to the influence of ground vibration. The Y direction is greatly affected by the transmission conductor (ground) line. Therefore, the following analysis mainly focuses on the X direction and Y direction ground acceleration time-history data and frequency domain data, as shown in Figures 22 and 23. When the vehicle speed is 250 km/h, the X-direction acceleration amplitude corresponding to the soft soil foundation is obviously larger than that of the hard soil foundation. However, with the increase in the distance to the track, the X-direction acceleration decay rate corresponding to the hard soil foundation is significantly larger than that of the soft soil foundation. As shown in the X-direction acceleration amplitude-frequency diagrams of the two sites, the acceleration frequency is mainly within 10 Hz, and the main frequencies of both are relatively close to the first-order natural vibration frequency (2.99 Hz) of the single transmission tower. However, in terms of the frequency domain energy distribution of soft and hard soils, the vibration acceleration of the hard soil foundation accounts for a significant proportion near the fundamental frequency of the transmission tower, which is larger than that of the soft soil type. Due to the presence of wheels, the peaks of the acceleration time history undergo periodic changes. At a speed of 250 km/h, the peak value of the X direction appears at a distance of 13.5 m from the track in Figure 22, indicating that the energy in the soft soil foundation is highest here, while the peak value of hard soil occurs at 4.5 m. The peak values of both soil types under the Y direction appear at 4.5 m, indicating that the highest energy of the Y component is at 4.5 m for both soil types. In Figure 23, the peak values of soft soil and hard soil appear at 13.5 m and 4.5 m, respectively, corresponding to Figure 22. From the acceleration time history data and amplitude frequency data in the Y direction, there is a certain difference between the Y direction and the X direction. As far as the acceleration amplitude level is concerned, there is a slight difference between the two, and the acceleration response amplitude under soft soil is also significantly greater than that of hard soil. The frequency domain energy distribution in the Y direction is basically the same as that in the X direction, but at high frequencies, the energy is slightly larger than that in the X direction.

Incorporating the time history, consider the effective displacement of the tower's top in the X and Y directions. It can be determined by Formula (12):

$$u_{\text{rms}} = \sqrt{\overline{u^2(t)}} = \sqrt{\frac{\int_0^T u^2(t)dt}{T}} \tag{12}$$

where u_{rms} is the effective displacement, $u(t)$ is the displacement at different times, and T is the duration of vibration action. By Formula (12), the effective displacement values of the tower top with different soil qualities and different distances from the track can be obtained, as shown in Table 11 and Figure 24.

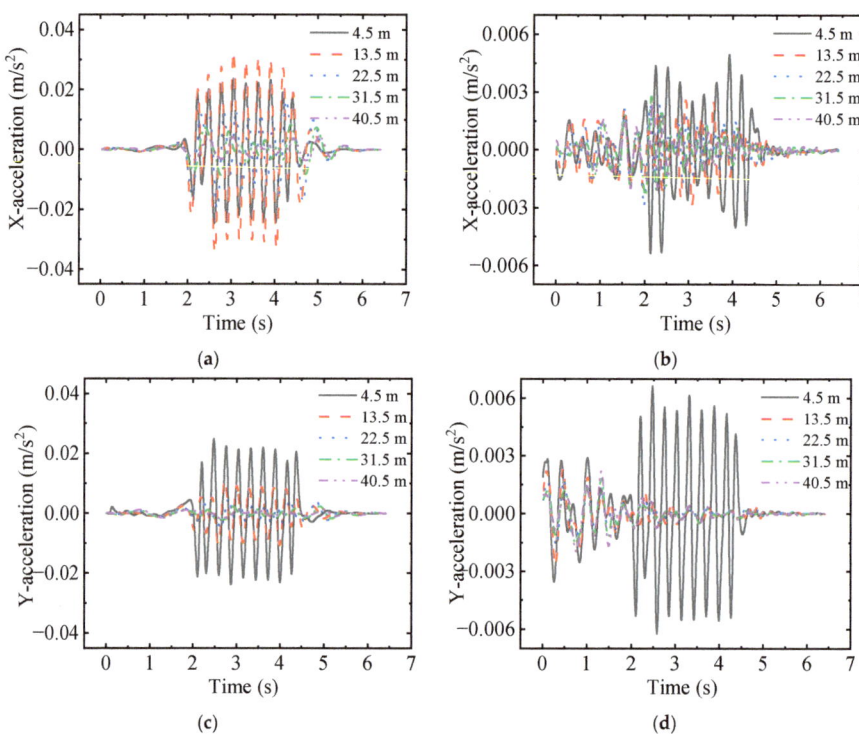

Figure 22. The X- and Y-direction acceleration time history diagram from 250 km/h to different distances from the track center. (**a**) Soft soil foundation. (**b**) Hard soil foundation. (**c**) Soft soil foundation. (**d**) Hard soil foundation.

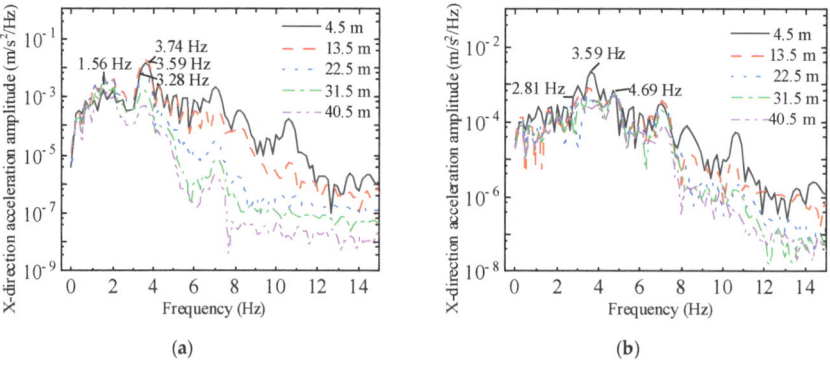

Figure 23. Cont.

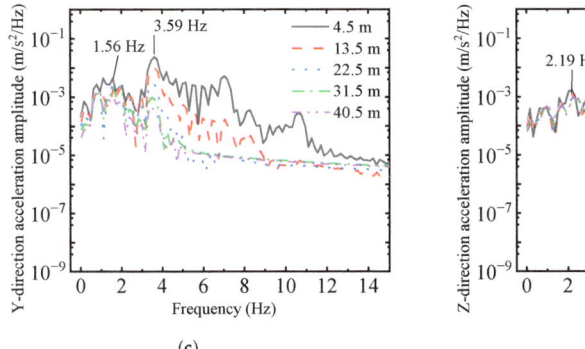
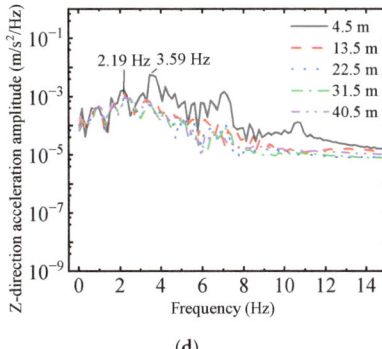

(c)   (d)

Figure 23. The X- and Y-direction acceleration amplitude-frequency diagram at different distances from the track center at 250 km/h. (**a**) Soft soil foundation. (**b**) Hard soil foundation. (**c**) Soft soil foundation. (**d**) Hard soil foundation.

Table 11. Effective displacement of the tower top at different distances.

Distance to Track (m)	Soft Soil X-Direction u_{rms} (mm)	Soft Soil Y-Direction u_{rms} (mm)	Hard Soil X-Direction u_{rms} (mm)	Hard Soil Y-Direction u_{rms} (mm)
4.5 m	4.07	0.56	3.49	2.46
13.5 m	4.25	0.66	4.06	2.16
22.5 m	4.25	0.84	4.17	1.75
31.5 m	4.23	0.98	3.51	1.57
40.5 m	3.03	0.90	3.44	1.48

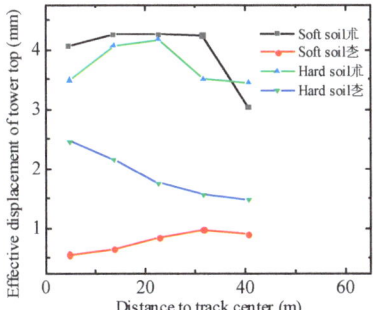

Figure 24. Effective displacement value of the tower top.

From the change trend of the effective displacement of the tower top with the distance from the track, the change trend of the effective displacement in the X direction is basically the same as the change law of the effective value of the ground vibration acceleration in the X direction, which increases first and then decreases, which is particularly evident in the case of soft soil foundation soil. The change trend of the tower top displacement in the X direction under the hard soil type foundation soil is in poor agreement with the change trend of the effective value of the ground X-direction vibration acceleration, which is mainly due to the use of a three-way ground vibration input in the excitation of the transmission tower-line system in this section. The ground vibration input in the direction is affected by the ground vibration in other directions, so it is different. From the overall displacement response, when the distance from the track center is greater than 13.5 m, all the effective displacements decrease except the Y-direction displacement of the soft soil type foundation. From the point of view of the effective displacement value of the tower top, at a close distance (4.5 m), the acceleration amplitude of hard soil is greater than that of soft soil, so the effective displacement of the top of the tower under hard soil is also greater than that of soft soil. However, due to the different Rayleigh wave velocities of the two soils, the Rayleigh wave velocity corresponding to the soft soil foundation soil is obviously smaller than that of the hard soil foundation soil; therefore, at a relatively long distance (40.5 m), the effective displacement value of the tower top under soft soil is significantly greater than that of hard soil.

To compare the stress response changes of the transmission tower under different working conditions, the stress data of the main material unit marked in Figure 21a are extracted, and Figures 25 and 26 draw the maximum axial tension and compression stress diagram of the unit with the height of the tower. The maximum tensile stress of the main material is mainly distributed in the tower body 12 m height, and the maximum compressive stress is mainly distributed in the tower body 8 m. The maximum tensile stress of the main material in the hard soil type foundation soil decreases with the increase in the orbital distance as a whole, which is almost consistent with the change law of ground vibration acceleration, which corresponds to the decrease in the consistency of soft soil type foundation soil.

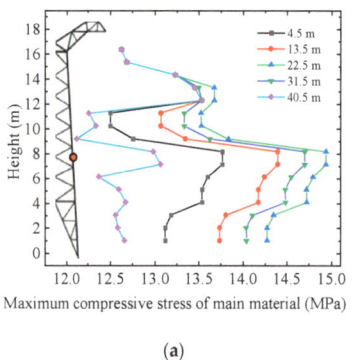
(a)

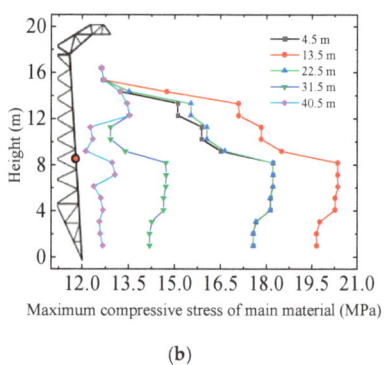
(b)

Figure 25. The maximum compressive stress distribution of the main material at different distances from the track center at 250 km/h. (**a**) Soft soil. (**b**) Hard soil.

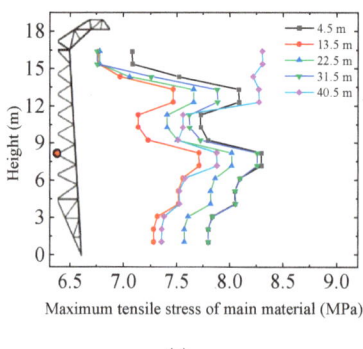
(a)

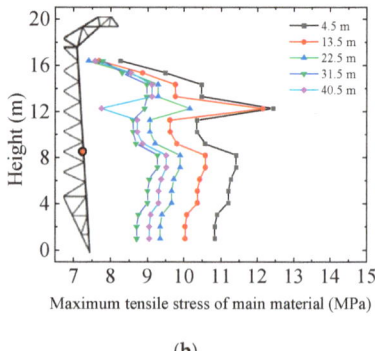
(b)

Figure 26. The maximum tensile stress distribution of the main material at different distances from

3.4. Influence of Different Train Speeds on the Vibration Response of the Transmission Tower-Line System

From the conclusion of the previous section, when the train speed is 250 km/h, the dynamic response of the transmission tower-line structure is the largest at a distance of 13.5 m from the center of the track. The following is to study the influence of different vehicle speeds on the dynamic response of the transmission tower-line structure under the environmental vibration caused by the train. In this section, the distance to the track is used as the control variable, and the 13.5 m influence of different train speeds on the dynamic response of the structure. Figures 27 and 28 show the time-history and amplitude-frequency diagrams of the ground X-direction acceleration under the two soil conditions at different train speeds down to track 13.5 m.

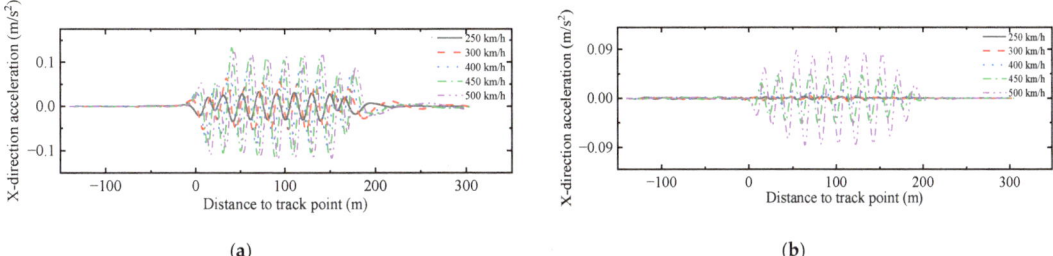

Figure 27. X-direction acceleration time history response at 13.5 m under soft and hard soil at different speeds. (**a**) Soft soil foundation. (**b**) Hard soil foundation.

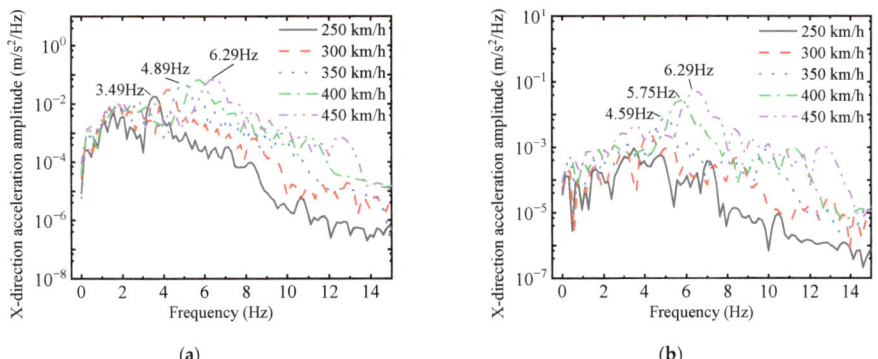

Figure 28. X-direction acceleration amplitude-frequency diagram at 13.5 m under soft and hard soil at different speeds. (**a**) Soft soil foundation. (**b**) Hard soil foundation.

Judging from the acceleration time-history data in the X direction, the acceleration amplitude corresponding to the soft soil type foundation soil is slightly larger than that of the hard soil type foundation soil; the acceleration amplitude in the X direction under both soil conditions increases with increasing vehicle speed, but the acceleration amplitude under the two soil conditions varies with the speed of the vehicle. The acceleration amplitude corresponding to the soft soil is faster at first and then slower, while the acceleration amplitude under the hard soil is gradually accelerated. In addition, from the analysis of the acceleration amplitude-frequency data in the two fields, with increasing vehicle speed, the frequency component of the acceleration gradually approaches the high-frequency section, which gradually moves away from the fundamental frequency of the transmission tower. The frequency component of the acceleration at the minimum vehicle speed is close to the high frequency of the transmission tower and gradually moves away as the vehicle speed increases.

Previous results showed that the Z-direction displacement fluctuation of the top of the transmission tower is small, and the response difference under different working conditions is also small. Therefore, only details of the X- and Y-direction displacements of the transmission tower are analyzed and compared in this section. The effective displacement of the tower top in the X and Y directions under different soil qualities corresponding to the vehicle speed needs to be calculated by using Formula (12). The results are shown in Table 12 and Figure 29.

Table 12. Effective displacement of the tower top under different train speeds.

Train Speed (km/h)	Soft Soil X-Direction u_{rms} (mm)	Soft Soil Y-Direction u_{rms} (mm)	Hard Soil X-Direction u_{rms} (mm)	Hard Soil Y-Direction u_{rms}
250	4.25	0.70	3.37	0.92
300	3.08	0.72	3.21	1.00
350	10.10	2.22	10.00	3.47
400	3.16	0.70	3.02	1.15
450	3.37	0.66	4.06	2.16

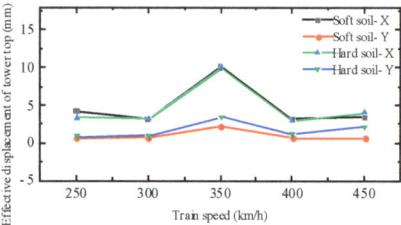

Figure 29. Effective tower top displacement at different speeds up to 13.5 m on the track.

As shown in Figure 29, when the speed is less than 450 km/h, the displacement in the X direction of the transmission tower is always greater than the displacement in the Y direction, which is mainly due to the weak stiffness of the transmission tower in the X direction. Under the ground vibration caused by the vehicle, the transmission tower does not increase linearly with increasing train speed but has a speed that has the greatest influence (350 km/h), mainly because the factors affecting the dynamic response of the structure are not only the amplitude of the time history acceleration but also the duration and frequency components of the ground vibration. Figure 28 shows from the amplitude-frequency diagram of the Y-direction acceleration of the middle ground that with increasing train speed, the main frequency range of the ground vibration gradually approaches the higher frequency band, and the difference between this and the fundamental frequency of the transmission tower will gradually increase. The ground vibration duration acting on the transmission tower-line structure will gradually decrease, so even if the vehicle-induced ground vibration acceleration amplitude will increase with increasing vehicle speed, the dynamic response of the structure will not show a linear increasing trend. Overall, due to the duration and frequency distribution of ground vibration, the effective displacement values of the tower top in the X direction and Y direction under the two kinds of soil are the largest at 350 km/h, which are 10.10 mm and 10.00 mm, respectively.

To compare the stress response changes of the main material of the transmission tower under different vehicle speeds, the main material element stress data marked in Figure 21a are extracted, and Figures 30 and 31 show the maximum axial tensile and compressive stresses of the element with the change in tower height.

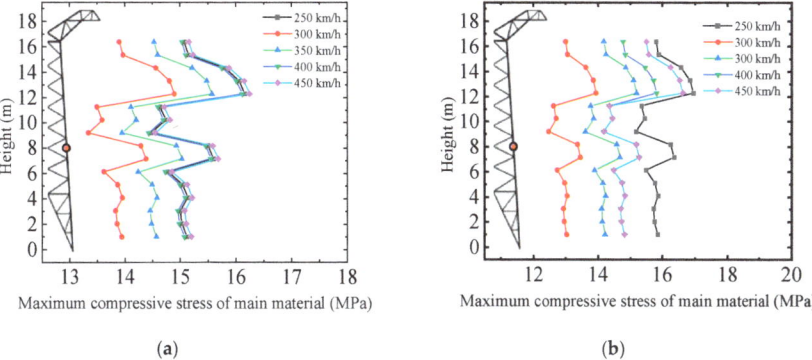

Figure 30. The maximum compressive stress of the main material is distributed along the height at different speeds. (**a**) Soft soil foundation. (**b**) Hard soil foundation.

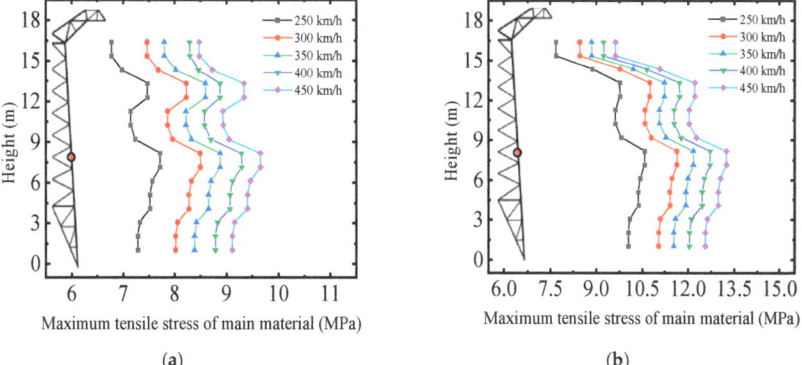

Figure 31. The maximum tensile stress of the main material at different speeds was distributed along the height. (**a**) Soft soil foundation. (**b**) Hard soil foundation.

From the data of the maximum tensile and compressive stress along the height of the main material of the transmission tower in Figures 30 and 31, the trend of the stress along the height is basically the same, and the maximum value of the tensile stress gradually increases with the speed of the vehicle, but the growth rate is first fast and then slow, while the compressive stress is the highest. The value as a whole satisfies the linear increasing trend with the vehicle speed.

The above control variables are used to compare the dynamic response differences of the top displacement of the transmission tower and the stress of the main material of the tower body at different vehicle speeds, soil qualities, and different distances from the track. To provide a more detailed analysis of the response changes of transmission towers at different soil types, vehicle speeds, and distances from the track, taking the tower top displacement that is greatly affected by ground vibration as an example, Formula (12) is used to analyze its effective values throughout the time domain, and the effective displacements of the tower top in the X and Y directions are fitted to obtain a three-dimensional curved surface, as shown in Figure 32.

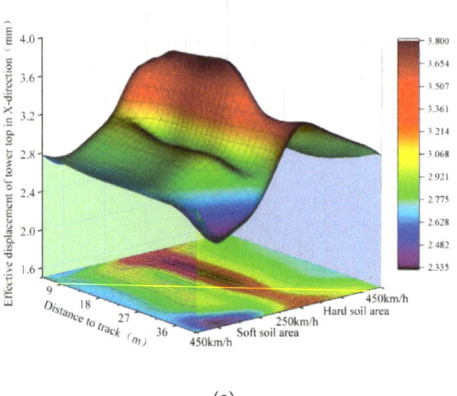
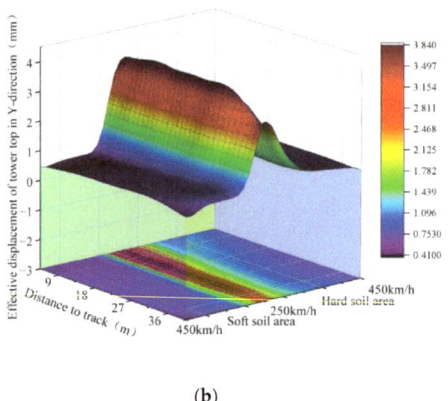

Figure 32. Transmission tower top displacement effective value fitting diagram. (**a**) Data fitting of effective displacement of tower top in X direction. (**b**) Data fitting of effective displacement of tower top in Y direction.

Figure 32 indicates that under the ground vibration caused by the train, the displacement of the transmission tower in the Y direction is affected by the speed of the train more than the displacement in the X direction. At 250 km/h, the dynamic response of the tower top displacement in the X and Y directions is very small. With increasing train speed, the change in the tower top displacement in the X direction is small, while the tower top displacement in the Y direction is almost exhausted quickly. This explains why the stiffness of the transmission tower-line system in-plane is small and is greatly affected by ground vibration. In addition, the displacement of the top of the tower at different distances from the track at the same speed is generally attenuated, but there is a trend of increasing first and then decreasing. From the three-dimensional surface map, it can be seen that the transmission tower line is greatly affected by the ground vibration in the range of 4.5 m~30 m from the track.

The effective value of ground vibration acceleration caused by high-speed trains increases with increasing vehicle speed and decreases with increasing distance to the track. However, the dynamic response corresponding to the transmission tower-line system is not so, which is affected by the frequency distribution, acceleration amplitude and vibration holding time of ground vibration. From the three-dimensional surface diagram, it can be concluded that the load holding time and frequency distribution occupy the main influence, and from the figure can be a more intuitive conclusion. The train speed with a greater influence is 250 km/h~350 km/h, and the transmission tower-line system has the most obvious response in the range of 4.5~30 m to the track.

For further analysis, the transfer function of the displacement of the foundation soil to the displacement of the tower top in the X direction and Y direction at 22.5 m under soft soil at 250 km/h are presented in Figure 33. As seen from Figure 33, the overall amplitude levels of the X direction and Y direction are similar, which can also be observed from the effective displacement value of the top of the tower. In addition, the frequency range of higher amplitude is mainly within 2~4 Hz, which is close to the first two natural frequencies of the transmission tower. The peaks of the transfer function in the X and Y directions occur at 2.99 Hz and 3.73 Hz, which is consistent with the first two mode shapes of the transmission tower shown in Figure 18. Overall, the amplitude of the transfer function in the fundamental frequency range of the tower is relatively high, showing an obvious

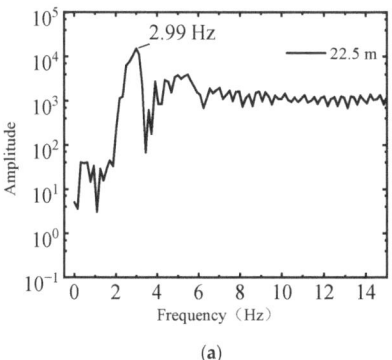 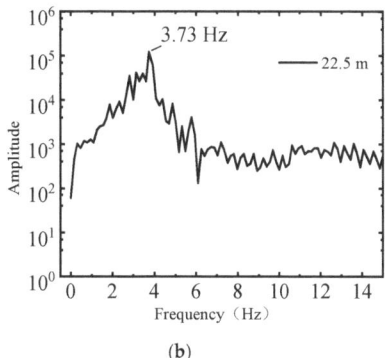

Figure 33. Transfer function of the displacement at the top of the tower response under a soft soil foundation. (**a**) Transfer function of X-direction displacement response at the top of the tower. (**b**) Transfer function of Y-direction displacement response at the top of the tower.

4. Conclusions

In this study, the commonly used ICE3 train and 220 kV typical transmission tower-line structure are used as the research objects, and the method of numerical simulation is used to study the impact of ground vibration generated by the train on the structure of the transmission tower line when the high-speed train crosses the transmission line. The influence of the train running speed, soil conditions and transmission tower-to-track distance on the dynamic response of the tower-line structure under the action of vehicle-induced ground vibration is analyzed and discussed. The main conclusions are as follows:

(1) The ground vibration characteristics of trains are mainly influenced by factors such as track irregularity, soil quality, and train speed. The irregularity of the track has a significant impact on the vibration response of structures near the track, and considering the irregularity of the track, the high-frequency components in the roadbed response are significantly higher than those in the smooth state. The roadbed structure also has a great inhibitory effect on high-frequency vibration at the vibration source, and the attenuation of vibration waves through the roadbed structure to the ground surface vibration beyond 4.5 m of the track can be ignored due to the influence of track irregularity. The soil quality of a free field has a significant impact on vehicle-induced surface vibration: the amplitude of the vehicle-induced vibration response on the surface corresponding to a soft soil foundation is significantly greater than that of hard soil, while the frequency distribution of ground vibration on a hard soil foundation is wider than that on soft soil. The vibration response amplitude of the ground surface increases significantly with increasing vehicle speed, but with increasing vehicle speed, the impact effect of wheel sets on the ground surface near the source gradually weakens;

(2) The predominant frequency of the acceleration responses of the transmission tower under soft and hard soil foundations is mainly within 10 Hz, and the main frequency of both is close to the first-order natural frequency (2.99 Hz) of a single transmission tower. The tower-line structure vibrates mainly in the low-frequency range, and the vibration of trains is distributed in a wide frequency range. The amplitude of the high-level displacement response transfer function of the X-direction and Y-direction tower tops is concentrated in the range of 2~4 Hz. This indicates that the vibration of the tower is sensitive in this frequency range. In addition, the effective displacement along the top of the tower (X direction) is greater than the dynamic response in the vertical direction (Y direction);

(3) The effective value of ground vibration acceleration caused by trains will increase with increasing train speed and decrease with increasing distance to the track. Due

to the influence of various factors, such as the frequency distribution, acceleration amplitude, and vibration duration of ground vibration, the dynamic response of the transmission tower-line system is not the same. From the point of view of the effective displacement value of the tower top, when the speed is 350 km/h, the effective displacement value of the tower top under the two kinds of soil is the largest. At a close distance (4.5 m), the acceleration amplitude of hard soil is greater than that of soft soil, so the effective displacement of the top of the tower under hard soil is also greater than that of soft soil. However, due to the different Rayleigh wave velocities of the two soils, the Rayleigh wave velocity corresponding to the soft soil foundation soil is obviously smaller than that of the hard soil foundation soil; therefore, at a relatively long distance (40.5 m), the effective displacement value of the tower top under soft soil is significantly greater than that of hard soil. Overall, for the crossing areas of the soft soil foundation soil in this article, the vibration response of the transmission tower is the highest when the train speed is 250~400 km/h and the distance to the track is within 40 m. The transmission tower in the crossing section of hard soil foundation soil has the highest vibration response when the train speed is 250~350 km/h and the distance to the track is within 30 m.

Author Contributions: Conceptualization, G.Z. and M.Z.; methodology, G.Z., M.W. and Y.L.; software, G.Z. and M.Z.; validation, M.W., Y.L. and G.Z.; formal analysis, M.W. and Y.L.; investigation, Y.L. and M.Z.; resources, M.Z.; data curation, M.W.; writing—original draft preparation, G.Z., M.W. and Y.L.; writing—review and editing, G.Z., Y.L. and M.Z. All authors have read and agreed to the published version of the manuscript.

Funding: This work was sponsored by the Natural Science Foundation of Henan (grant no. 222300420549) and the Cultivating Fund Project for Young Teachers of Zhengzhou University (grant no. JC21539028).

Data Availability Statement: Data are contained within this article.

Conflicts of Interest: The authors declare no conflict of interest.

References

1. We, Y. Preface: Current research progress on mechanics of high-speed rail. *J. Mech. Engl. Ed.* **2014**, *30*, 846–848. [CrossRef]
2. Jia, Y.Y. *Research on Vibration Source Model and Ground Vibration Response of Subway Train*; Science and Technology Literature Publishing House: Beijing, China, 2016.
3. Timoshenko, S.P. Method of analysis of statical and dynamical stresses in rail. *J. Phys. Soc. Jpn.* **1927**. [CrossRef]
4. Jenkins, H.H.; Stephenson, J.E.; Clayton, G.A.; Morland, G.W.; Lyon, D. The effect of track and vehicle parameters on wheel/rail vertical dynamic loads. *J. Railw. Eng. Soc.* **1974**, *3*, 2–16.
5. Ditzel, A.; Herman, G.C.; Drijkoningen, G.G. Seismograms of moving trains: Comparison of theory and measurements. *J. Sound Vibr.* **2001**, *248*, 635–652. [CrossRef]
6. Kaynia, A.M.A.; Madshus, C.A.; Zackrisson, P.B. Ground vibration from high-speed trains: Prediction and countermeasure. *J. Geotech. Geoenviron. Eng.* **2000**, *126*, 531–537. [CrossRef]
7. Sheng, X.S.X.; Jones, C.J.C.; Thompson, D.T.D. A theoretical model for ground vibration from trains generated by vertical track irregularities. *J. Sound Vibr.* **2004**, *272*, 937–965. [CrossRef]
8. Hayakawa, K.; Sawatake, M.; Gotoh, R.; Matsui, T. Reduction effect of ballast mats and eps blocks on ground vibration caused by trains and its evaluation. In INTER-NOISE and NOISE-CON Congress and Conference Proceedings; Institute of Noise Control Engineering: Washington, DC, USA, 1992; pp. 593–596.
9. Xia, H.; Cao, Y.M.; De Roeck, G. Theoretical modeling and characteristic analysis of moving-train induced ground vibrations. *J. Sound Vibr.* **2010**, *329*, 819–832. [CrossRef]
10. Erkal, A.; Kocagoz, M.S. Interaction of vibrations of road and rail traffic with buildings and surrounding environment. *J. Perform. Constr. Facil.* **2020**, *34*, 04020038. [CrossRef]
11. Motazedian, D.; Hunter, J.A.; Sivathayalan, S.; Pugin, A.; Pullan, S.; Crow, H.; Banab, K.K. Railway train induced ground vibrations in a low V_s soil layer. *Soil Dyn. Earthq. Eng.* **2012**, *36*, 1–11. [CrossRef]
12. Niu, D.X.; Deng, Y.H.; Mu, H.D.; Chang, J.; Xuan, Y.; Cao, G. Attenuation and propagation characteristics of railway load-induced vibration in a loess area. *Transp. Geotech.* **2022**, *37*, 100858. [CrossRef]

high-speed trains. In Proceedings of the 2010 WASE International Conference, USA, 14–15 August 2010.

14. Hesami, S.; Ahmadi, S.; Ghalesari, A.T. Numerical modeling of train-induced vibration of nearby multi story building: A case study. *KSCE J. Civ. Eng.* **2016**, *20*, 1701–1713. [CrossRef]
15. Zhou, Y.; Wang, B. Dynamic analysis of building vibration induced by train along railways. *J. Vib. Shock* **2006**, *25*, 36–41.
16. Celebi, E.; Kirtel, O.; Istegun, B.; Zulfikar, A.C.; Goktepe, F.; Faizan, A.A. In situ measurements and data analysis of environmental vibrations induced by high-speed trains: A case study in northwestern turkey. *Soil Dyn. Earthq. Eng.* **2022**, *156*, 107211. [CrossRef]
17. Schaumann, M.; Gamba, D.; Guinchard, M.; Scislo, L.; Wenninger, J. Effect of ground motion introduced by hl-lhc ce work on lhc beam operation. In Proceedings of the 10th International Particle Accelerator Conference, Melbourne, Australia, 19–24 May 2019; p. THPRB116.
18. Schaumann, M.; Gamba, D.; Garcia Morales, H.; Corsini, R.; Guinchard, M.; Scislo, L.; Wenninger, J. The effect of ground motion on the lhc and hl-lhc beam orbit. *Nucl. Instrum. Methods Phys. Res. Sect. A Accel. Spectrometers Detect. Assoc. Equip.* **2023**, *1055*, 168495. [CrossRef]
19. Farahani, M.V.; Sadeghi, J.; Jahromi, S.G.; Sahebi, M.M. Modal based method to predict subway train-induced vibration in buildings. *Structures* **2023**, *47*, 557–572. [CrossRef]
20. Davenport, A.G. Gust response factors for transmission line loading. In *Wind Engineering*; Cermak, J.E., Ed.; Elsevier: Pergamon, Turkey, 1980; pp. 899–909.
21. Zhao, S.; Yan, Z.; Li, Z.; Dong, J.; Zhong, Y. Investigation on wind tunnel tests of an aeroelastic model of 1000 kv sutong long span transmission tower-line system. *Proc. Chin. Soc. Electr. Eng.* **2018**, *38*, 5257–5265.
22. Li, Z.L.; Hu, Y.J.; Tu, X. Wind-induced response and its controlling of long-span cross-rope suspension transmission line. *Appl. Sci.* **2022**, *12*, 1488. [CrossRef]
23. Yang, F.; Zhang, H.; Wang, F.; Huang, G. Gust response coefficients of transmission line conductor. *J. Vib. Shock* **2021**, *40*, 85–91.
24. Wang, Y.; Yao, Y.; Zhang, Y.; Huang, J. Ice-shedding and breaking force analysis of transmission tower line system under different icing conditions. *Ind. Constr.* **2019**, *49*, 7–12.
25. Xue, J.Y.; Mohammadi, F.; Li, X.; Sahraei-Ardakani, M.; Ou, G.; Pu, Z.X. Impact of transmission tower-line interaction to the bulk power system during hurricane. *Reliab. Eng. Syst. Saf.* **2020**, *203*, 107079. [CrossRef]
26. Feng, H.F.; Wang, Y.R. Research on safety and reliability of overhead transmission lines across high-speed railway lines design. *Electr. Plat. Railr.* **2020**, *31*, 30–32.
27. Yin, M.; He, Z.H.; Zhang, Y.C.; Liu, Y.C.; Yang, J.; Zhang, Z.Y. Dynamic response of transmission lines crossing high-speed railway. *IEEE Trans. Power Deliv.* **2022**, *37*, 2143–2152. [CrossRef]
28. Zhang, M.; Liu, Y.; Liu, H.; Zhao, G.F. Dynamic response of an overhead transmission tower-line system to high-speed train-induced wind. *Wind Struct.* **2022**, *34*, 335–353.
29. Liu, H. Analysis of the Influence of Train Wind on the Transmission Tower-Line Structure across High-Speed Railway. Master's Thesis, Zhengzhou University, Zhengzhou, China, 2021.
30. Wang, F.X. Numerical Simulation and Vibration Characteristics Analysis of Environmental Vibration of High-Speed Railway Based on ABAQUS. Master's Thesis, Beijing Jiaotong University, Beijing, China, 2014.
31. Chen, X.M.; Wang, L.; Tao, X.X. Research on the evaluation method of track regularity of trunk railway in China. *China Railw. Sci.* **2008**, *29*, 21–27.
32. Xu, J.; Chen, X.M.; Jia, H.B. Power spectrum analysis of track irregularity for 350 km/h Shanghai-Nanjing passenger dedicated line. *J. Civ. Eng. Manag.* **2013**, *30*, 22–26.
33. Qiu, G.D. Research on Train Vibration Response Based on Track Random Irregularity. Master's Thesis, South China University of Technology, Guangzhou, China, 2014.
34. Chen, G.; Zhai, W.M. Numerical simulation of random process of railway track irregularity. *J. Southwest Jiaotong Univ.* **1999**, *34*, 13–17.
35. TB10621-2014; Code for Design of High-Speed Railway. China Railway Publishing House: Beijing, China, 2019.
36. Liu, Y. *Basic Principles of Soil Dynamics*; Tsinghua University Press: Beijing, China, 2019.
37. Clough, R.W.; Penzien, J. *Dynamics of Structures*; McGraw-Hill: New York, NY, USA, 1975.
38. Han, H.Y. Numerical Simulation Analysis of Ground Vibration Caused by High-Speed Trains on Soft Soil Subgrade. Master's Thesis, Southwest Jiaotong University, Chengdu, China, 2012.
39. Zhang, L.L. Random Wind Field Research and Wind-Resistant Reliability Analysis of High-Rise and High-Rise Structures. Doctoral Thesis, Tongji University, Shanghai, China, 2006.
40. Zhang, M.; Zhao, G.; Wang, L.; Li, J.; Ma, X. Wind-induced coupling vibration effects of high-voltage transmission tower-line systems. *Shock Vib.* **2017**, *2017*, 1205976. [CrossRef]

Disclaimer/Publisher's Note: The statements, opinions and data contained in all publications are solely those of the individual author(s) and contributor(s) and not of MDPI and/or the editor(s). MDPI and/or the editor(s) disclaim responsibility for any injury to people or property resulting from any ideas, methods, instructions or products referred to in the content.

Article

Experimental Study on Bearing Behavior and Soil Squeezing of Jacked Pile in Stiff Clay

Banglu Xi [1,2,*], Guangzi Li [2] and Xiaochuan Chen [1]

1. Anhui Province Key Laboratory of Green Building and Assembly Construction, Anhui Institute of Building Research & Design, Hefei 230031, China
2. School of Civil Engineering, Hefei University of Technology, Hefei 230009, China
* Correspondence: xibanglu@126.com; Tel.: +86-18-3217-00065

Abstract: In order to study the bearing behavior and soil-squeezing of jacked piles in stiff clay, two groups of pile penetration tests were performed, with a rough pile that can reproduce the quick-shear behavior of the pile–soil interface, i.e., group 1 in stiffer clay, and group 2 in softer clay for comparison. For each group, the adjacent pile was additionally penetrated at different pile spacings to study the soil-squeezing effect on an adjacent pile. The results show that the penetration resistance increased rapidly at the beginning and then increased at a lower rate. This is because the resistance at the pile end increased rapidly at the beginning and then kept stable with fluctuations, whereas the resistance at the pile side continually increased due to the increasing contact area. Therefore, the ratio of the resistance at the pile end to the total penetration resistance exhibited a softening behavior, which first increased to a peak and then gradually decreased. In addition, there was soil-squeezing stress and soil-squeezing displacement in the ground and adjacent piles due to pile penetration. In stiffer clay, the soil-squeezing stress was larger than that in softer clay due to the higher strength, whereas the soil-squeezing displacement was smaller than that in softer clay due to the low compressibility. In addition, the nonlinear equation form $y = ae^{-bx}$ can be employed to describe the effect of pile spacing on the vertical flotation, horizontal deviation, and pile strain of the adjacent pile.

Keywords: jacked pile; soil squeezing effect; bearing behavior; model test

Citation: Xi, B.; Li, G.; Chen, X. Experimental Study on Bearing Behavior and Soil Squeezing of Jacked Pile in Stiff Clay. *Buildings* **2023**, *13*, 2609. https://doi.org/10.3390/buildings13102609

Academic Editor: Yong Tan

Received: 4 September 2023
Revised: 25 September 2023
Accepted: 29 September 2023
Published: 16 October 2023

Copyright: © 2023 by the authors. Licensee MDPI, Basel, Switzerland. This article is an open access article distributed under the terms and conditions of the Creative Commons Attribution (CC BY) license (https:// creativecommons.org/licenses/by/ 4.0/).

1. Introduction

Jacked piles have been widely employed in civil engineering due to the advantages of low cost, strong quality controllability, fast construction speed, and convenient quality testing [1–4]. In the design and construction of jacked piles, the soil-squeezing effect is essential to prevent engineering accidents such as pile breakage and pile floating [5–7].

Great efforts have been made to study the soil-squeezing effect during pile penetration by theoretical methods [8–10], numerical simulations [11–14], field tests [15–17], and indoor model tests [18–22]. The theoretical methods, e.g., cavity expansion theory [23], can be employed to predict the soil displacement and ground deformations for jacked pile penetration, but the pile–soil frictional behavior is ignored and more reliable experimental data are necessary for validation. The numerical simulation can provide a unique view of the soil deformation and stress evolution, but large soil deformation and soil failure appear during jacked pile penetration, which can hardly be well simulated by the finite element method [24,25]. The discrete element method, which can capture the soil's large deformation and failure process well, is limited by its huge computation cost [14,26,27], especially when the water effect and unique particle shape should be taken into consideration for clays [28,29]. The field test can provide the most reliable data about the soil-squeezing effect, but it is limited by its huge cost and complex geological conditions [15,30]. In recent decades, the physical model test has been quite popular at investigating the bearing and soil-squeezing effect. With a focus on the penetration resistance of jacked piles, previous research has investigated the effects of time [31], soil plugging [32], friction fatigue [33], the pile-bearing layer [34], pile diameters [16], etc. With a focus on the soil-squeezing effect, various test methods and techniques have been employed, including the half-model test [35,36], X-ray radiography tomography [32]), computed tomography (CT) [19], and transparent soils [37–39], which are quite expensive when applied to monitoring soil deformation.

However, most indoor model tests of jacked piles focus on the penetration resistance and soil-squeezing effect in sands [33,35,40] and soft clays [30,34] with restively low shear strength. Little attention has been paid to those in stiffer clays with low compressibility and high shear strength, especially the soil-squeezing effect on the adjacent pile in stiffer clay. As a result, the design method and experimental findings on sands and soft clays can hardly be applied to the analysis of the soil–pile interaction in stiffer clays of Central China, which constitutes the main motivation for this paper.

In addition, the piles in most experimental tests are usually smooth, simulated by aluminum piles [6,19,34], steel piles [31,33,40], polymethyl methacrylate piles [27], etc., which cannot capture the quick-shear behavior between concrete piles and soil. Therefore, a rough pile, which can reproduce the quick-shear behavior between a concrete pile and stiffer clay, was employed here to study the penetration resistance and soil-squeezing effect in stiffer and softer clays. After briefly introducing the stiffer clay and rough pile, the closed-end rough piles were penetrated into softer and stiffer clays to analyze the penetration resistance and soil-squeezing effect. After this, the adjacent pile was penetrated at different pile spacings, focusing on the soil-squeezing effect on the adjacent pile. The experimental data can provide advice for better designing and constructing jacked piles in stiffer clay areas of Central China.

2. Model Setup
2.1. Experimental Apparatus

Figure 1 provides the experimental apparatus used in the test, including the soil bin, bracket, two railways, and penetration system. The soil bin was carefully designed and manufactured to remove the boundary effects on experimental results. Gui et al. [41] stated that the boundary effect was neglectable when the distance between the penetration point and the side boundary wall was 10 times larger than the cone diameter. Here, the pile diameter D was scaled to be 40 mm in the test, and the square soil bin was designed to be 1.0 m in side length L. Thus, the ratio of the soil bin side length to the pile diameter was 25, which is larger than suggested in previous research [16,41,42], and the boundary effect can be ignored. The bracket (i.e., 400 mm in height) is employed to provide enough space to allow a stable penetration resistance to occur. The railways can move freely on the bracket and the penetration system can freely move on the railways, allowing the pile to penetrate at different positions. The penetration system consists of the motor and shaft. The pile can be fixed on the shaft and forced into the stiffer clay by the motor with a velocity ranging from 0.2 to 2.0 mm/s.

2.2. Test Material

In China, stiffer clay is generally distributed in the middle areas (e.g., Anhui province), whereas softer clay is generally distributed in the east–south areas (e.g., Shanghai City and Guangdong province), where jacked piles were first widely employed in civil engineering. As a result, the abundant engineering experience and design method for softer clays leads to frequent engineering accidents in stiffer clay, e.g., pile breakage and pile floating [43]. Thus, two clays were employed here to perform a pile penetration test, one simulating natural stiff clay with high strength and low compressibility and the other simulating soft and saturated clay, aiming to show the difference in the penetration resistance and soil-squeezing effect of jacked piles. The stiffer clay in Hefei, Anhui province, was employed to prepare these two clays. In light of the remolded clay exhibiting a lower strength and higher compressibility than the natural stiff clay, the remolded clay treated with 1% chunam in mass was employed to reproduce the mechanical and compressive behavior of natural stiff clay here. Note that a similar technique has been employed to mimic the structure effect of natural clay [21,44]. Figure 2 presents the relationships between the deviatoric stress and the axial strain of the stiffer and softer clays employed in the test. Figure 2 shows that the stiffer clay (i.e., remolded clay treated with 1% chunam in mass) exhibited a larger shear strength than the softer clay, which represents a constitutive model for cohesive soils, as portrayed in [45,46]. The mechanical strength of the stiffer clay (i.e., c = 82.3 kPa, φ = 13.8°) was quite similar to those of the natural stiff clay in Hefei (i.e., c = 82.8 kPa, φ = 14.1°). The saturated softer clay exhibited a much lower strength, i.e., c = 46.2 kPa, φ = 10.6°. The related physical and mechanical parameters are provided in Table 1.

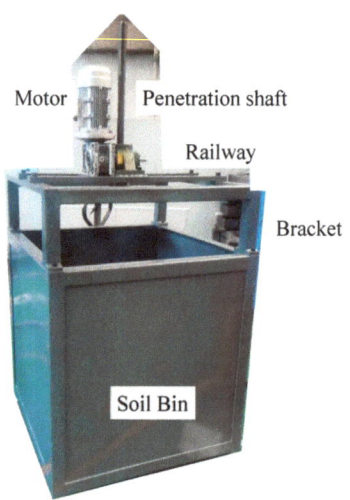

Figure 1. Experimental apparatus.

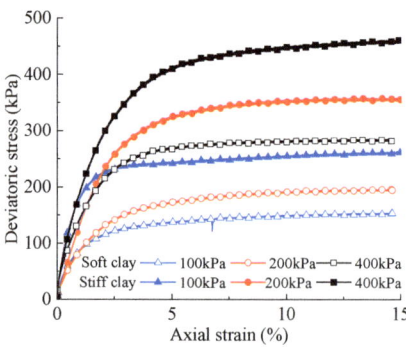

Figure 2. Stress–strain relationships of the stiffer and softer clays.

Table 1. Physical and mechanical parameters of stiffer and softer clays.

Soil	Specific Gravity	Internal Friction Angle	Cohesion	Bulk Density	Water Content
Stiffer clay	2.73	13.8°	82.3 kPa	1.902 g/cm^3	20.2 (%)
Softer clay	2.73	10.6°	46.2 kPa	2.275 g/cm^3	40.1 (%)

Concrete jacked pile has a rough surface significantly different from smooth aluminum piles [6,19,34], steel piles [31,33,40], polymethyl methacrylate piles [27], etc. To obtain reasonable penetration resistance and the soil-squeezing effect of a jacked pile, the model pile should be rough to reproduce the inter-surface quick-shear behavior between the concrete pile and natural stiff clay. Here, an acrylic plexiglass pile with a surface stuck by sand uniformly was employed to capture the frictional behavior of the concrete pile, as shown in Figure 3a. Figure 3b provides the relationships between shear stress τ and normal force σ using sands with diameters of <1 mm, 1–2 mm, and 2–3 mm by performing direct shear tests at a rate of 0.8 mm/min. Note that the quick-shear behavior for the concrete–soil interface was also examined for comparison. It shows that the interface frictional angle and cohesion increased with the sand size, and the acrylic plexiglass sample stuck uniformly by sands with sizes < 1 mm reproduced the shear behavior of the concrete–clay interface well. Therefore, an acrylic plexiglass pile uniformly stuck by sand with a size of < 1 mm (Figure 4) was employed in the physical model test to simulate a concrete pile. Note that the pile end also plays an important role in pile penetration. Therefore, a steel pile end (Figure 4) in the shape of a cross was employed in the test, which has been widely employed in jacked pile engineering in China. The characteristic of the model pile is provided in Table 2.

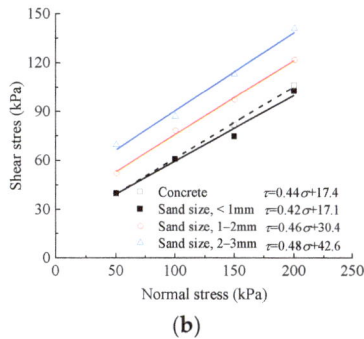

Figure 3. Acrylic plexiglass sample stuck uniformly by sands for direct shear test: (**a**) sample; (**b**) shear–stress relationship.

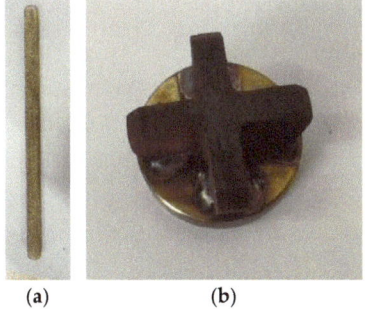

Figure 4. Model pile (**a**) and pile end (**b**).

Table 2. Model pile characteristics.

Material	Sand Size	Outside Diameter (m)	Inside Diameter (m)	Pile Length (m)	Elastic Modulus (GPa)
Acrylic plexiglass	<1 mm	0.04 m	0.02	0.7	3.16

2.4. Monitoring System

Figure 5 provides the arrangements of the monitoring sensors in the tests. A force sensor was installed between the shaft and the pile to monitor the total penetration resistance. In the ground, three earth pressure cells and pore pressure cells were buried at depths of 50 mm, 100 mm, and 150 mm to monitor the squeezing stress and pore pressure, respectively. On the surface, three displacement sensors were placed with a spacing of 50 mm to monitor the soil-squeezing displacement. Note that the resistance at the pile end could not be monitored directly because the actual shaped pile end was employed instead of an earth pressure cell in the test. Therefore, two strain gauges were attached to the pile top and end to monitor the pile strain, which could be used to calculate the total penetration and end resistances. It is worth mentioning that the monitored data in the reduced-scale model tests could hardly be scaled to the real-life jacked pile accurately because some important factors (e.g., soil particles and stress level, cohesion) were not perfectly scaled using the in situ clays instead of similar material, which has been widely employed in geotechnical engineering [6,34].

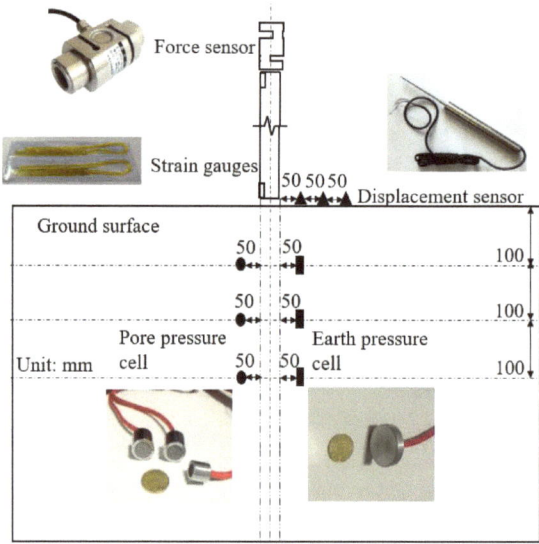

Figure 5. Test schematic diagram.

2.5. Test Preparation

The testbed was filled with five layers, with each layer measuring 160 mm. For stiffer clay, the dried and crushed clay particles were mixed with 1% chunam by mass uniformly by a mixer. Then, the mixture was wetted to a water content of 20.2%, which was the same as the natural stiff clay. Afterwards, the stiffer clay was poured into the soil bin and compacted to the target dry density layer by layer to a final thickness of 800 mm. As for the softer clay, the dried clay particles were wetted to a water content of 40.1% and then poured into the soil bin layer by layer. Note that more water was sprayed on the ground to ensure saturation after compacting each layer to the target density. After finishing the soil filling, the testbed was consolidated for one week.

The jacked pile penetration showed a significant squeezing effect on the adjacent pile, leading to engineering accidents, including breakage and floating of the adjacent pile. Therefore, after the jacked pile was forced at the center to study the penetration resistance and soil-squeezing effect, another pile was forced at different pile spacings from the center pile to investigate the soil-squeezing effect on the adjacent pile. Two groups of tests were performed, i.e., group 1 in stiffer clay and group 2 in softer clay, as shown in Table 3. In light

were chosen to be 2.5 D, 3.5 D, 4.5 D, and 5.5 D in the test.

Table 3. Test scheme.

Test Group	Soil	Pile Spacing	Penetration Velocity
1	Stiffer clay	2.5 D	0.5 mm/s
		3.5 D	0.5 mm/s
		4.5 D	0.5 mm/s
		5.5 D	0.5 mm/s
2	Softer clay	2.5 D	0.5 mm/s
		3.5 D	0.5 mm/s
		4.5 D	0.5 mm/s
		5.5 D	0.5 mm/s

3. Results
3.1. Bearing Behavior and Soil Squeezing Effect in the Ground
3.1.1. Total Penetration Resistance

Figure 6 presents the monitored total penetration resistance F_t monitored by the force sensor in stiffer and softer clays. It shows that F_t increased rapidly at the beginning and then increased at a lower rate. This is because both the pile end and the side resistances increased significantly at the beginning, which led to a relatively high increasing rate. As the pile penetrated, the resistance at the pile end reached the peak and then tended to be stable, whereas the resistance at pile side increased continually due to the increasing contacting side area, which contributed to the increasing total penetration resistance at a lower increasing rate. As expected, the stiffer clay exhibited a much larger penetration resistance due to the higher shear strength and a higher increasing rate due to the lower compressibility compared with the softer clay. Note that the penetration resistance in stiff clay seemed to increase more rapidly when the penetration depth exceeded 250 mm, which was probably caused by the inevitable inhomogeneity in the ground.

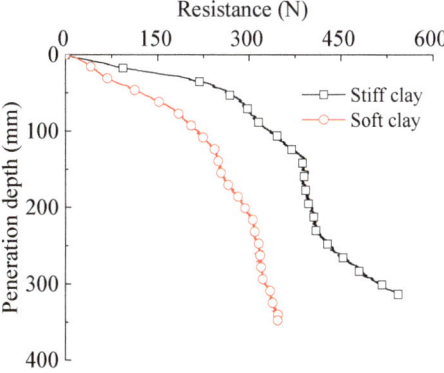

Figure 6. Total pile penetration resistances in the stiffer and the softer clay.

3.1.2. Resistance at Pile End

The monitored strain at the pile top and end was employed to calculate the axial force at the pile top and end using the following equation based on elastic mechanics [48]:

$$F = E\varepsilon_z \pi (r_o^2 - r_i^2) \tag{1}$$

where E is the pile modulus, ε_z is the measured strain, and r_o and r_i are the outside and inside radius of the pile, respectively. Note that the axial force at the pile end is similar in value to the total penetration resistance F_t and that the axial force at the pile end is similar to the resistance at the pile end F_e. Figure 7 provides the calculated F_t and F_e in stiffer and softer clays. It shows that the calculated F_t evolved in a similar trend and value as that monitored by the force sensor. Therefore, the strain at the top and end could be employed to show the evolution of F_e/F_t, as shown in Figure 8. Figure 8 shows that F_e/F_t exhibited a softening behavior, which first increased to a peak and then gradually decreased. It can be easily understood that the increase in F_e caused by soil strength was larger than that of the resistance at pile side F_s due to the increasing contact area at the beginning. After the soil beneath the pile end failed, F_e tended to be stable and F_s increased continually, which led to a decreasing F_e/F_t. It is worth mentioning that the peak and final F_e/F_t for stiffer and softer clays were quite similar in value, probably because pile–soil quick-shear behavior decreased with the soil's mechanical properties. However, due to lower compressibility, the penetration depth needed to reach the peak F_e/F_t was larger in stiffer clay. Nevertheless, the ratio of the resistance at the pile end to the total penetration resistance in stiffer clay can directly refer to that in softer clay.

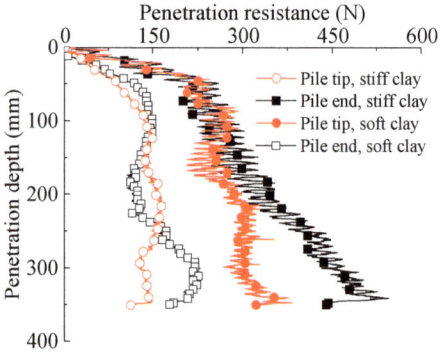

Figure 7. Pile end resistances in the stiffer and the softer clay.

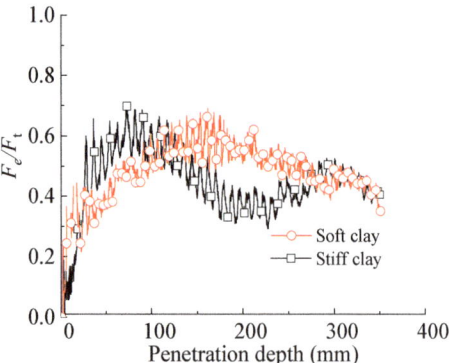

Figure 8. Contribution of the pile end resistances.

3.1.3. Earth Pressure

Figure 9 presents the evolutions of the horizontal pressures at different depths to show the soil-squeezing stress. Note that the initial earth pressure caused by consolidation was set to be 0 kPa, and only the pressure due to pile penetration was monitored here. Figure 9 shows that the horizontal earth pressure first increased to a peak with the increasing penetration depth and then decreased gradually. Note that the peak earth pressures in the stiffer clay were larger than those in the softer clay, indicating that larger squeezing stress

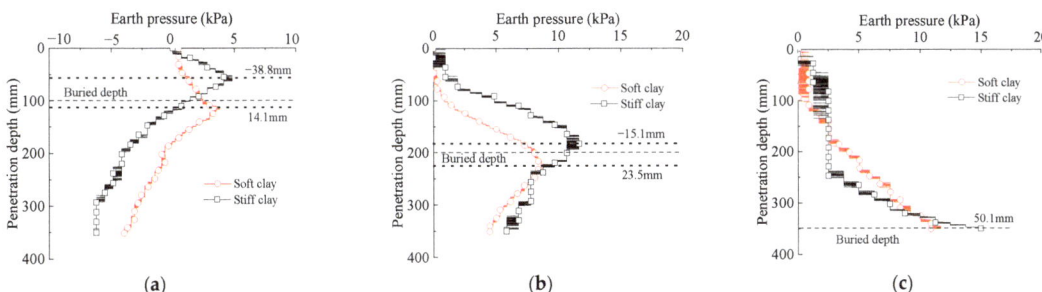

Figure 9. Soil-squeezing stress in the ground at a depth of (**a**) 100 mm, (**b**) 200 mm, and (**c**) 300 mm.

In cohesionless soil, the peak pressure increased to the peak value at the buried depth of the earth pressure, i.e., the distance between the penetrator cone and earth pressure cell d_p approached 0 mm when the earth pressure reached the peak [14]. However, d_p in the cohesive soil here was non-zero and increased with the buried depth of the earth pressure cell, as shown in Figure 10. Note that the d_p for the cell buried at a depth of 300 mm was chosen to be 50 mm, although the earth pressures were still increasing. In addition, d_p in the softer clay was much larger than that in the stiffer clay at the same buried depth. This is because both the resistances at the pile end and the side contributed to the increasing earth pressure. However, the resistance at the pile end increased rapidly to the peak and then tended to be stable, whereas the resistance at pile side continuously increased, as shown in Figure 5. Consequently, the resistance at the pile end contributed more to the shallow earth pressure cell, whereas the resistance at the pile side contributed more to the deep earth pressure cell.

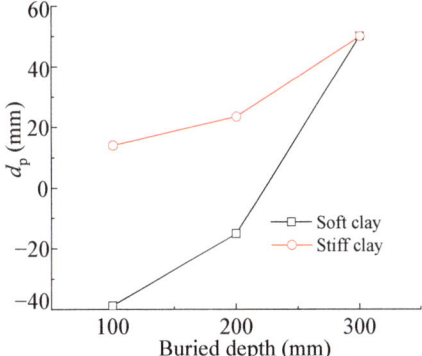

Figure 10. Relationships between d_p and buried depth.

3.1.4. Pore Pressure

Previous research has shown that the pore pressure in saturated clay increases significantly during pile penetration, which weakens the soil shear strength and leads to low pile resistance [15]. Figure 11 provides the pore pressure evolution in both the stiffer and the softer clay. It shows that the pore pressure increased rapidly to peak when the pile approached and then tended to be stable, which is different from the earth pressure, which decreased significantly after the peak value. This is because several days were needed for the pore pressure to dissipate due to clay's low permeability. In addition, the peak pore pressures in stiffer clay were much smaller than in softer clay, although the earth pressures were larger, as shown in Figure 9. It can be easily understood that the stiffer clay was unsaturated and the soil skeleton bore more squeezing stress from pile penetration.

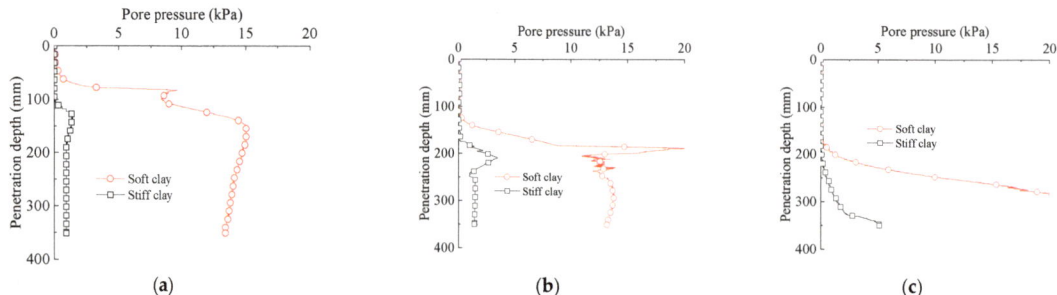

Figure 11. Pore pressure in the ground at a depth of (**a**) 100 mm, (**b**) 200 mm, and (**c**) 300 mm.

3.1.5. Soil-Squeezing Displacement

Figure 12 shows the ground heave in the stiffer and the softer clay, i.e., the soil-squeezing displacement on the ground surface. Note that no ground heave was observed at the point with a distance of 150 mm from the pile. Figure 12 shows that the ground heaved more significantly with the softer clay than with the stiffer clay. In particular, the ground heave at the point with a distance of 50 mm in the softer clay was nearly four times larger than that in the stiffer clay. This is because the softer clay was easier to compact and move due to its higher compressibility.

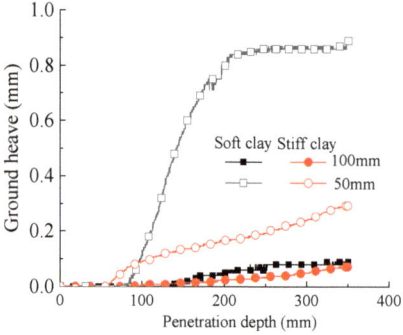

Figure 12. Ground heave.

Above all, it can be concluded that both soil-squeezing stress and soil-squeezing displacement exist in the ground due to pile penetration. Compared with the softer clay, the stiffer clay exhibited larger soil-squeezing stress due to the higher shear strength and smaller soil-squeezing displacement due to the low compressibility.

3.2. Soil-Squeezing Effect on the Adjacent Pile

3.2.1. Soil-Squeezing Displacement on the Adjacent Pile

The maximum soil-squeezing displacement on the adjacent pile was nearly 1 mm, which could barely be observed. Thus, the displacement sensors placed on the pile top were employed to monitor the soil-squeezing displacements. Figure 13 provides the center pile's vertical flotation and horizontal deviation caused by adjacent pile penetration at different pile spacings in the stiffer and the softer clay. Note that larger vertical flotation and horizontal pile deviation lead to pile floating in engineering, which affects the pile-bearing behavior. Figure 13 shows that the vertical pile flotation and horizontal pile deviation

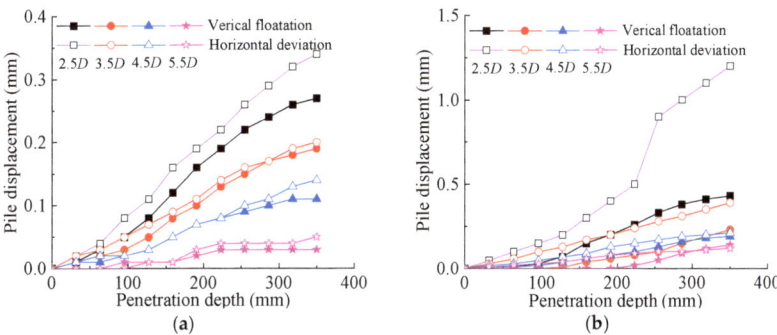

Figure 13. Vertical pile flotation and horizontal pile deviation due to adjacent pile penetration: (**a**) stiffer clay; (**b**) softer clay.

Figure 14 presents the relationships between vertical flotation/horizontal deviation and the pile spacing in the stiffer and the softer clay. Figure 14 shows that the vertical pile flotation and the horizontal pile deviation in the softer clay were much larger, indicating that the piles were easier to float and incline. This is because the soil squeezing displacement is larger and the pore pressure is higher in softer clay, leading to larger vertical pile flotation and horizontal pile deviation. Therefore, from the view of soil-squeezing displacement due to adjacent pile penetration, the spacing distance necessary to reduce the soil-squeezing effect is larger in softer clay. The linear, exponential, and logarithmic equations were employed to describe the relationships between vertical flotation/horizontal deviation and pile spacing. The fitting equations are provided in Table 4, which shows that the nonlinear equation form $y = ae^{-bx}$ could be employed to describe the effect of pile spacing on the vertical flotation/horizontal deviation of the adjacent pile. In other words, the vertical flotation/horizontal deviation of the adjacent pile could be predicted using the equation form $y = ae^{-bx}$ with the help of some filed test data.

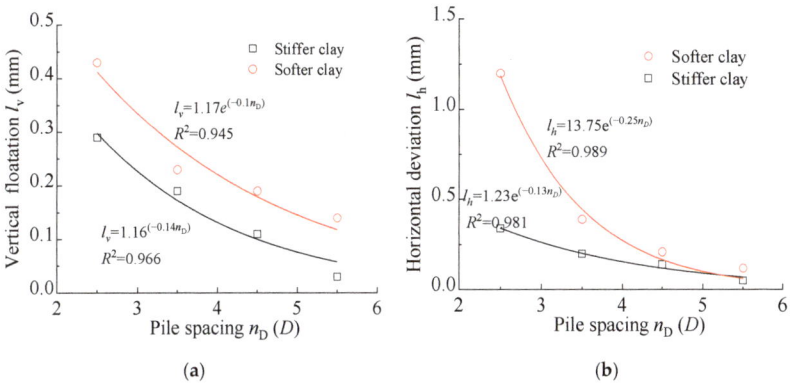

Figure 14. Relationships of vertical flotation (**a**)/horizontal deviation (**b**) and pile spacing.

Table 4. Fitting equations.

Soil	Equation	Horizontal Deviation	Vertical Flotation
Stiffer clay	Linear	$I_h = -0.093n_D + 0.5545$, $R^2 = 0.972$	$I_v = -0.086n_D + 0.499$, $R^2 = 0.996$
	Exponential	$I_h = 1.23e^{(-0.13nD)}$, $R^2 = 0.981$	$I_v = 1.16e^{(-0.14nD)}$, $R^2 = 0.966$
	Logarithmic	$I_h = -0.356\ln(n_D) + 0.66$, $R^2 = 0.979$	$I_v = -0.33\ln(n_D) + 0.6$, $R^2 = 0.992$
Softer clay	Linear	$I_h = -0.342n_D + 1.848$, $R^2 = 0.802$	$I_v = -0.091n_D + 0.611$, $R^2 = 0.854$
	Exponential	$I_h = 13.75e^{(-0.25nD)}$, $R^2 = 0.989$	$I_v = 1.17e^{(-0.1nD)}$, $R^2 = 0.945$
	Logarithmic	$I_h = -1.363\ln(n_D) + 2.319$, $R^2 = 0.884$	$I_v = -0.359\ln(n_D) + 0.73$, $R^2 = 0.920$

3.2.2. Soil-Squeezing Stress on the Adjacent Pile

Figure 15 presents the pile strain at the pile end due to adjacent pile penetration in the stiffer and the softer clay. Note that large pile strain due to soil-squeezing stress leads to pile breakage in engineering. Figure 15 shows that as the pile spacing increased, the strain at the pile end decreased significantly, which implies that the squeezing stress at the pile end decreased with the increasing pile spacing, as expected.

Figure 16 presents the relationships between the strain at the pile end and the pile spacing in the stiffer and the softer clay. Figure 16 shows the strain at the pile end in the stiffer clay was larger than that in the softer clay, indicating that larger squeezing stress appeared on the pile end in the stiffer clay. Therefore, from the view of soil-squeezing stress due to adjacent pile penetration, the spacing distance necessary to reduce the soil-squeezing effect is larger in softer clay. In addition, the nonlinear equation form $y = ae^{-bx}$ can also be employed to describe the effect of pile spacing on the strain at the adjacent pile end.

Above all, it can be concluded that the soil-squeezing effect due to adjacent pile penetration also leads to a smaller squeezing displacement but a larger squeezing force on the pile in stiffer clay than in softer clay due to the high shear strength and low compressibility of stiffer clay. Consequently, the pile spacing for stiffer clay should be determined mainly by controlling the pile strain caused by the squeezing stress to avoid pile breakage, whereas the pile spacing for softer clay should be determined mainly by controlling the pile vertical flotation and horizontal deviation to avoid pile flotation. In addition, high-strength PHC piles should be employed in stiff clay to avoid pile breakage.

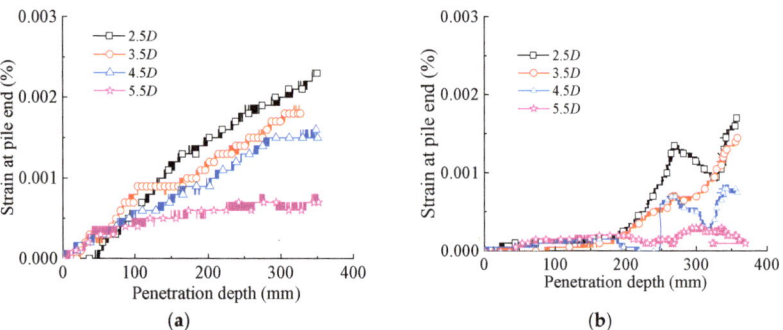

Figure 15. Strain at the pile end due to adjacent pile penetration: (**a**) stiffer clay; (**b**) softer clay.

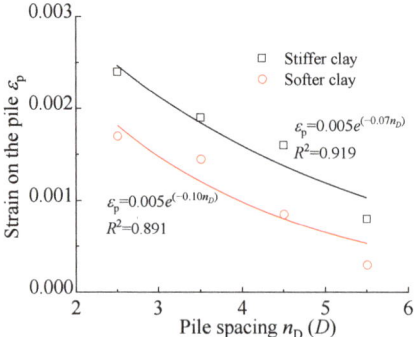

Figure 16. Relationships between the strain at the pile end and the pile spacing.

4. Conclusions

To better design and construct jacked piles in stiff clay areas of Central China, two groups of jacked pile penetration tests were performed with a rough pile that can reproduce the quick-shear behavior of the pile–soil interface, i.e., group 1 in stiffer clay and group 2 in softer clay for comparison, to study the bearing behavior and soil-squeezing effect in stiffer clay. For each group, the adjacent pile was additionally penetrated at different pile spacings to investigate the soil-squeezing effect due to adjacent pile penetration. The main conclusions are summarized as follows:

(1) During pile penetration, the resistance at the pile end increased rapidly at the beginning and then tended to be stable with fluctuations, whereas the resistance on the pile side continually increased due to the increasing contact area. As a result, the total penetration resistance increased rapidly at the beginning and then increased at a lower increasing rate.

(2) The ratio of the resistance at the pile end to the total penetration resistance exhibited a softening behavior, which first increased to a peak and then gradually decreased. The peak and final proportion for the stiffer and the softer clay were similar in value, whereas the penetration depth needed to reach the peak proportion was larger in the stiffer clay due to its low compressibility.

(3) There was soil-squeezing stress and soil-squeezing displacement in the ground due to pile penetration. In the stiffer clay, the soil-squeezing stress was larger due to the higher strength, whereas the soil-squeezing displacement was smaller due to the low compressibility.

(4) The nonlinear equation form $y = ae^{-bx}$ could be employed to describe the effect of pile spacing on the vertical flotation, horizontal deviation, and pile strain of the adjacent pile.

Author Contributions: Visualization, writing, funding acquisition, and formal analysis, B.X.; methodology, investigation, and data curation, G.L.; conceptualization, validation, and resources. X.C. All authors have read and agreed to the published version of the manuscript.

Funding: The research was funded by the Science and Technology Plan for Housing and Urban Rural Construction in Anhui Province (2021-YF27) and the Anhui Province Key Laboratory of Green Building and Assembly Construction (2021-JKYL1-002).

Data Availability Statement: Not applicable.

Conflicts of Interest: The authors declare no conflict of interest.

References

1. Doherty, P.; Gavin, K.; Gallagher, D. Field investigation of base resistance of pipe piles in clay. *Proc. Inst. Civ. Eng. Geotech. Eng.* **2010**, *163*, 13–22. [CrossRef]
2. Cooke, R.W.; Price, G.; Tarr, K. Jacked piles in London Clay: A study of load transfer and settlement under working conditions. *Geotechnique* **1979**, *29*, 113–147.
3. Zhang, Z.; Wang, Y.H. Examining setup mechanisms of driven piles in sand using laboratory model pile tests. *J. Geotech. Geoenviron. Eng.* **2015**, *141*, 04014114. [CrossRef]
4. Han, F.; Ganju, E.; Salgado, R.; Prezzi, M. Comparison of the load response of closed-ended and open-ended pipe piles driven in gravelly sand. *Acta Geotech.* **2019**, *14*, 1785–1803.
5. Liu, C.; Tang, X.; Wei, H.; Wang, P.; Zhao, H. Model tests of jacked-pile penetration into sand using transparent soil and incremental particle image velocimetry. *KSCE J. Civ. Eng.* **2020**, *24*, 1128–1145.
6. Wang, Y.; Sang, S.; Liu, X.; Huang, Y.; Zhang, M.; Miao, D. Model test of jacked pile penetration process considering influence of pile diameter. *Front. Phys.* **2021**, *9*, 616410.
7. Hwang, J.H.; Liang, N.; Chen, C.H. Ground response during pile driving. *J. Geotech. Geoenviron.* **2001**, *127*, 939–949. [CrossRef]
8. Baligh, M.M. Strain path method. *Int. J. Geotech. Eng.* **1985**, *111*, 1108–1136. [CrossRef]
9. Wang, Y.; Li, L.; Gong, W. An analytical solution for settlement of jacked piles considering pile installation effects. *Eur. J. Environ. Civ. Eng.* **2022**, *26*, 200–215. [CrossRef]
10. Zhang, Q.Q.; Liu, S.W.; Zhang, S.M.; Zhang, J.; Wang, K. Simplified non-linear approaches for response of a single pile and pile groups considering progressive deformation of pile–soil system. *Soils Found.* **2016**, *56*, 473–484.
11. Burd, H.J.; Abadie, C.N.; Byrne, B.W.; Houlsby, G.T.; Martin, C.M.; McAdam, R.A.; Andrade, M.P. Application of the PISA design model to monopiles embedded in layered soils. *Géotechnique* **2020**, *70*, 1067–1082. [CrossRef]
12. Chen, H.; Li, L.; Li, J. An elastoplastic solution for spherical cavity undrained expansion in overconsolidated soils. *Comput. Geotech.* **2020**, *126*, 103759.

13. Hu, P.; Stanier, S.A.; Cassidy, M.J.; Wang, D. Predicting peak resistance of spudcan penetrating sand overlying clay. *J. Geotech. Geoenviron.* **2014**, *140*, 04013009. [CrossRef]
14. Jiang, M.; Dai, Y.; Cui, L.; Shen, Z.; Wang, X. Investigating mechanism of inclined CPT in granular ground using DEM. *Granul. Matter* **2014**, *16*, 785–796. [CrossRef]
15. Lei, H.Y.; Li, X.; Lu, P.Y.; Huo, H.F. Field test and numerical simulation of squeezing effect of pipe pile. *Rock Soil Mech.* **2012**, *33*, 1006–1012.
16. Wang, Y.Y.; Sang, S.K.; Zhang, M.Y.; Jeng, D.S.; Yuan, B.X.; Chen, Z.X. Laboratory study on pile jacking resistance of jacked pile. *Soil Dyn. Earthq. Eng.* **2022**, *154*, 107070. [CrossRef]
17. Zhang, L.M.; Wang, H. Field study of construction effects in jacked and driven steel H-piles. *Géotechnique* **2009**, *59*, 63–69. [CrossRef]
18. Gill, D.R.; Lehane, B. An optical technique for investigating soil displacement patterns. *Geotech. Test J.* **2001**, *24*, 324–329.
19. Huang, B.; Zhang, Y.; Fu, X.; Zhang, B. Study on visualization and failure mode of model test of rock-socketed pile in softer rock. *Geotech. Test J.* **2018**, *42*, 1624–1639.
20. Lalicata, L.M.; Desideri, A.; Casini, F.; Thorel, L. Experimental observation on laterally loaded pile in unsaturated silty soil. *Can. Geotech. J.* **2019**, *56*, 1545–1556. [CrossRef]
21. Liu, E.L.; Shen, Z.J. Experimental study on mechanical properties of artificially structured soils. *Rock Soil Mech.* **2007**, *28*, 679–683.
22. White, D.J.; Bolton, M.D. Displacement and strain paths during plane-strain model pile installation in sand. *Géotechnique* **2004**, *54*, 375–397. [CrossRef]
23. Randolph, M.F.; Wroth, C.P. An analytical solution for the consolidation around a driven pile. *Int. J. Numer. Anal. Met.* **1979**, *3*, 217–229. [CrossRef]
24. Henke, S.; Grabe, J. Simulation of pile driving by 3-dimensional Finite-Element analysis. In Proceedings of the 17th European Young Geotechnical Engineers' Conference, Zagreb, Croatia, 20–22 July 2006; pp. 215–233.
25. Su, D.; Wu, Z.; Lei, G.; Zhu, M. Numerical study on the installation effect of a jacked pile in sands on the pile vertical bearing capacities. *Comput. Geotech.* **2022**, *145*, 104690.
26. Li, L.; Wu, W.; Liu, H.; Lehane, B. DEM analysis of the plugging effect of open-ended pile during the installation process. *Ocean Eng.* **2021**, *220*, 108375.
27. Liu, J.; Duan, N.; Cui, L.; Zhu, N. DEM investigation of installation responses of jacked open-ended piles. *Acta Geotech.* **2019**, *14*, 1805–1819.
28. Katti, D.R.; Matar, M.I.; Katti, K.S.; Amarasinghe, P.M. Multiscale modeling of swelling clays: A computational and experimental approach. *KSCE J. Civ. Eng.* **2009**, *13*, 243–255. [CrossRef]
29. Bayesteh, H.; Hoseini, A. Effect of mechanical and electro-chemical contacts on the particle orientation of clay minerals during swelling and sedimentation: A DEM simulation. *Comput. Geotech.* **2021**, *130*, 103913.
30. Wen, L.; Kong, G.; Li, Q.; Zhang, Z. Field tests on axial behavior of grouted steel pipe micropiles in marine softer clay. *Int. J. Geomech.* **2020**, *20*, 06020006. [CrossRef]
31. Lim, J.K.; Lehane, B.M. Set-up of pile shaft friction in laboratory chamber tests. *Int. J. Phys. Model. Geo.* **2014**, *14*, 21–30. [CrossRef]
32. Paniagua, P.; Andó, E.; Silva, M.; Emdal, A.; Nordal, S.; Viggiani, G. Soil deformation around a penetrating cone in silt. *Geotech. Lett.* **2013**, *3*, 185–191. [CrossRef]
33. White, D.J.; Lehane, B.M. Friction fatigue on displacement piles in sand. *Géotechnique* **2004**, *54*, 645–658.
34. Ye, Z.H.; Zhou, J.; Tang, S.D. Model test on pile bearing behaviors in clay under different pile tip conditions. *J. Tongji Univ. (Nat. Sci.)* **2009**, *37*, 733–737.
35. Arshad, M.I.; Tehrani, F.S.; Prezzi, M.; Salgado, R. Experimental study of cone penetration in silica sand using digital image correlation. *Géotechnique* **2014**, *64*, 551–569. [CrossRef]
36. Mo, P.Q.; Marshall, A.M.; Yu, H.S. Layered effects on soil displacement around a penetrometer. *Soils Found.* **2017**, *57*, 669–678. [CrossRef]
37. Ni, Q.C.C.I.; Hird, C.C.; Guymer, I. Physical modelling of pile penetration in clay using transparent soil and particle image velocimetry. *Géotechnique* **2010**, *60*, 121–132. [CrossRef]
38. Qi, C.G.; Zheng, J.H.; Zuo, D.J.; Chen, G. Measurement on soil deformation caused by expanded-base pile in transparent soil using particle image velocimetry (PIV). *J. Mt. Sci.* **2017**, *14*, 1655–1665. [CrossRef]
39. Chen, Z.; Omidvar, M.; Iskander, M.; Bless, S. Modelling of projectile penetration using transparent soils. *Int. J. Phys. Model. Geo.* **2014**, *14*, 68–79. [CrossRef]
40. Paik, K.; Salgado, R. Determination of bearing capacity of open-ended piles in sand. *J. Geotech. Geoenviron.* **2003**, *129*, 46–57. [CrossRef]
41. Gui, M.W.; Bolton, M.D.; Garnier, J.; Corte, J.F.; Bagge, G.; Laue, J.; Renzi, R. Guidelines for cone penetration tests in sand. In *Centrifuge*; Balkema: Rotterdam, The Netherlands, 1998; pp. 155–160.
42. Mo, P.Q.; Marshall, A.M.; Yu, H.S. Centrifuge modelling of cone penetration tests in layered soils. *Géotechnique* **2015**, *65*, 468–481. [CrossRef]
44. Suebsuk, J.; Horpibulsuk, S.; Liu, M.D. Modified Structured Cam Clay: A generalised critical state model for destructured, naturally structured and artificially structured clays. *Comput. Geotech.* **2010**, *37*, 956–968. [CrossRef]

45. Amorosi, A.; Kavvadas, M. A constitutive model for structured soils. *Géotechnique* **2000**, *50*, 263–273.
46. Savvides, A.A.; Papadrakakis, M. A computational study on the uncertainty quantification of failure of clays with a modified cam-clay yield criterion. *SN Appl. Sci.* **2021**, *3*, 659. [CrossRef]
47. *JGJ 94-2008*; Ministry of Construction of the People's Republic of China. Technical Code for Building Pile Foundations. China Architecture & Building Press: Beijing, China, 2008.
48. Wu, J.L. *Elasticity*; Higher Education Press: Beijing, China, 2001.

Disclaimer/Publisher's Note: The statements, opinions and data contained in all publications are solely those of the individual author(s) and contributor(s) and not of MDPI and/or the editor(s). MDPI and/or the editor(s) disclaim responsibility for any injury to people or property resulting from any ideas, methods, instructions or products referred to in the content.

MDPI AG
Grosspeteranlage 5
4052 Basel
Switzerland
Tel.: +41 61 683 77 34

Buildings Editorial Office
E-mail: buildings@mdpi.com
www.mdpi.com/journal/buildings

Disclaimer/Publisher's Note: The title and front matter of this reprint are at the discretion of the Guest Editors. The publisher is not responsible for their content or any associated concerns. The statements, opinions and data contained in all individual articles are solely those of the individual Editors and contributors and not of MDPI. MDPI disclaims responsibility for any injury to people or property resulting from any ideas, methods, instructions or products referred to in the content.